Sequence Transformations
and Their Applications

This is Volume 154 in
MATHEMATICS IN SCIENCE AND ENGINEERING
A Series of Monographs and Textbooks
Edited by RICHARD BELLMAN, *University of Southern California*

The complete listing of books in this series is available from the Publisher upon request.

SEQUENCE TRANSFORMATIONS AND THEIR APPLICATIONS

Jet Wimp

DEPARTMENT OF MATHEMATICAL SCIENCES
DREXEL UNIVERSITY
PHILADELPHIA, PENNSYLVANIA

1981

ACADEMIC PRESS
A Subsidiary of Harcourt Brace Jovanovich, Publishers
New York London Toronto Sydney San Francisco

ACADEMIC PRESS, INC.
111 Fifth Avenue, New York, New York 10003

United Kingdom Edition published by
ACADEMIC PRESS, INC. (LONDON) LTD.
24 28 Oval Road, London NW1 7DX

Library of Congress Cataloging in Publication Data

Wimp, Jet.
Sequence transformations and their applications.

(Mathematics in science and engineering)
Bibliography: p.
Includes index.
1. Sequences (Mathematics) 2. Transformations (Mathematics) 3. Numerical analysis. I. Title. II. Series.
QA292.W54 515'.24 80-68564
ISBN 0-12-757940-0

PRINTED IN THE UNITED STATES OF AMERICA

81 82 83 84 9 8 7 6 5 4 3 2 1

Contents

Chapter 3 Linear Lozenge Methods

Chapter 4 Optimal Methods and Methods Based on Power Series

Chapter 5 Nonlinear Lozenges; Iteration Sequences

Chapter 6 The Schmidt Transformation; The ε-Algorithm

Chapter 7 Aitken's δ^2-Process and Related Methods

Chapter 8 Lozenge Algorithms and the Theory of Continued Fractions

Chapter 9 Other Lozenge Algorithms and Nonlinear Methods

Chapter 10 The Brezinski–Håvie Protocol

Chapter 11 The Brezinski–Håvie Protocol and Numerical Quadrature

Chapter 12 Probabilistic Methods

Chapter 13 Multiple Sequences

Appendix

Preface

In this book we shall be concerned with the practical aspects of sequence transformations. In particular, we shall discuss transformations T mapping sequences in a Banach space $\mathscr{B}$ (often, but not always, the complex field) into sequences in $\mathscr{B}$. Certain practical requirements are ordinarily made of T: that its domain $\mathscr{D}$ contain an abundance of "interesting" sequences and for $\mathbf{s} \in \mathscr{D}$ also $\alpha\mathbf{s} + \mathbf{c} \in \mathscr{D}$, $\mathbf{c}$ being any constant sequence; further, we shall usually require that T satisfy the following requirements:

(i) T is *homogeneous*: $T(\alpha\mathbf{s}) = \alpha T(\mathbf{s})$ for any scalar α;

(ii) T is *translative*: $T(\mathbf{s} + \mathbf{c}) = T(\mathbf{s}) + T(\mathbf{c})$ for any constant sequence $\mathbf{c}$;

(iii) T is *regular for* $\mathbf{s}$: if $\mathbf{s}$ converges, then $T(\mathbf{s})$ converges to the same limit.

Often more than (iii) is required, namely,

(iii′) T is *accelerative for* $\mathbf{s}$: $T(\mathbf{s})$ converges more rapidly than $\mathbf{s}$.

This requirement sometimes takes the form that

$$\overline{\lim_{n\to\infty}} \frac{\|\{T(\mathbf{s})\}_n - s\|}{\|s_n - s\|^p} = \beta < 1$$

for some index $p \geq 1$, where $\{T(\mathbf{s})\}_n$ and s_n are the nth components of $T(\mathbf{s})$ and $\mathbf{s}$, respectively, and s is the limit of $\mathbf{s}$.

Historically, most of the work done in this area up to 1950 focused on transformations that are also linear: $T(\mathbf{s} + \mathbf{t}) = T(\mathbf{s}) + T(\mathbf{t})$. Such transformations have a very simple structure, namely, the components of $T(\mathbf{s})$

can be characterized by weighted scalar means of the components of **s** (at least when $\mathscr{B}$ is separable), and such transformations have beautiful theoretical properties. [The classical work in this area is the book "Divergent Series" (Hardy, 1956), and more modern developments are discussed in Cooke (1950), Zeller (1958), Petersen (1966), and Peyerimhoff (1969).] However, linear methods are distinctly limited in their usefulness primarily because *the class of sequences for which the methods are regular is too large.* In defense of this somewhat paradoxical statement, I only remark that experience indicates the size of the domain of regularity of a transformation and its efficiency (i.e., the sup of p values in the foregoing equation) seem to be inversely related. Furthermore, linear transformations whose domains of regularity are all convergent sequences (called regular transformations) generally accelerate convergence at most linearly, i.e., $p = 1$, $0 < \beta < 1$. Obviously, for safety's sake, when one uses a nonregular method, one wants a criterion for deciding when **s** belongs to its domain of regularity. This, however, is not the problem it might seem to be.

Linear regular transformations are discussed (at length, in fact) in this book, but primarily those transformations whose application can be effected through a certain simple computational procedure called a *lozenge method.*

As the reader will find, the subject touches virtually every area of analysis, including interpolation and approximation, Padé approximation, special functions, continued fractions, and optimization methods, to name a few; and the proofs of the theorems draw their techniques from all these disciplines. Incidentally, I have included a proof only if it is either short or conceptually important for the discussion at hand. It was simply not feasible to include very detailed and computational proofs, e.g., estimates for the Lebesgue constants for various transformations (Section 2.4), or inequalities satisfied by the iterates in the ε-algorithm, or long proofs whose flavor was totally that of another discipline—results on Padé theory, for instance, or results requiring the theory of Hilbert subspaces. In such cases, I have always indicated where the proof can be found.

The techniques given will, I hope, be useful in any practical problem that requires the evaluation of the limit of a sequence: the summation of series, numerical quadrature, the solution of systems of equations. Particularly welcome should be the discussion of methods to accelerate the convergence of sequences arising from Monte Carlo statistical experiments. Since the convergence of Monte Carlo computations is so poor, $O(n^{-1/2})$, n being the number of trials, techniques for enhancing convergence are highly desirable.

A closely related subject is the iterative solution of (operator) equations. In fact, any sequence transformation can be used to define such an iterative method (cf. Chapter 5). However, this is not the subject proper of this book, there being available already several excellent works in this area. I have, in fact, restricted myself mostly to material which has not appeared in book form in English. Some of the material is available in French [any numerical analyst will have on his shelf C. Brezinski's two important volumes (Brezinski, 1977, 1978)], but much of the material has never appeared in book form, some has not appeared in published papers [the thesis work of Higgins (1976) and Germain-Bonne (1978) for instance], and much is new altogether.

I have not usually opted for abstraction. In most instances the transformations can be generalized from complex sequences to Banach-space-valued sequences, and often I have indicated how this can be done and have established appropriate convergence results. But where abstraction can confuse rather than elucidate, I have left well enough alone. For instance, I believe that the theory of Padé approximants, at least for my purposes, is most firmly at home in classical function theory.

My notation may at times seem idiosyncratic, but it is one I have found necessary to diminish clutter and bring some focus to the development. Before the reader gets into the book, I strongly advise him to read the section on notation. Otherwise, certain unfamiliar conventions—for instance, $x_n R\colon y_n$, which I have found most useful—may well render the material completely opaque. The notation for special functions is, by and large, as in the Bateman manuscript volumes. Ad hoc notation is explained in Notation or as needed.

I have provided many numerical examples, but these are illustrative only, not exhaustive. The reader interested in further numerical examples and applications should consult C. Brezinski's (1978) book, and, for a comparison of methods, the survey of Smith and Ford (1979).

The problem of rounding is always an annoying one in a book dealing with numerical methods. Generally speaking, all numbers free from decimal points or occurring in definitions may be considered exact. Others, particularly those occurring in tables, have been rounded to the number of places given. However, I should be surprised if I have been consistent.

Acknowledgments

Several people have contributed to this book. John Quigg has read and commented on some of the material. Bob Higgins, my former student, has provided most of the theory in Chapters 12 and 13. Steve Yankovich and Stanley Dunn have contributed their programming and analytical skills for the preparation of numerical examples. Drexel University has been generous in its support and encouragement. I am grateful to Alison Chandler, whose combined typing and mathematical skills led to such a beautifully prepared manuscript, and to Don Johnson and Harold Schwalm, Jr., who assisted in the proofreading.

Finally, I consider myself fortunate to be working in a field where friends are so easily made. My colleagues have proved to be warm and enthusiastic. I have enjoyed thoroughly meeting and exchanging ideas with Bernard Germain-Bonne and Florent Cordellier. I am particularly indebted to correspondence and discussions with Claude Brezinski. He has generously provided me with unpublished results (Chapter 10). Some of the ideas in the book originated in a lengthy afternoon discussion with Claude and other colleagues. That meeting demonstrated to me the delights of the mutual, as opposed to solitary, quest.

Notation

Spaces

$\mathscr{M}$	metric or pseudometric space
$\mathscr{V}$	linear space
$\mathscr{E}$	topological vector space over real or complex field
$\mathscr{B}$	Banach space
—*	dual space
$B(\mathscr{B}, \mathscr{B}')$	space of all bounded linear mappings of one Banach space into another

$\|T\| = \sup_{\|x\| \leq 1} \|T(x)\|$, $T \in B$, $x \in \mathscr{B}$

Ω cone in $\mathscr{E}$ (Ω contains a nonzero vector and if $x \in \Omega$, $\lambda x \in \Omega$, $\lambda > 0$).

for any matrix $A = [a_{ij}]$, $1 \leq i \leq n$, $1 \leq j \leq m$, first subscript of a_{ij} denotes row position, the second column position, of the element

Real and Complex Numbers

$\mathscr{C}^p$	space of ordered complex p-tuples, $p > 1$
$\mathscr{C}$	complex numbers
$\mathscr{R}^p$	space of ordered real p-tuples, $p > 1$

$\mathscr{R}$ real numbers

$\mathscr{R}^0$ nonnegative real numbers

$\mathscr{R}^+$ positive reals

J integers

J^0 nonnegative integers

J^+ positive integers

m, n, k, r, i, j generally denote integers

$d(A, B) = \inf_{x \in A, y \in B} |x - y|$

$D(A, B) = \sup_{x \in A, y \in B} |x - y|$

$N_\rho(a) = \{z \,|\, |z - a| < \rho\}$

$\partial N_\rho(a) = \{z \,|\, |z - a| = \rho\}$

$\overline{N}_\rho(a) = \{z \,|\, |z - a| \leq \rho\}$

$N_\rho(0) = N_\rho$

$N_1 = N$ the unit circle

Sequences

boldface letters denote sequences, **s**, **t**, etc.

for any space $\mathscr{A}$, $\mathscr{A}_S$ denotes the space of sequences with elements in $\mathscr{A}$; $\mathbf{s} = \{s_n\} \in \mathscr{A}_S$, $s_n \in \mathscr{A}$

$\mathscr{A}_C$ space of convergent sequences

$\mathscr{A}_N$ space of null sequences, e.g., $\mathscr{A}$ a metrizable t.v.s.

$\mathscr{C}_\sigma, \mathscr{C}_l, \mathscr{C}_{l'}, \mathscr{R}_{\mathrm{TM}}, \mathscr{R}_{\mathrm{TO}}, \mathscr{C}_{\mathrm{E^m}}(\Gamma)$ special real and complex sequence spaces (see Sections 1.4, 1.5, 2.2)

related sequences (the space $\mathscr{A}$ must be such that the definitions make sense)

a: $a_n = s_n - s_{n-1}, \quad n \geq 1; \quad a_0 = s_0, \quad \text{so} \quad s_n = \sum_{k=0}^{n} a_k$

r: $r_n = s_n - s, \quad s = \lim_{n \to \infty} s_n$

$\boldsymbol{\rho}$: $\rho_n = a_{n+1}/a_n$

h: $h_n = r_{n+1}/r_n = (s_{n+1} - s)/(s_n - s)$

$\Delta^k\mathbf{s}$: $\{\Delta^k\mathbf{s}\}_n = \Delta^k s_n, \qquad k \geq 1$

$\sum'$ indicates first term of sum is to be halved

$\sum''$ indicates first and last term of (finite) sum are to be halved

$$T_n(f) = \frac{1}{n}\sum_{k=0}^{n}{}'' f(k/n), \qquad n \geq 1 \qquad \text{(trapezoidal sum)}$$

sequence relationships: let R be a binary relationship between members of two sequences **x**, **y**

$x_n R\!: y_n$ means $x_n R y_n$ holds for an infinite number of values of n

$x_n R. y_n$ means $x_n R y_n$ holds for all n sufficiently large

this notation is used only when the sequence variable is n

Example

$|\lambda_{nk}| \leq . \, 1$ means: for some n_0, $|\lambda_{nk}| \leq 1, 0 \leq k \leq n, n > n_0$

Functions

Ψ class of real nondecreasing bounded functions on $[0, \infty)$ having infinitely many points of increase

Ψ^* subclass of Ψ such that $\int_0^\infty t^n \, d\psi < \infty, n \geq 0; \psi \in \Psi^*$

support of ψ is the set of points of increase

$(\alpha)_k = \alpha(\alpha + 1) \cdots (\alpha + k - 1)$ (Pochhammer's symbol)

$$\begin{vmatrix} 1 & a_1 & \cdots & a_1^{k-1} \\ 1 & a_2 & \cdots & a_2^{k-1} \\ \vdots & \vdots & & \vdots \\ 1 & a_k & \cdots & a_k^{k-1} \end{vmatrix} = V_k(a_1, a_2, \ldots, a_k)$$

$$= \prod_{j=1}^{k-1} \prod_{r=j+1}^{k} (a_r - a_j) \qquad \text{(van der Monde determinant)}$$

$$H_n^{(k)}(\mathbf{s}) = \begin{vmatrix} s_n & \cdots & s_{n+k-1} \\ \vdots & & \vdots \\ s_{n+k-1} & \cdots & s_{n+2k-2} \end{vmatrix}, \qquad n \geq 0, \qquad k \geq 1 \qquad \text{(Hankel determinant)}$$

$R_n^{(\alpha,\beta)}(x) = P_n^{(\alpha,\beta)}(2x-1)$, Jacobi polynomial shifted to [0, 1]

All other special functions are as defined in the Erdélyi volumes (1953).

Special Sequences

$$s_n = \sum_{k=0}^{n} a_k$$

$(\mathrm{LN}\ 2)_n$: $a_k = \dfrac{(-1)^k}{k+1}$, $s = \ln 2 = 0.693147180559945$

$(\mathrm{RT})_n$: $a_k = \dfrac{(-1)^k}{\sqrt{k+1}}$, $s = 0.604898643421630$

$(\mathrm{PI}^2)_n$: $a_k = \dfrac{1}{(k+1)^2}$, $s = \dfrac{\pi^2}{6} = 1.644934066846559$

$(\mathrm{GAM})_n$: $a_0 = 1$, $a_k = \dfrac{1}{k+1} + \ln\left(\dfrac{k}{k+1}\right)$, $k > 0$,

$s = \gamma = 0.577215664901533$

$(\mathrm{EX}\ 1)_n$: $a_k = (k+1)(0.8)^k$, $s = 25$

$(\mathrm{EX}\ 2)_n$: $a_k = (0.8)^k/(k+1)$, $s = 1.25 \ln 5$

$(\mathrm{EX}\ 3)_n$: $a_k = (k+1)(1.2)^k$, **s** divergent

$(\mathrm{FAC})_n$: $a_k = (-1)^k k!$, **s** divergent, but generated by

$$\int_0^\infty \frac{e^{-t}}{t+1} = 0.5963473611$$

$(\mathrm{IT}\ 1)_n$: generated by $s_{n+1} = 20[s_n^2 + 2s_n + 10]^{-1}$; $s_0 = 1$;

$s = 1.368808107$

$(\mathrm{IT}\ 2)_n$: generated by $s_{n+1} = (20 - 2s_n^2 - s_n^3)/10$; $s_0 = 1$;

s divergent

$(\mathrm{LUB})_n$: $a_k = \dfrac{(-1)^{\langle k/2 \rangle}}{k+1}$, $\langle \cdot \rangle$ = greatest integer contained in

$s = 1.131971754$

Numerics

Generally, in tables n SF representing a number is a rounded value; for instance,

$$\pi = 3$$
$$\pi = 3.1$$
$$\pi = 3.14$$
$$\pi = 3.142, \ldots .$$

For rational numbers, it may occasionally be important to know that the given value is exact. If that is the case, we write

$$\tfrac{3}{2} = 1.5 \qquad \text{(exact).}$$

In definitions, all numbers are exact, e.g., $s_n = (1.18)^n$, or it is indicated by ... that the number has been truncated.

Chapter 1 Sequences and Series

1.1. Order Symbols and Asymptotic Scales, Continuous Variables

Let $\mathscr{V}$ and $\mathscr{V}'$ (see Notation) be equipped with pseudometrics d and d', respectively; let Ω be a cone in $\mathscr{V}$ and $\phi, \psi \in \mathscr{T}(\mathscr{V}, \mathscr{V}')$.

$$\phi = O(\psi) \qquad \text{in} \quad \Omega \tag{1}$$

means for some $M > 0$ there is an $R(M) > 0$ such that

$$d'(\phi, 0)/d'(\psi, 0) < M, \qquad x \in \Omega, \qquad d(x, 0) > R. \tag{2}$$

Further,

$$\phi = o(\psi) \qquad \text{in} \quad \Omega \tag{3}$$

means for any $\varepsilon > 0$ there is an $R(\varepsilon) > 0$ such that

$$d'(\phi, 0)/d'(\psi, 0) < \varepsilon, \qquad x \in \Omega, \qquad d(x, 0) > R. \tag{4}$$

If ϕ, ψ depend parametrically on $a \in \mathscr{A}$ and (2) holds for all $a \in \mathscr{A}$, then we shall write "$\phi = O(\psi)$ in Ω *uniformly* in $\mathscr{A}$," and similarly for (3).

For the foregoing definitions to apply, the *implicit* assumption is made that denominators are never zero; for example, there must be some R such that $d'(\psi, 0) \neq 0$, $x \in \Omega$, $d(x, 0) > R$. Thus anytime an order symbol is used, an implicit statement is being made about the zeros of $d'(\psi, 0)$.

The concept of *asymptotic equivalence* is often useful. This is written

$$\phi \approx \psi \qquad \text{in} \quad \Omega \tag{5}$$

and means both $\phi - \psi = o(\psi)$ and $\psi - \phi = o(\phi)$ in Ω.

Now let $\boldsymbol{\phi} \in \mathscr{T}_S(\mathscr{V}, \mathscr{V}')$, where $\mathscr{V}$ and $\mathscr{V}'$ are linear spaces with pseudometrics d and d'. $\boldsymbol{\phi}$ is called an *asymptotic scale in* Ω if, for every $k \geq 0$,

$$\phi_{k+1} = o(\phi_k) \qquad \text{in} \quad \Omega, \tag{6}$$

and if this holds uniformly in k or uniformly in some parameter space, we speak of a uniform asymptotic scale (properly qualified). See Erdélyi (1956) for many examples.

Let $f \in \mathscr{T}(\mathscr{V}, \mathscr{V}')$, $\mathbf{A} \in \mathscr{C}_S$ and $\boldsymbol{\phi}$ be an asymptotic scale in Ω. The statement

$$f \sim \sum_{k=0}^{\infty} A_k \phi_k \qquad \text{in} \quad \Omega \tag{7}$$

is to be read "f has the right-hand side as an asymptotic expansion in Ω with respect to the scale $\boldsymbol{\phi}$" and means, for every $k \geq 0$,

$$f - \sum_{r=0}^{k} A_r \phi_r = o(\phi_k) \qquad \text{in} \quad \Omega. \tag{8}$$

Often $\boldsymbol{\phi}$ is understood from context, so "with respect to the scale $\boldsymbol{\phi}$" may be deleted from the definition.

Note that o, O, and $\sim$ are transitive and $\approx$ is symmetric. Clearly no asymptotic scale can contain the zero vector or two identical vectors. If d' is a *metric* induced by a *norm* $\|\cdot\|$, the asymptotic expansion (7) is unique (but not otherwise). This is a simple consequence of the fact that $\phi = o(\eta)$, $\psi = o(\eta)$ then imply $\phi + \psi = o(\eta)$. Thus assume another expansion (7) with coefficients $\mathbf{A}'$ holds. Setting $k = 0$ in (8) and its analog and subtracting the two gives

$$(A_0 - A_0')\phi_0 = o(\phi_0), \tag{9}$$

or $|A_0 - A_0'| < \varepsilon$ for all ε, so that $A_0 = A_0'$, similarly, $A_j = A_j'$, $j > 0$.

1.2. Integer Variables

In discussions of sequences, the relevant variable x in ϕ or ψ takes values in J^0. We write ϕ_n or ψ_n for ϕ or ψ, respectively, or, when there is a possibility of confusion with the index of an asymptotic scale, $\phi(n)$ or $\psi(n)$. 1.1(1) is then written

$$\phi_n = O(\psi_n) \qquad \text{in} \quad J^0 \tag{1}$$

and means that for some $M > 0$, there is an $N > 0$ such that

$$d'(\phi_n, 0)/d'(\psi_n, 0) < M \qquad \text{for} \quad n > N.$$

A similar modification is made of 1.1(3). An additional complexity occurs when ϕ and ψ depend on a p-tuple with elements in J^0, say, $n = (m_1, m_2, \ldots, m_p)$. It is usually important to know exactly how the elements m_j become infinite, and it is hardly ever sufficient to say, for instance, that $m_1 + m_2 + \cdots + m_p > N$. In fact, the concept of a path in n-space becomes important (see Section 1.3).

1.3. Sequences and Transformations in Abstract Spaces

In this book we shall be concerned with two kinds of sequence transformations. The first is the transformation of a given sequence $\mathbf{s} \in \mathscr{A}_S$ into a sequence $\bar{\mathbf{s}} \in \mathscr{A}'_S$ with, generally, a formula given to compute $\bar{s}_n$ in terms of elements of $\mathbf{s}$. (In some situations there is no explicit formula.)

The other case is where the given sequence $\mathbf{s}$ is mapped into a countable set of sequences $\mathbf{s}^{(k)}$, $k \geq 0$, with a formula given (called a *lozenge algorithm*) for filling out the array $\{s_n^{(k)}\}$, $n, k \geq 0$.

The whole point is to compare the convergence of the transformed sequence(s) with that of the original sequence. The most useful concepts are formulated in the definitions that follow.

Definition 1. Let $\mathbf{s}, \mathbf{t} \in \mathscr{M}_C$, a metric space.

(i) $\mathbf{t}$ converges *as* $\mathbf{s}$ means $d(s_n, s) = O(d(t_n, t))$ and $d(t_n, t) = O(d(s_n, s))$.
(ii) $\mathbf{t}$ converges *more rapidly* than $\mathbf{s}$ means $d(t_n, t) = o(d(s_n, s))$.
(iii) The convergence of $\mathbf{s}$ is pth order if, for some $p \in J^+$,

$$d(s_{n+1}, s) = O(d(s_n, s)^p)$$

and (1)

$$d(s_n, s)^p = O(d(s_{n+1}, s)).$$

It is easy to show that p, if it exists, is unique.

Definition 2. Let $T \in \mathscr{T}(\mathscr{A}, \mathscr{M}_S)$ where $\mathscr{A} \subset \mathscr{M}_C$ and $T(\mathbf{s}) = \bar{\mathbf{s}}$.

(i) T is *regular* for $\mathscr{A}$ if $\mathbf{s} \in \mathscr{A} \Rightarrow \bar{\mathbf{s}} \in \mathscr{M}_C$ and $s = \bar{s}$.
(ii) T is *accelerative* for $\mathscr{A}$ (or accelerates $\mathscr{A}$) if T is regular for $\mathscr{A}$ and $\bar{\mathbf{s}}$ converges more rapidly than $\mathbf{s}$, $\mathbf{s} \in \mathscr{A}$.

Definition 3. Let $T \in \mathscr{T}(\mathscr{A}, \mathscr{M}_S)$ where $\mathscr{A} \subset \mathscr{M}_S$. T *sums* $\mathscr{A}$ if $T(\mathbf{s}) \in \mathscr{M}_C$, $\mathbf{s} \in \mathscr{A}$.

$\mathbf{P} = \{(i_m, j_m) | i_m, j_m \in J^0\}$ is called a *path* if $i_0 = j_0 = 0$, $i_{m+1} \geq i_m$, $j_{m+1} \geq j_m$, and either $i_{m+1} = i_m + 1$ or $j_{m+1} = j_m + 1$ (or both). Increasing m assigns a direction along $\mathbf{P}$. Obviously, $i_m + j_m \to \infty$. Paths where j_m is

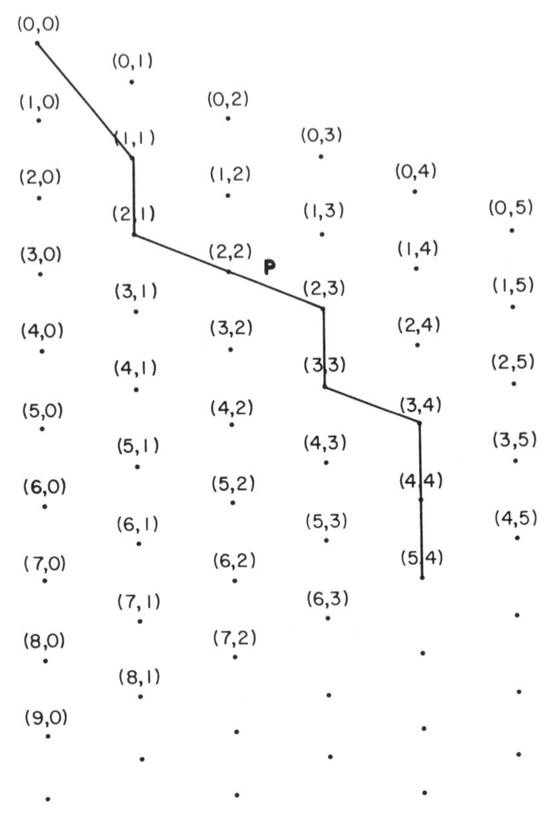

Fig. 1.

ultimately constant are called *vertical paths*, paths with i_m ultimately constant *diagonal paths*. Figure 1 shows how the (n, k) position on $\mathbf{P}$ is labeled for illustrative purposes. Generally the (n, k) position of the diagram itself will be occupied by the $(n + 1)$th component of the $(k + 1)$th member of the set $\mathbf{s}^{(k)}$, i.e., $s_n^{(k)}$. $s_n^{(k)}$ may converge as $n + k \to \infty$ along certain paths but not along others. The following definitions contain the key ideas.

Let $\mathbf{P}$ be a path and $\phi(n, k), \psi(n, k) \in \mathscr{T}(\mathbf{P}, \mathscr{V})$ where $\mathscr{V}$ is equipped with a pseudometric d.

$$\phi = O(\psi) \quad \text{in} \quad \mathbf{P} \tag{2}$$

means for some $M > 0$ there is an $N > 0$ such that $d(\phi, 0)/d(\psi, 0) < M$ for $(n, k) \in \mathbf{P}$, $n + k > N$. A similar interpretation is made of o.

Definition 4. Let $\mathscr{M}$ be a metric space, $\mathbf{T} \in \mathscr{T}_S(\mathscr{A}, \mathscr{M}_S)$ where $\mathscr{A} \subset \mathscr{M}_C$ and let $T_k(\mathbf{s}) = \mathbf{s}^{(k)}$, $k \geq 0$.

(i) **T** is called *regular for* $\mathscr{A}$ *on* **P** if $\mathbf{s} \in \mathscr{A} \Rightarrow d(s_n^{(k)}, s) = o(1)$ in **P**.
(ii) **T** is called *accelerative for* $\mathscr{A}$ *on* **P** if **T** is regular for $\mathscr{A}$ on **P** and if

$$d(s_n^{(k)}, s)/d(s_n, s) = o(1) \qquad \text{in} \quad \mathbf{P}, \quad \mathbf{s} \in \mathscr{A}. \tag{3}$$

If, in the foregoing definitions, $\mathscr{A} \equiv \mathscr{M}_C$, we shall omit the words *for* $\mathscr{A}$ and say simply that $\mathscr{T}$ is *regular*, etc.

We now discuss certain computational aspects of the foregoing definitions. Usually $T_0 = I$, the identity transformation, so $\mathbf{s}^{(0)} = \mathbf{s}$ and an algorithm that is *computationally feasible* for filling out the array $\{s_n^{(k)}\}$ will start with the values $s_n^{(0)} = s_n$ and assign one and only one value to each (n, k) position in the array. There seems to be no easy characterization of those algorithms that are feasible in this sense. However, several important ones have been discovered recently. Among these are formulas of the kind

$$s_n^{(k+1)} = G(s_{n+1}^{(k)}, s_n^{(k)}), \qquad s_n^{(0)} = s_n, \quad n, k \geq 0, \tag{4}$$

called a *deltoid*; and

$$s_n^{(k+1)} = H(s_{n+1}^{(k-1)}, s_{n+1}^{(k)}, s_n^{(k)}), \quad n, k \geq 0, \quad s_n^{(-1)} = 0, \quad s_n^{(0)} = s_n, \quad n \geq 0; \tag{5}$$

$$s_n^{(k+1)} = H'(s_{n+1}^{(k)}, s_n^{(k-1)}, s_n^{(k)}), \quad n, k \geq 0, \quad s_n^{(-1)} = 0, \quad s_n^{(0)} = s_n, \quad n \geq 0, \tag{6}$$

called *rhomboids*.

There is as yet no general theory for constructing such algorithms. Those that are known have been derived using ad hoc arguments from diverse areas of analysis: Lagrangian interpolation, the theory of orthogonal polynomials, and the transformation theory of continued fractions. Much work remains to be done in this area.

For transformations in vector spaces, there are several important concepts that involve the linearity of the underlying space.

Definition 5. Let $T \in \mathscr{T}(\mathscr{A}, \mathscr{V}_S)$ where $\mathscr{A} \subset \mathscr{V}_S$. T is *linear* if, for all $\mathbf{x}, \mathbf{y} \in \mathscr{A}$ and $c_1, c_2 \in \mathscr{C}$, $T(c_1\mathbf{x} + c_2\mathbf{y}) = c_1 T(\mathbf{x}) + c_2 T(\mathbf{y})$; otherwise, T is nonlinear. T is *homogeneous* if $T(c\mathbf{x}) = cT(\mathbf{x})$ for $\mathbf{x} \in \mathscr{A}$, $c \in \mathscr{C}$. T is *translative* if $T(\mathbf{d} + \mathbf{x}) = \mathbf{d} + T(\mathbf{x})$, where $\mathbf{d}$ is a constant sequence ($d_n \equiv d$) whenever $\mathbf{d} + \mathbf{x}, \mathbf{x} \in \mathscr{A}$.

1.4. Properties of Complex Sequences

When the metric space of the previous sections is the complex field, its sequence space possesses elegant properties. Some of these have been long known, and others are surprisingly recent. This section contains a discussion of some of these results.

Definition 1. Let $\mathbf{s} \in \mathscr{C}_C$ and

$$r_{n+1}/r_n = (s_{n+1} - s)/(s_n - s) = \rho + o(1). \tag{1}$$

(i) If $0 < |\rho| < 1$, $\mathbf{s}$ converges *linearly* and we write $\mathbf{s} \in \mathscr{C}_l$.
(ii) If $\rho = 1$, $\mathbf{s}$ converges *logarithmically* and we write $\mathbf{s} \in \mathscr{C}_{l'}$.

Definition 2. $\mathbf{s} \in \mathscr{C}_\sigma$ means $\overline{\lim}_{n\to\infty} |a_n|^{1/n} < 1$.

Theorem 1. Let $|\rho| \neq 0, 1$. Then

$$\lim_{n\to\infty} \frac{s_{n+1} - s}{s_n - s} = \rho \qquad \text{iff} \quad \lim_{n\to\infty} \frac{a_{n+1}}{a_n} = \rho. \tag{2}$$

Remark. For the divergent case $|\rho| > 1$, s can be any number.

Proof. The validity of either limit implies $a_n \neq. 0$. Assume, without loss of generality, $a_n \neq 0$ for any n; otherwise delete the finite number of a_ns that are zero and relabel the members of $\mathbf{a}$ and $\mathbf{s}$.

$\Rightarrow$: We have

$$[a_{n+1} + (s_n - s)]/(s_n - s) \approx \rho \tag{3}$$

or

$$a_{n+1} \approx (\rho - 1)(s_n - s), \qquad a_n = (\rho - 1)(s_{n-1} - s). \tag{4}$$

Dividing the former by the latter shows

$$a_{n+1}/a_n \approx (s_n - s)/(s_{n-1} - s) \approx \rho. \tag{4'}$$

Note that for this part of the theorem ρ *can* be zero.

$\Leftarrow$: We do only the convergent case $0 < |\rho| < 1$. The other is similar. Since $\sum a_n$ converges,

$$s_n - s = -\sum_{k=n+1}^{\infty} a_k. \tag{5}$$

We can write

$$a_{n+1}/a_n = \rho(1 + \varepsilon_n), \qquad \boldsymbol{\varepsilon} \in \mathscr{C}_N. \tag{6}$$

Let

$$g_n = \sup_{j \geq n} |\varepsilon_j|. \tag{7}$$

Then $\mathbf{g} \in \mathscr{C}_N$. Taking products in (6) gives

$$a_n = a_0 \rho^n \prod_{j=0}^{n-1} (1 + \varepsilon_j), \tag{8}$$

empty products interpreted as 1. Define

$$\begin{aligned} F_n &= \left| \sum_{k=n}^{\infty} a_k \right| \Big/ |a_n| = \left| \sum_{k=0}^{\infty} \rho^k \prod_{j=n}^{k+n-1} (1 + \varepsilon_j) \right| \\ &\geq. \left| \sum_{k=0}^{r} \rho^k \prod_{j=n}^{k+n-1} (1 + \varepsilon_j) \right| - \sum_{k=r+1}^{\infty} |\rho|^k (1 + g_n)^k \\ &= \left| \sum_{k=0}^{r} \rho^k \prod_{j=n}^{k+n-1} (1 + \varepsilon_j) \right| - \frac{[|\rho|(1 + g_n)]^{r+1}}{1 - |\rho|(1 + g_n)}. \end{aligned} \tag{9}$$

Thus

$$\begin{aligned} \varliminf_{n \to \infty} F_n &\geq \left| \sum_{k=0}^{r} \rho^k \right| - \frac{|\rho|^{r+1}}{1 - |\rho|} \\ &\geq \frac{1}{|1 - \rho|} - |\rho|^{r+1} \left(\frac{1}{|1 - \rho|} + \frac{1}{1 - |\rho|} \right). \end{aligned} \tag{10}$$

For r sufficiently large, the right-hand side is >0. Thus $\varliminf F_n > 0$ and $1/F_n$ is bounded.

Now

$$\begin{aligned} \frac{s_{n+1} - s}{s_n - s} &=. \sum_{k=n+1}^{\infty} a_{k+1} \Big/ \sum_{k=n+1}^{\infty} a_k = \sum_{k=n+1}^{\infty} a_k \rho (1 + \varepsilon_k) \Big/ \sum_{k=n+1}^{\infty} a_k \\ &=. \rho + u_n, \end{aligned} \tag{11}$$

$$u_n =. \sum_{k=n+1}^{\infty} a_k \varepsilon_k \Big/ \sum_{k=n+1}^{\infty} a_k. \tag{12}$$

(The foregoing operations are valid since it will turn out that $s_n \neq. s$.) Thus

$$\begin{aligned} |u_n| &\leq. \sum_{k=n+1}^{\infty} |a_k \varepsilon_k| / |a_{n+1}| F_{n+1} \\ &\leq. C g_{n+1} \sum_{k=0}^{\infty} |\rho|^k (1 + g_{n+1})^k \\ &= \frac{C g_{n+1}}{1 - \rho(1 + g_{n+1})}, \end{aligned} \tag{13}$$

which actually shows a bit more, namely,

$$\frac{s_{n+1} - s}{s_n - s} = \rho + O\left(\sup_{k>n} \left| \frac{a_{k+1}}{\rho a_k} - 1 \right| \right). \quad \blacksquare \tag{14}$$

Corollary. $\mathscr{C}_\sigma \supset \mathscr{C}_l$.

Proof. This is true since

$$\overline{\lim_{n\to\infty}} |a_n|^{1/n} \leq \lim_{n\to\infty} |a_{n+1}/a_n|. \quad \blacksquare \tag{15}$$

Another useful result has to do with the order of growth of partial products.

Theorem 2. If

$$v_n = \prod_{j=0}^{n-1} (1+\varepsilon_j), \qquad n \geq 1, \quad v_0 = 1 \tag{16}$$

for some $\boldsymbol{\varepsilon} \in \mathscr{C}_N$, $\varepsilon_j \neq -1, j \geq 0$, then there is an $\boldsymbol{\varepsilon}^* \in \mathscr{C}_N$ such that

$$v_n = e^{n\varepsilon_n^*}. \tag{17}$$

Proof. We have

$$v_n = C \prod_{j=n_0}^{n-1} (1+\varepsilon_j), \qquad n > n_0, \tag{18}$$

where $|\varepsilon_j| < 1, j \geq n_0$. Thus

$$v_n = C \exp\left\{n\left[\left(\sum_{j=n_0}^{n-1} \ln(1+\varepsilon_j)\right)\Big/ n\right]\right\} \tag{19}$$

but the quantity in square brackets is the Cesaro means of a null sequence, and hence the nth term of a null sequence, say, δ_n. So

$$v_n = Ce^{n\delta_n} = e^{n(\delta_n + \ln C/n)} = e^{n\varepsilon_n^*}, \qquad n > n_0, \quad \boldsymbol{\varepsilon}^* \in \mathscr{C}_N, \tag{20}$$

and this may be extended to all $n \geq 0$. ($\boldsymbol{\varepsilon}^*$, because of the multiple valuedness of log, is not unique.) $\blacksquare$

1.5. Further Properties of Complex Sequences

Some unusual convergence properties have recently been demonstrated for complex sequences. These properties are a help in determining whether important sequence transformations are regular or accelerative. Sources for this material are Tucker (1967, 1969).

In what follows let **s**, **a** $\in \mathscr{C}_S$ and be related in the usual way. For all $a_n \neq 0$ and $n \geq 0$, define ρ_n by

$$\rho_n = a_{n+1}/a_n, \tag{1}$$

and if **s** is convergent, τ_n by

$$\tau_n = (s - s_n)/a_n. \tag{2}$$

Otherwise ρ_n, τ_n are undefined.

Since we shall in general be concerned only with members of a sequence with large index, the notation "$x_n R.y_n$" (see Notation) will be employed constantly.

Lemma 1. Let $\mathbf{s} \in \mathscr{C}_C$ and $a_n \neq .\, 0$. Let $c \in \mathscr{C}$, and define

$$c_n = c + (s_n - s), \qquad n \geq 0; \tag{3}$$

then

$$1 + c\left(\frac{1 - \rho_n}{a_{n+1}}\right) + \frac{c_n}{a_n} - \frac{c_{n+1}}{a_{n+1}} =. \frac{1 - \rho_n}{a_{n+1}}(s - s_n). \tag{4}$$

Proof

$$1 + c\left(\frac{1 - \rho_n}{a_{n+1}}\right) + \frac{c_n}{a_n} - \frac{c_{n+1}}{a_{n+1}}$$

$$=. 1 + c\left(\frac{1}{a_{n+1}} - \frac{1}{a_n}\right) + \frac{c + s_n - s}{a_n} - \frac{c + s_{n+1} - s}{a_{n+1}}$$

$$=. 1 + \frac{s - s_{n+1}}{a_{n+1}} - \frac{s - s_n}{a_n} =. \frac{s - s_n}{a_{n+1}} - \frac{s - s_n}{a_n}$$

$$=. \left(\frac{1}{a_{n+1}} - \frac{1}{a_n}\right)(s - s_n) =. \left(\frac{1 - \rho_n}{a_{n+1}}\right)(s - s_n). \quad \blacksquare \tag{5}$$

Theorem 1. Let

$$\frac{1 - \rho_n}{a_{n+1}} = \frac{1}{a_{n+1}} - \frac{1}{a_n} = O(1). \tag{6}$$

Then $\mathbf{s}$ diverges.

Proof. Assume $\mathbf{s}$ converges. Since $(1 - \rho_n)/a_{n+1}$ is bounded, there is an $\varepsilon > 0$ such that $|\varepsilon(1 - \rho_n)/a_{n+1}| <. \frac{1}{4}$.

Let c be any complex number satisfying $|c| = \varepsilon$, so that

$$-\text{Re}[c(1 - \rho_n)/a_{n+1}] <. \tfrac{1}{4}. \tag{7}$$

Set $c_n = c + (s_n - s)$. From the previous lemma

$$\text{Re}\left[1 + \frac{c(1 - \rho_n)}{a_{n+1}} + \frac{c_n}{a_n} - \frac{c_{n+1}}{a_{n+1}}\right] =. \text{Re}\left[\frac{1 - \rho_n}{a_{n+1}}(s - s_n)\right], \tag{8}$$

so

$$1 + \text{Re}\left[\frac{c(1 - \rho_n)}{a_{n+1}}\right] + \text{Re}\,\frac{c_n}{a_n} - \text{Re}\,\frac{c_{n+1}}{a_{n+1}} <. \frac{1}{4}. \tag{9}$$

Using (7) and (9) shows

$$\frac{1}{2} + \operatorname{Re}\frac{c_n}{a_n} <. \operatorname{Re}\frac{c_{n+1}}{a_{n+1}} - \operatorname{Re}\left[\frac{c(1-\rho_n)}{a_{n+1}}\right] - \frac{1}{4} <. \operatorname{Re}\frac{c_{n+1}}{a_{n+1}} \tag{10}$$

from which it follows that $\operatorname{Re} c_n/a_n \to \infty$ and so $\operatorname{Re} c_n/a_n >. 0$. Since $c_n \to c$,

$$a_n \notin. \{z \mid \arg c + 3\pi/4 \leq \arg z \leq \arg c + 5\pi/4\}. \tag{11}$$

Choosing $\arg c$ to be successively 0, $\pi/2$, π, $3\pi/2$ shows that **a** cannot be a complex sequence, a contradiction. ■

[This beautiful proof is due to Tucker (1967).]

We state without proof a similar result for infinite products.

Theorem 2. Let

$$1/a_{n+1} - 1/a_n = O(1). \tag{12}$$

Then $\prod_{n=0}^{\infty} (1 + a_n)$ diverges.

Lemma 2. Let ρ_n be defined ultimately and

$$|\rho_n| \leq. \rho < \tfrac{1}{2}. \tag{13}$$

Then

$$0 < (1 - 2\rho)/(1 - \rho) \leq. |\tau_n/\rho_n| \leq. 1/(1 - \rho). \tag{14}$$

Proof. Note that τ_n is, ultimately, defined and that

$$\tau_n =. \rho_n + \rho_n\rho_{n+1} + \rho_n\rho_{n+1}\rho_{n+2} + \cdots, \tag{15}$$

since $\tau_{n+1} + 1 =. \tau_n/\rho_n$ and the above series converges. We have

$$|\tau_n| \leq. |\rho_n| + |\rho_n\rho_{n+1}| + \cdots \leq. |\rho_n|/(1 - \rho) \leq. \rho/(1 - \rho) < 1. \tag{16}$$

Thus $|\tau_n/\rho_n| \leq. 1/(1 - \rho)$ and

$$\begin{aligned} |\tau_n/\rho_n| &=. |1 + \tau_{n+1}| \geq. |1 - |\tau_{n+1}|| =. 1 - |\tau_{n+1}| \\ &\geq. 1 - \rho/(1 - \rho) = (1 - 2\rho)/(1 - \rho) > 0. \quad ■ \end{aligned} \tag{17}$$

Theorem 3. Let **s**, **s*** be two sequences such that $a_n^*/a_n = o(1)$ and $|\rho_n| \leq. \rho < \frac{1}{2}$, $|\rho_n^*| \leq. \rho^* < 1$ for some numbers ρ, ρ^*. Then **s*** converges more rapidly than *s*.

Proof. An implication of the hypothesis is $a_n \neq. 0$, $a_n^* \neq. 0$. The previous lemma shows

$$0 < (1 - 2\rho)/(1 - \rho) \leq. |\tau_n/\rho_n| \tag{18}$$

and

$$|\tau_n^*/\rho_n^*| \leq 1/(1 - \rho^*). \tag{19}$$

One concludes that

$$\left|\frac{s_n^* - s^*}{s_n - s}\right| =. \left|\frac{a_{n+1}^*}{a_{n+1}}\right| \left|\frac{\tau_n^*/\rho_n^*}{\tau_n/\rho_n}\right|$$

$$\leq. \left|\frac{a_{n+1}^*}{a_{n+1}}\right| [(1 - \rho^*)(1 - 2\rho)(1 - \rho)]^{-1} = o(1). \quad \blacksquare \tag{20}$$

Tucker gives an example (1967, p. 358) to show that $\frac{1}{2}$ cannot be replaced by a larger number.

Lemma 3. Let **b**, **s**, $\mathbf{s}^* \in \mathscr{C}_S$ with

$$s_n^* = s_n + b_{n+1}. \tag{21}$$

Then $\mathbf{s}^*$ converges more rapidly than **s** to the same limit if and only if

$$b_{n+1} \approx s - s_n = o(1). \tag{22}$$

Proof. Either hypothesis implies the convergence of **s** and $s - s_n \neq. 0$. In either case, therefore,

$$b_{n+1}/(s - s_n) + (s - s_n^*)/(s - s_n) =. 1, \tag{23}$$

and from this the lemma follows. $\blacksquare$

Theorem 4. Let **t**, $\mathbf{s} \in \mathscr{C}_C$ and

$$t_n = s_n + a_{n+1}\alpha_{n+1}, \qquad u_n = s_n + a_{n+1}\beta_{n+1}, \tag{24}$$

and suppose **t** converges more rapidly than **s** to the same limit. Then **u** converges more rapidly than **s** to the same limit if and only if $\beta_n \approx \alpha_n$.

Proof. From the previous theorem, $a_{n+1}\alpha_{n+1} \approx s - s_n = o(1)$. Also **u** converges more rapidly than **s** to the same limit if and only if $a_{n+1}\beta_{n+1} \approx s - s_n$. Since $s - s_n \neq. 0$, we have $a_{n+1}\alpha_{n+1} \neq. 0$, $a_{n+1}\beta_{n+1} \neq. 0$, and so $a_n, \alpha_n, \beta_n \neq. 0$. By transitivity of $\approx$ we conclude $a_{n+1}\alpha_{n+1} \approx a_{n+1}\beta_{n+1}$ or $\alpha_n \approx \beta_n$. This step is reversible, so the theorem follows. $\blacksquare$

Theorem 5. Let **s** be convergent and

$$s_n^* = s_n + a_{n+1}\alpha_{n+1}. \tag{25}$$

Then the three conditions below are all equivalent:

(i) $\mathbf{s}^*$ converges more rapidly than **s** to the same limit;
(ii) $\alpha_{n+1} \approx \tau_n/\rho_n$;
(iii) $\alpha_n \approx 1 + \tau_n$.

Proof. From Lemma 3, $\mathbf{s}^*$ converges more rapidly than $\mathbf{s}$ iff $a_{n+1}\alpha_{n+1} \approx -r_n \to 0$; this is equivalent to $\alpha_{n+1} \approx -r_n/a_{n+1} = \tau_n/\rho_n$. Moreover, $\alpha_{n+1} \approx \tau_n/\rho_n$ is equivalent to $\alpha_n \approx 1 + \tau_n$ since $\tau_n/\rho_n = 1 + \tau_{n+1}$. ■

1.6. Totally Monotone and Totally Oscillatory Sequences

Definition. $\mathbf{s}$ is totally monotone (written $\mathbf{s} \in \mathscr{R}_{\mathrm{TM}}$) if

$$(-1)^k \Delta^k s_n \geq 0, \qquad n, k \geq 0. \tag{1}$$

$\mathbf{s}$ is totally oscillatory (written $\mathbf{s} \in \mathscr{R}_{\mathrm{TO}}$) if $\{(-1)^n s_n\} \in \mathscr{R}_{\mathrm{TM}}$.

Here

$$\Delta v_n = v_{n+1} - v_n, \qquad n \geq 0,$$

$$\Delta^k v_n = \begin{cases} \Delta^{k-1}(\Delta v_n), & k \geq 1 \\ (-1)^k \sum_{r=0}^{k} \binom{k}{r} (-1)^r v_{n+r}, & n, k \geq 0. \end{cases}$$

If $\mathbf{s} \in \mathscr{R}_{\mathrm{TM}}$, one has

$$0 \leq s_{n+1} \leq s_n \leq s_0, \tag{2}$$

so $\mathbf{s}$ converges since it is monotone decreasing and bounded.

On the other hand, if $\mathbf{s} \in \mathscr{R}_{\mathrm{TO}}$, $\mathbf{s}$ is alternating and so converges to 0.

Examples. The sequences $\mathbf{s}^{(k)} \in \mathscr{R}_{\mathrm{TM}}$, where

$$s_n^{(1)} = 1/(n+2), \qquad s_n^{(2)} = 1/(n+1)^2, \qquad s_n^{(3)} = n!/(\alpha)_{n+1}, \qquad \alpha > 0, \tag{3}$$

since

$$s_n^{(k)} = \int_0^1 t^n \, d\psi_k(t) \tag{4}$$

for ψ_k bounded and nondecreasing

$$\psi_1 = \tfrac{1}{2}t^2, \qquad \psi_2 = (1 - t \ln t), \qquad \psi_3 = -(1-t)^\alpha/\alpha \tag{5}$$

(see Theorem 3).

$\mathscr{R}_{\mathrm{TM}}$ is an important regularity space for certain nonlinear transformations in $\mathscr{T}(\mathscr{R}_S, \mathscr{R}_S)$.

Theorem 1. Let $\mathbf{s} \in \mathscr{R}_{\mathrm{TM}}$. Then the sequences whose nth elements are given below are also $\in \mathscr{R}_{\mathrm{TM}}$. (Empty products are interpreted as 1.)

(i) $(1 - s_n)^{-1} \qquad (s_0 < 1)$;

(ii) $\prod_{j=0}^{n-1}(1-s_j) \qquad (s_0 \le 1)$;

(iii) $\lambda^{(-1)^{k+1}\Delta^k s_n} \qquad (0 < \lambda \le 1, \quad k \ge 0, a, k \text{ fixed})$;

(iv) $\lambda^{\Sigma_{j=0}^{n-1} s_j} \qquad (0 \le \lambda \le 1)$.

Proof. The proofs are straightforward. We prove here only (i); the reader is referred to Wynn's paper (1972) for the others. Write

$$1/(1 - s_n) = t_n. \tag{6}$$

Multiplying both sides by $1 - s_n$ and using the difference formula 1.6(40) gives

$$(-\Delta)^k t_n = (1 - s_n)^{-1} \sum_{j=1}^{k} \binom{k}{j} (-\Delta)^{k-j} t_{n+j} (-\Delta)^j s_n. \tag{7}$$

Now $s_n - s_{n+1} \ge 0$, so $0 \le s_n < 1$ for all n, and thus Eq. (7) provides an immediate induction argument on k. ∎

Theorem 2. Let $\mathbf{s}, \mathbf{t} \in \mathscr{R}_{\mathrm{TM}}$. Then $\{s_n t_n\}$, $\{as_n + bt_n\} \in \mathscr{R}_{\mathrm{TM}}$, $a, b > 0$.

Proof. Obvious. ∎

Theorem 3. $\mathbf{s}$ is totally monotone if and only if there is a function $\psi(t)$ bounded and nondecreasing on [0, 1] that satisfies

$$s_n = \int_0^1 t^n \, d\psi(t), \qquad n \ge 0. \tag{8}$$

Proof

$\Leftarrow$: Write

$$(-1)^k \Delta^k s_n = \int_0^1 (1 - t)^k t^n \, d\psi(t) \ge 0. \tag{9}$$

$\Rightarrow$: For all k and $0 \le n \le k$,

$$(-1)^{k-n} \Delta^{k-n} s_n \ge 0, \tag{10}$$

or

$$\sum_{r=0}^{k-n} \binom{k-n}{r} (-1)^r s_{n+r} = r_n, \qquad 0 \le n \le k, \tag{11}$$

where $\mathbf{r} \in \mathscr{R}_S^0$. It is easily seen that this system of equations has the solution

$$s_n = \sum_{m=n}^{k} \binom{k-n}{m-n} r_m = \sum_{m=n}^{k} \frac{m(m-1)\cdots(m-n+1)}{k(k-1)\cdots(k-n+1)} L_m, \qquad 0 \le n \le k, \tag{12}$$

where

$$L_m = \binom{k}{m} r_m \geq 0. \tag{13}$$

This can be written

$$s_n = \int_0^1 \Phi_{k,n}(t)\, d\psi_k(t), \qquad 0 \leq n \leq k, \tag{14}$$

where

$$\begin{aligned}\Phi_{k,n}(t) &= \frac{t(t-1/k)(t-2/k)\cdots[t-(n-1)/k)}{1(1-1/k)(1-2/k)\cdots[1-(n-1)/k]} \\ &= t^n + O(k^{-1})\end{aligned} \tag{15}$$

uniformly in t, and $\psi_k(t)$ is the step function defined by

$$\psi_k(t) = \begin{cases} 0, & t \leq 0, \\ L_0, & 0 < t \leq 1/k, \\ L_0 + L_1, & 1/k < t \leq 2/k, \\ \vdots & \vdots \\ L_0 + L_1 + \cdots + L_{k-1}, & (k-1)/k < t < 1, \\ L_0 + L_1 + \cdots + L_k, & 1 \leq t. \end{cases} \tag{16}$$

Because of (12) with $n = 0$,

$$s_0 = \psi_k(1) \geq \psi_k(t) \geq \psi_k(0) = 0, \qquad 0 \leq t \leq 1. \tag{17}$$

But any sequence of bounded nondecreasing functions on [0, 1] contains a subsequence converging to a bounded nondecreasing function [see Wall (1948, p. 246)]. It is easy to justify taking the limit over this subsequence inside the integral sign (Wall, 1948, p. 245), so for some bounded nondecreasing $\psi(t)$,

$$s_n = \int_0^1 t^n\, d\psi(t), \qquad n \geq 0. \quad \blacksquare \tag{18}$$

For $\mathbf{s} \in \mathscr{R}_S$, the determinants

$$H_n^{(k)}(\mathbf{s}) = \begin{vmatrix} s_n & s_{n+1} & \cdots & s_{n+k-1} \\ s_{n+1} & s_{n+2} & \cdots & s_{n+k} \\ \vdots & \vdots & & \vdots \\ s_{n+k-1} & s_{n+k} & \cdots & s_{n+2k-2} \end{vmatrix}, \qquad n, k \geq 0, \tag{19}$$

are called *Hankel determinants.*

Theorem 4. If $\mathbf{s} \in \mathscr{R}_{\mathrm{TM}}$, $H_n^{(k)}(\mathbf{s}) \geq 0$. If $\mathbf{s} \in \mathscr{R}_{\mathrm{TO}}$, $(-1)^{nk} H_n^{(k)}(\mathbf{s}) \geq 0$.

Proof. Let

$$Q_n^{(k)} = \sum_{i,j=0}^{k-1} s_{n+i+j} \xi_i \xi_j = \int_0^1 t^n(\xi_0 + \xi_1 t + \cdots + \xi_{k-1} t^{k-1})^2 \, d\psi(t)$$
$$\geq 0. \tag{20}$$

But this means by a known result on quadratic forms (Bellman, 1970, p. 75) that the determinant of the coefficients of $Q_k^{(n)}$ must be nonnegative, which gives the first part of the theorem. The second is similar. ■

Theorem 5 (Brezinski). Let $f(x) = \sum_{k=0}^{\infty} c_k x^k$ be a power series with nonnegative coefficients and radius of convergence $\rho > 0$. Let $\mathbf{s} \in \mathscr{R}_{\mathrm{TM}}$ and $s_0 < \rho$. Then $\{f(s_n)\} \in \mathscr{R}_{\mathrm{TM}}$.

Proof. Obvious, since the sequence $\mathbf{s}^{(k)} \in \mathscr{R}_{TM}$ when

$$s_n^{(k)} = c_0 + c_1 s_n + \cdots + c_k s_n^k$$

by Theorem 2, and any limit of totally monotone sequences is totally monotone. [This also provides a proof of Theorem 1(i).] ■

Theorem 6. If $\mathbf{s} \in \mathscr{R}_{\mathrm{TM}}$, then

$$H_n^{(k)}(\Delta^{2r}\mathbf{s}) \geq 0 \qquad \text{and} \qquad (-1)^k H_n^{(k)}(\Delta^{2r+1}\mathbf{s}) \geq 0.$$

If $\mathbf{s} \in \mathscr{R}_{\mathrm{TO}}$, then

$$(-1)^{kn} H_n^{(k)}(\Delta^{2r}\mathbf{s}) \geq 0 \qquad \text{and} \qquad (-1)^{k(n+1)} H_n^{(k)}(\Delta^{2r+1}\mathbf{s}) \geq 0.$$

Proof. Obvious. ■

1.7. Birkhoff–Poincaré Logarithmic Scales

Let $\rho \in J^+, \mu_1, \mu_2, \ldots, \mu_\rho, \theta$ be complex constants, μ_0 an integral multiple (positive or negative) of $1/\rho$. Define

$$Q(\omega) = \mu_0 \omega \ln \omega + \mu_1 \omega + \mu_2 \omega^{(\rho-1)/\rho} + \cdots + \mu_\rho \omega^{1/\rho}, \qquad \omega \in \mathscr{R}^+. \tag{1}$$

Consider the sequence of functions

$$\psi_{i,j}(\omega) = e^{Q(\omega)} \omega^{\theta - j/\rho} (\ln \omega)^i, \qquad i, j = 0, 1, 2, \ldots, \quad \omega \in \mathscr{R}^+, \tag{2}$$

and let $F = \{\psi_{i,j}\}$. It is easily verified that F is strictly (nonreflexively) well ordered under the operation written "$\phi = o(\psi)$ in $\mathscr{R}^+$." Even so, one may not be able to rearrange the elements of F into a scale. However, if i is bounded, $i = 0, 1, 2, \ldots, p$, then one may define a unique asymptotic scale on F, say, $\{\phi_n\}$, with

$$\phi_n = \psi_{i_n, k_n}. \tag{3}$$

This scale is called the *Birkhoff–Poincaré logarithm scale* (B–P log scale); p is called the index of the scale. If $p = 0$, it is called simply the *Birkhoff–Poincaré scale*. The special case $\mu_j = \theta = 0$, $\rho = 1$ is called the *Poincaré scale*.

Any function satisfying a fairly general difference equation or differential equation is known to possess an asymptotic expansion in a B–P log scale, or, more precisely, the function can be written as a linear combination of such expansions, once it is decided how to interpret sums of asymptotic expansions. [In fact, this can easily be done; see Wimp (1974b).] For difference equations, this is called the Birkhoff–Trjitzinsky theory (1930, 1932), and for differential equations, the theory of subnormal forms. [See Wasow (1965) and the references given there.]

Theorem 1 (Birkhoff–Trjitzinsky). Consider the difference equation

$$A_0(\omega)y(\omega) + A_1(\omega)y(\omega + 1) + \cdots + A_m(\omega)y(\omega + m) = 0, \tag{4}$$

where A_i is defined for $\omega \in \mathscr{R}^0$ and $A_m(\omega) \neq 0$.

Let A_i have an asymptotic expansion with respect to some B–P scale F with $Q = \mu_j = \theta = 0$ in $\mathscr{R}^+$. Then there is a B–P log scale G and a basis of solutions $y_1, y_2, \ldots, y_m$ of the equation such that y_j has an asymptotic expansion g_i with respect to G in $\mathscr{R}^+$.

The general form of these solutions is

$$y(\omega) = e^{Q(\omega)}\omega^\theta[(a_{00} + a_{01}\omega^{-1/\rho} + \cdots) + (a_{10} + a_{11}\omega^{-1/\rho} + \cdots)\ln\omega + \cdots + (a_{m0} + a_{m1}\omega^{-1/\rho} + \cdots)(\ln\omega)^m]. \tag{5}$$

Proof. The proof is the subject of two papers. The first (Birkhoff, 1930) treats the formal (constructive) theory of the question; the second (Birkhoff and Trjitzinsky, 1932) treats the analytic theory. ■

While the theorem is simple to state, in the construction of the asymptotic expansions there are many complexities. For instance, ρ for G and F need not be the same. Further, once certain expansions g_i are obtained others may be found from these by formal manipulations, thus vastly simplifying the work involved.

For example, the difference equation

$$y(\omega) - \frac{(\omega + 1)[(2\omega + b + c + 1) + \lambda]}{(\omega + b)(\omega + c)} y(\omega + 1) + \frac{(\omega + 1)(\omega + 2)}{(\omega + b)(\omega + c)} y(\omega + 2) = 0, \qquad b, c > 0 \tag{6}$$

has solutions

$$h_1(\omega) = \frac{\Gamma(b+\omega)\Gamma(c+\omega)}{\Gamma(\omega+1)}\Psi(b+\omega, b+1-c; \lambda),$$
$$h_2(\omega) = \frac{\Gamma(\omega+b)}{\Gamma(\omega+1)}\Phi(b+\omega, b+1-c; \lambda) \tag{7}$$

(see Wimp, 1974a).

There is a formal basis of solutions

$$g_i(\omega) = \exp[(-1)^{i+1}\lambda^{1/2}\omega^{1/2}]\omega^{\theta}\sum_{k=0}^{\infty} B_k(-1)^{ik}\omega^{-k/2}, \qquad \theta = \tfrac{1}{2}(b+c) - \tfrac{5}{4} \tag{8}$$

and

$$h_1(\omega) \sim \sqrt{\pi}\,\lambda^{(c-b)/2-1/4}e^{\lambda/2}g_1(\omega) \qquad \text{in} \quad \mathcal{R}^+. \tag{9}$$

Here the A_j are rational; $\rho = 1$ for F; for G, $\rho = 2$.

Let us see how the construction proceeds for the important case $m = 1$. For the formal computations, assume all series are convergent for $\omega > R$, say. Then Eq. (4) can be written

$$y(\omega+1) = \omega^{\mu/\rho}(a_0 e^{\mu/\rho} + a_1\omega^{-1/\rho} + \cdots)y(\omega), \qquad \mu \in J, \quad a_0 \neq 0. \tag{10}$$

Let

$$y(\omega) = \omega^{\mu\omega/\rho}a_0^{\omega}z(\omega). \tag{11}$$

Then

$$z(\omega+1) = b(\omega)z(\omega), \qquad b(\omega) = 1 + b_1\omega^{-1/\rho} + \cdots. \tag{12}$$

Finally, let $u(\omega) = \ln z(\omega)$. Then u satisfies

$$u(\omega+1) - u(\omega) = \ln b(\omega) = b_1\omega^{-1/\rho} + (2b_2 - b_1^2)\frac{\omega^{-2/\rho}}{2} + \cdots$$
$$= c_1\omega^{-1/\rho} + c_2\omega^{-2/\rho} + \cdots. \tag{13}$$

In this equation write

$$u(\omega) = d_{1-\rho}\omega^{1-1/\rho} + d_{2-\rho}\omega^{1-2/\rho} + \cdots + d_{-1}\omega^{1/\rho}$$
$$+ d_c \ln\omega + d_1\omega^{-1/\rho} + d_2\omega^{-2/\rho} + \cdots, \tag{14}$$

and it is immediately found that $d_{1-\rho}, d_{2-\rho}, \ldots, d_0$ are uniquely determined, with

$$d_{1-\rho} = \frac{c_1}{1-\rho^{-1}}, \qquad d_{2-\rho} = \frac{c_2}{1-2\rho^{-1}}, \qquad \ldots,$$
$$d_{-1} = \frac{c_{\rho-1}}{1-(\rho-1)\rho^{-1}}, \qquad d_0 = c_\rho, \tag{15}$$

by comparison of terms $\omega^{-1/\rho}, \omega^{-2/\rho}, \ldots, \omega^{-1}$. On comparison of terms $\omega^{-k/\rho}$, $(k > \rho)$, one gets equations of the form

$$[(-k + \rho)/\rho]\, d_{k-\rho} + \alpha_k = c_k, \qquad k \geq \rho + 1, \tag{16}$$

in which α_k is a known polynomial in $d_{1-\rho}, d_{2-\rho}, \ldots, d_{k-1-\rho}$. Thus all the d_i are determined in succession and uniquely. Then, writing

$$y(\omega) = \omega^{\mu\omega/\rho} a_0^{\omega} e^{u(\omega)} \tag{17}$$

and exponentiating the series for $u(\omega)$ gives the desired formal series. This construction, of course, shows that a unique formal asymptotic series always exists, and also that, for the first-order case, ρ for $F = \rho$ for G and the index of G is zero.

The next result shows how the partial sums of a series grow when the general term of the series has an asymptotic expansion of the form $e^{Q}s$.

Theorem 2. Let

$$s_n = \sum_{k=0}^{n} a_k \tag{18}$$

with $\mathbf{s} \in \mathscr{C}_C$. Let

$$a_n \sim e^{Q(\omega)} \omega^{\theta} p(\omega), \qquad \omega = n + \zeta \quad \text{in} \quad J^+ \tag{19}$$

where ζ is arbitrary and complex and where

$$p(\omega) = \alpha_0 + \alpha_1 \omega^{-1/\rho} + \alpha_2 \omega^{-2/\rho} + \cdots. \tag{20}$$

Then

$$s_n - s \sim e^{Q(\omega)} \omega^{\theta^*} p^*(\omega) \qquad \text{in} \quad J^+ \tag{21}$$

where θ^*, α_0^* are as follows:

Case I. $Q \not\equiv 0$. Denote the first nonzero μ_j in the sequence

$$\{\mu_0, \mu_1, \ldots, \mu_\rho\}$$

by μ_τ. Then

$$\theta^* = \begin{cases} \theta + \mu_0, & \tau = 0 \\ \theta + (\tau - 1)/\rho, & 1 \leq \tau \leq \rho; \end{cases} \tag{22}$$

$$\alpha_0^* = \begin{cases} -\alpha_0 e^{\mu_0 + \mu_1}, & \tau = 0 \\ \alpha_0/(1 - e^{-\mu_1}), & \tau = 1; \\ \alpha_0 \rho/\mu_\tau(\rho + 1 - \tau), & 2 \leq \tau \leq \rho. \end{cases} \tag{23}$$

Case II. $Q \equiv 0$.

$$\theta^* = \theta + 1, \qquad \alpha_0^* = \alpha_0/(\theta + 1). \tag{24}$$

Proof. A straightforward application of Theorem 1; see Wimp (1974b) (whose $\tau = 0$ values are incorrect). ■

Also of interest is a related result for the partial sums of divergent series.

Theorem 3. Let $\mathbf{s} \notin \mathscr{C}_C$ and let a_n, p, ω be as in (19), for some constant c. Then

$$s_n - c \sim e^{Q(\omega)}\omega^{\theta^*}p^*(\omega) \qquad \text{in} \quad J^+ \tag{25}$$

where θ^*, α_0^* are as in (22) and (23) except in the following cases:

Case I. $\mu_0 \neq 0$. Then $\alpha_0^* = \alpha_0$, $\theta^* = \theta$.

Case II. $Q \equiv 0$ and $p(\omega)$ contains a term ω^{-1}. Then for some c, d,

$$s_n - c + d \ln \omega \sim \omega^{\theta+1}p^*(\omega) \qquad \text{in} \quad J^+. \tag{26}$$

Corollary. Let

$$a_n \sim \lambda^n n^\theta\left(\alpha_0 + \frac{\alpha_1}{n} + \frac{\alpha_2}{n^2} + \cdots\right), \qquad \alpha_0 \neq 0. \tag{27}$$

Then for some constants β_j, γ_j, δ_j,

$$s_n - s \sim \begin{cases} \dfrac{\lambda^{n+1}n^\theta}{(\lambda-1)}\left(\alpha_0 + \dfrac{\beta_1}{n} + \dfrac{\beta_2}{n^2} + \cdots\right), & |\lambda| < 1 \\[2ex] \dfrac{-n^{\theta+1}}{(\theta+1)}\left(\alpha_0 + \dfrac{\gamma_1}{n} + \dfrac{\gamma_2}{n^2} + \cdots\right), & \lambda = 1, \quad \operatorname{Re}\theta < -1, \end{cases} \tag{28}$$

and

$$s_n \sim \alpha_0 \ln n + \delta_1/n + \delta_2/n^2 + \cdots, \qquad \lambda = 1, \quad \theta = -1. \tag{29}$$

As examples of the use of these formulas, consider the computation of e^x and $\zeta(s)$ from their defining series. Let

$$\begin{aligned} e^x &= s_n - r_n, \quad s_n = \sum_{k=0}^{n} \frac{x^k}{k!} \\ \zeta(\sigma) &= s'_n - r'_n, \quad s'_n = \sum_{k=1}^{n} \frac{1}{k^\sigma}, \qquad \operatorname{Re}\sigma > 1. \end{aligned} \tag{30}$$

For s_n,

$$\rho = 1, \qquad \mu_0 = -1, \qquad \mu_1 = \ln ex, \qquad \theta = -\tfrac{1}{2}, \qquad \alpha_0 = 1/\sqrt{2\pi}, \tag{31}$$

and for s'_n,

$$\rho = 1, \qquad \mu_0 = \mu_1 = 0, \qquad \theta = -\sigma, \qquad \alpha_0 = 1, \qquad \alpha_j = 0, \qquad j > 0, \tag{32}$$

and ζ for both may be taken to be 0. We have

$$e^x - s_n = e^x - \sum_{k=0}^{n} \frac{x^k}{k!} \sim \frac{(ex)^n}{n^{n+3/2}} \left(\frac{x}{\sqrt{2\pi}} + \frac{\alpha_1}{n} + \cdots \right), \tag{33}$$

and the series on the right can be rewritten

$$\frac{x^{n+1}}{(n+1)!} \left(1 + \frac{\beta_1}{n} + \frac{\beta_2}{n^2} + \cdots \right). \tag{34}$$

Higher coefficients are easily determined by formal series manipulations [see Smith (1978)]:

$$\beta_1 = 1, \qquad \beta_2 = x, \qquad \beta_3 = x^2 - 2x, \qquad \ldots. \tag{35}$$

For the Riemann zeta function,

$$\zeta(\sigma) - \sum_{k=1}^{n} \frac{1}{k^\sigma} \sim n^{1-\sigma} \left(\alpha_0 + \frac{\alpha_1}{n} + \frac{\alpha_2}{n^2} + \cdots \right), \tag{36}$$

where

$$\alpha_0 = 1/(\sigma - 1), \qquad \alpha_1 = -\tfrac{1}{2}, \tag{37}$$

and

$$\begin{aligned} \alpha_{2k+1} &= 0, & k \geq 1, \\ \alpha_{2k} &= \frac{(\sigma)_{2k-1}}{(2k)!} B_{2k}, & k \geq 1, \end{aligned} \tag{38}$$

the higher coefficients being obtained by the formula in Wimp (1974b, 2.42) or from the integral representation for $\zeta(\sigma)$. Equation (36) is known to hold for all complex $\sigma \neq 1$ [see Olver (1974, p. 292)].

In what follows let

$$y(n) = \lambda^n n^\theta (\alpha_0 + \alpha_1/n + \cdots), \qquad \alpha_0 \neq 0 \tag{39}$$

be a formal asymptotic series. In future sections we shall need to know the effect of certain difference operators on this series.

Lemma 1

$$\Delta^j u(n) v(n) = \sum_{r=0}^{j} \binom{j}{r} \Delta^r u(n) \Delta^{j-r} v(n+r). \tag{40}$$

Proof. See Milne-Thomson (1960, p. 35). ■

Lemma 2

$$\Delta^p y(n) = \begin{cases} \lambda^n(\lambda-1)^p n^\theta(\alpha_0+\beta_1/n+\cdots), & \lambda\neq 1, \\ (-\theta)_p(-1)^p n^{\theta-p}(\alpha_0+\gamma_1/n+\cdots), & \lambda=1, \quad \theta\neq 0, 1, 2, \ldots, p-1, \end{cases} \tag{41}$$

for some β_i, γ_i.

Proof. Left to the reader. ∎

Theorem 4

$$\sum_{r=0}^{j} d_{j,r}(\lambda)\Delta^{p+r}y(n) = \lambda^n(\lambda-1)^p(-1)^j(-\theta)_j n^{\theta-j}(\alpha_0+\alpha_1'/n+\cdots),$$

$$\lambda \neq 1, \quad \theta \neq 0, 1, 2, \ldots, j-1, \tag{42}$$

where α_1', α_2', ..., depend on j and p and

$$d_{j,r}(\lambda) = \binom{j}{r}\lambda^{-j}(1-\lambda)^{j-r}. \tag{43}$$

Proof. The first part of the previous lemma gives

$$\lambda^{-n}\Delta^p y(n) = (\lambda-1)^p n^\theta(\alpha_0+\beta_1/n+\cdots). \tag{44}$$

Then letting $v(n) = \lambda^{-n}$ and $u(n) = \Delta^p y(n)$ and using the two previous lemmas gives the theorem. ∎

The paper by Wimp (1974b) includes a number of applications of the previous results, particularly to the problem of finding asymptotic formulas for the remainder terms in expansions in orthogonal polynomials.

Brezinski has shown that something like Theorems 2 and 3 is true for series of arbitrary real terms provided the terms are ultimately positive and that their differences are ultimately of one sign.

First, note that $u_n \approx v_n$ is equivalent to the statement that $u_n \neq. 0$, $v_n \neq. 0$, and

$$u_n/v_n = 1 + o(1). \tag{45}$$

Lemma 3. Let $a_n >. 0$, **s** diverge to $+\infty$, and $b_n = o(1)$. Then

$$\sum_{k=0}^{n} a_k b_k = o(s_n). \tag{46}$$

Proof

$$\sum_{k=0}^{n} \mu_{nk} b_k = o(1), \tag{47}$$

where

$$\mu_{nk} = a_k/s_n, \tag{48}$$

by the Toeplitz limit theorem, Theorem 2.1(3). Note **b** may be a complex sequence. ■

Theorem 5. Let $\mathbf{s} \in \mathscr{R}_S$, $a_n >. 0$ and $h_n = a_n/\Delta a_n$ with $\Delta h_n = o(1)$.

Case I. $\Delta a_n <. 0$. Then **s** converges and

$$r_n = s_n - s \approx a_{n+1}^2/\Delta a_{n+1}. \tag{49}$$

Case II. $\Delta a_n >. 0$. Then **s** diverges and

$$s_n \approx a_n a_{n+1}/\Delta a_n. \tag{50}$$

Proof

Case I.

$$h_n - h_0 = \sum_{k=0}^{n-1} \Delta h_k = o\left(\sum_{k=0}^{n-1} 1\right) = o(n), \tag{51}$$

by the lemma (with $\mathbf{s} = \{n\}$). This means

$$[n(a_{n+1}/a_n - 1)]^{-1} = o(1), \tag{52}$$

or, since $a_{n+1}/a_n <. 1$, the sequence $\{n(a_{n+1}/a_n - 1)\}$ is definitely divergent to $-\infty$. By Raabe's test **s** is convergent. Thus

$$\sum_{k=n}^{\infty} a_k = \sum_{k=n}^{\infty} h_k \Delta a_k = -h_n a_n - \sum_{k=n+1}^{\infty} a_k \Delta h_{k-1}. \tag{53}$$

Now

$$\left| \sum_{k=n+1}^{\infty} a_k \Delta h_{k-1} \right| \le \sup_{k \ge n} |\Delta h_k|(s - s_n) \le. \sup_{k \ge n} |\Delta h_k|(s - s_{n-1}). \tag{54}$$

Thus

$$\sum_{k=n+1}^{\infty} a_k \Delta h_{k-1} = (s - s_{n-1})\xi_n, \tag{55}$$

where $\boldsymbol{\xi} \in \mathscr{R}_N$, so (53) may be written

$$s - s_{n-1} = -h_n a_n - (s - s_{n-1})\xi_n, \tag{56}$$

or, since $s - s_n \ne. 0$,

$$-h_n a_n/(s - s_{n-1}) =. 1 + \xi_n \tag{57}$$

and letting $n \to \infty$ gives the result.

Case II. Since $a_{n+1} >. a_n$, $s_n \to +\infty$. Also,

$$s_n = \sum_{k=0}^{n} h_k \Delta a_k = h_n a_{n+1} - h_0 a_0 - \sum_{k=1}^{n} a_k \Delta h_{k-1}, \tag{58}$$

or

$$a_0 + \sum_{k=1}^{n} a_k(1 + \Delta h_{k-1}) = h_n a_{n+1} - h_0 a_0. \tag{59}$$

But, by the lemma,

$$\sum_{k=1}^{n} a_k(1 + \Delta h_{k-1}) = s_n(1 + \xi_n) \tag{60}$$

where $\xi \in \mathscr{R}_N$. Since $s_n \neq. 0$,

$$h_n a_{n+1}/s_n =. 1 + a_0(1 + h_0)/s_n + \xi_n, \tag{61}$$

and letting $n \to \infty$ gives the result. ■

Sometimes the variable appearing in a Poincaré series is $n + \beta$ rather than n. This is immaterial, however, as the following result indicates.

Theorem 6. Let

$$v_n \sim \sum_{r=0}^{\infty} c_r(n + \alpha)^{\theta - r}, \qquad \theta, \alpha \in \mathscr{C}. \tag{62}$$

Then, for any $\beta \in \mathscr{C}$,

$$v_n \sim \sum_{r=0}^{\infty} c'_r(n + \beta)^{\theta - r}, \qquad c'_r = \sum_{m=0}^{r} \frac{c_m(\beta - \alpha)^{r-m}(m - \theta)_{r-m}}{(r - m)!}. \tag{63}$$

Proof.

$$\begin{aligned}
\sum_{r=0}^{k} c_r(n + \alpha)^{\theta - r} &=. \sum_{r=0}^{k} c_r(n + \beta)^{\theta - r}\left(1 + \frac{\alpha - \beta}{n + \beta}\right)^{\theta - r} \\
&=. \sum_{r=0}^{k} c_r(n + \beta)^{\theta} \sum_{s=r}^{\infty} \frac{(\beta - \alpha)^{s-r}(r - \theta)_{s-r}}{(s - r)!(n + \beta)^{s}} \\
&=. \sum_{r=0}^{k} c_r(n + \beta)^{\theta} \sum_{s=r}^{k} \frac{(\beta - \alpha)^{s-r}(r - \theta)_{s-r}}{(s - r)!(n + \beta)^{r}} + O(n^{\theta - k - 1}) \\
&=. \sum_{s=0}^{k} c'_s(n + \beta)^{\theta - s} + O(n^{\theta - k - 1}),
\end{aligned} \tag{64}$$

and from this the theorem follows immediately. ■

Chapter 2 Linear Transformations

2.1. Toeplitz's Theorem in a Banach Space

The most famous result dealing with the regularity of linear transformations is the Toeplitz limit theorem. In its classical guise, this concerns the convergence of transformations of $\mathscr{C}_S$ where the $(n+1)$th member of the transformed sequence is a weighted mean of the first $n+1$ members of the original sequence:

$$\bar{s}_n = \sum_{k=0}^{n} \mu_{nk} s_k. \tag{1}$$

The theory of this transformation is covered quite adequately in the existing literature (Knopp, 1947, Hardy, 1956; Petersen, 1966, Peyerimhoff, 1969).

For what follows, we shall need an abstract version of the theorem. This, in a way, is fortunate, since the proof is cleaner than the proof of $\mathscr{C}_S$, which is rather computational. However, we shall have to begin with two lemmas concerning linear operators in Banach spaces.

Let $\mathscr{B}^j$, $j = 1, 2, 3$, be Banach spaces and μ_{nk}, ζ_{nk} be sequences of linear operators,

$$\mu_{nk} \in B(\mathscr{B}^1, \mathscr{B}^2), \qquad \zeta_{nk} \in B(\mathscr{B}^3, \mathscr{B}^2), \quad n, k \geq 0. \tag{2}$$

Lemma 1 (Uniform Boundedness Principle). If

$$\lim_{n\to\infty} \lim_{k\to\infty} \zeta_{nk}(y) \tag{3}$$

exists for every $y \in \mathscr{B}^3$, then the linear operators

$$\zeta_n(y) = \lim_{k\to\infty} \zeta_{nk}(y) \tag{4}$$

satisfy

$$\|\zeta_n\| \le M, \qquad n \ge 0. \tag{5}$$

Lemma 2. If $\|\zeta_{nk}\| \le M$, n, $k \ge 0$, and

$$\lim_{n\to\infty} \lim_{k\to\infty} \zeta_{nk}(y) = \zeta(y) \tag{6}$$

exists for all y in some dense subset $W \subset \mathscr{B}^3$, then $\zeta(y)$ exists for all $y \in \mathscr{B}^3$ and $\zeta \in B(\mathscr{B}^3, \mathscr{B}^2)$, with $\|\zeta\| \le M$.

We omit the proofs of these lemmas, which are straightforward double applications of standard functional analysis results [see Banach (1932, Theorems 3 and 5); Zeller (1952)]. (All limits above, of course, are in the norm topology.)

Now let $\mathbf{s}$ be a convergent sequence in $\mathscr{B}^1$, i.e., $\mathbf{s} \in \mathscr{B}^1_C$ and let $\bar{\mathbf{s}} \in \mathscr{B}^2_S$ with

$$\bar{s}_n = \sum_{k=0}^{\infty} \mu_{nk}(s_k) \tag{7}$$

subject to certain convergence considerations, which we shall examine presently.

$T(\mathbf{s}) = \bar{\mathbf{s}}$ is called a *generalized Toeplitz transformation.*

Theorem 1 (Toeplitz Limit Theorem). The sum (7) converges, $n \ge 0$, and $\bar{\mathbf{s}} \in \mathscr{B}^2_C$ iff

(i) $\|\sum_{j=0}^{k} \mu_{nj}(s_j)\| \le M$ for $\|s_j\| \le 1$, $j \ge 0$, and $n, k \ge 0$;
(ii) $\sum_{k=0}^{\infty} \mu_{nk}(y)$, $n \ge 0$, and $\lim_{n\to\infty} \sum_{k=0}^{\infty} \mu_{nk}(y)$ exist for $y \in \mathscr{B}^1$;
(iii) $\lim_{n\to\infty} \mu_{nk}(y)$ exists for $y \in \mathscr{B}^1$, $k \ge 0$.

If (i)–(iii) are satisfied,

$$\bar{s} = \lim_{n\to\infty} \sum_{k=0}^{\infty} \mu_{nk}(s) + \sum_{k=0}^{\infty} \lim_{n\to\infty} \mu_{nk}(s_k - s). \tag{8}$$

Proof. To apply the lemmas, put $\mathscr{B}^3 = \mathscr{B}^1_C$ with the usual norm

$$\|\mathbf{x}\| = \sup_n \|x_n\|, \tag{9}$$

and

$$\zeta_{nk}(\mathbf{s}) = \sum_{j=0}^{k} \mu_{nj}(s_j). \tag{10}$$

For W, the dense subset of $\mathscr{B}^3$, we pick all finite linear combinations of sequences of the form

$$\{c, c, c, \ldots\} \qquad \text{and} \qquad \{0, 0, 0, \ldots, 0, b, 0, \ldots\}. \tag{11}$$

This works since for any $\mathbf{x} \in \mathscr{B}_C^1$,

$$\{x, x, x, \ldots\} + \{x_0 - x, x_1 - x, x_2 - x, \ldots, x_k - x, 0, 0, \ldots\} \to \mathbf{x}, \quad k \to \infty. \tag{12}$$

$\Rightarrow$: The necessity of (ii) and (iii) is immediate. For (i) note that each sequence of the form

$$(s_0, s_1, s_2, \ldots, s_k, 0, 0, \ldots) \qquad \text{with} \quad \|s_j\| \leq 1 \tag{13}$$

belongs to $\mathscr{B}_C^1$ and has norm $\|\mathbf{s}\| \leq 1$.

$\Leftarrow$: Consider $\mathbf{c} \in W$, $c_j = c$. Statement (i) shows that condition 1 of Lemma 2 is satisfied and (ii) guarantees the second. For $(0, 0, \ldots, 0, b, 0, \ldots) \in W$, condition 1 is clearly satisfied and (iii) guarantees the second. Thus $\bar{\mathbf{s}}$ converges and $T(\mathbf{s}) = \bar{\mathbf{s}}$ defines a continuous linear transformation from $\mathscr{B}_C^1$ to $\mathscr{B}_C^2$. (8) is obvious. Equation (7) shows

$$\|T\| = \sup_{\|\mathbf{s}\| = 1} \sup_n \left\| \sum_{k=0}^{\infty} \mu_{nk}(s_k) \right\|. \quad \blacksquare \tag{14}$$

Theorem 2 (Toeplitz Regularity). $T(\mathbf{s}) = \bar{\mathbf{s}}$ is regular for $\mathscr{B}_C$ iff

(i) $\|\sum_{j=0}^{k} \mu_{nj}(s_j)\| \leq M$ for $\|s_j\| \leq 1, j \geq 0, n, k \geq 0$;
(ii) $\sum_{k=0}^{\infty} \mu_{nk}(y) = y + o(1), n \geq 0, y \in \mathscr{B}$;
(iii) $\mu_{nk}(y) = o(1)$ in n, $k \geq 0$, $y \in \mathscr{B}$.

Proof

$\Leftarrow$: Obvious.
$\Rightarrow$: We determine the limits $\bar{u}, \bar{v}$ of the two sequences

$$\mathbf{u} = (u, u, u, u, \ldots), \qquad \mathbf{v} = (0, 0, \ldots, 0, v, 0, \ldots) \tag{15}$$

where v in the second is in the $(K + 1)$th position. Applying Eq. (8) yields

$$\bar{u} = \lim_{n \to \infty} \sum_{k=0}^{\infty} \mu_{nk}(u), \tag{16}$$

$$\bar{v} = \lim_{n \to \infty} \mu_{nK}(v). \tag{17}$$

Requiring $\bar{u} = u$ and $\bar{v} = 0$ finishes the proof. $\blacksquare$

Any transformation (7) defined by a matrix $\{\mu_{nk}\}$ of linear operators is called a *generalized Toeplitz summation method.*

2.2. Complex Toeplitz Methods

The only possible linear transformation is

$$\mu_{nk}(z) = \mu_{nk} z, \qquad \mu_{nk} \in \mathscr{C}. \tag{1}$$

In all practical situations $\mu_{nk} = 0$, $k > n$, and so the matrix

$$U = [\mu_{nk}] = \begin{bmatrix} \mu_{00} & 0 & 0 & \cdots \\ \mu_{10} & \mu_{11} & 0 & \cdots \\ \mu_{20} & \mu_{21} & \mu_{22} & \cdots \\ \vdots & \vdots & \vdots & \ddots \end{bmatrix} \tag{2}$$

is lower triangular. If rows sum to 1,

$$\sum_{k=0}^{n} \mu_{nk} = 1, \tag{3}$$

then U, or the transformation defined by U, $T(\mathbf{s}) = \bar{\mathbf{s}}$,

$$\bar{s}_n = \sum_{k=0}^{n} \mu_{nk} s_k, \qquad n = 0, 1, 2, \ldots \tag{4}$$

is called a *triangle*.

We now restate the Toeplitz limit theorem in a form suitable for U in Eq. (2).

Theorem 1. U is regular iff

(i) $\sum_{k=0}^{n} |\mu_{nk}| \leq M$;
(ii) $\sum_{k=0}^{n} \mu_{nk} = 1 + o(1)$;
(iii) $\mu_{nk} = o(1)$, $\quad k$ fixed.

Proof. (ii) and (iii) are obvious. Condition (i) of Theorem 2.1(2) produces $|\sum_{j=0}^{k} \mu_{nj} s_j| \leq M$ for $|s_j| \leq 1$, but for n fixed, there is an $\mathbf{s}$ for which this maximum is attained, namely, $s_j = \operatorname{sgn} \mu_{nj}$. The smallest M that will do is, in fact, the norm of T, and

$$\|T\| = \sup_n M_n, \qquad M_n = \sum_{k=0}^{n} |\mu_{nk}|. \quad \blacksquare \tag{5}$$

Of course, if U is a triangle, condition (ii) can be deleted.

A method U satisfying (i)–(iii) is called a *Toeplitz method*.

If $\mu_{nk} \geq 0$, U is called *positive*. Complex Toeplitz methods are very useful and, when applied to the right sequences, can greatly enhance convergence. Because of their numerical stability, positive methods are the most frequently used ones.

Real positive Toeplitz triangles (even triangles that are "nearly" positive) have an important limit-preserving property; i.e., *negative elements appear in only a finite number of columns of U iff*

$$\varliminf_{n\to\infty} \bar{s}_n \geq \varliminf_{n\to\infty} s_n, \qquad \varlimsup_{n\to\infty} \bar{s}_n \leq \varlimsup_{n\to\infty} s_n \tag{6}$$

for all real bounded sequences **s** [see Cooke (1955, p. 160)].

One cannot expect too much from any linear summability method. The improvement in convergence is, in general, no greater than exponential; in other words, $(\bar{s}_n - s)/(s_n - s) = O(\gamma^n)$, $0 < \gamma < 1$, and one cannot find a method, at least a positive triangle, that is accelerative for all convergent sequences. To see this, let U be such a method and **s** a monotone decreasing null sequence.

Then

$$\frac{\bar{s}_n - s}{s_n - s} = \sum_{k=0}^{n} \frac{\mu_{nk} s_k}{s_n} > \sum_{k=0}^{n} \frac{\mu_{nk} s_n}{s_n} = 1. \tag{7}$$

Pennacchi (1968) has shown that no method of the form

$$\bar{s}_n = \sum_{j=0}^{p} \mu_j s_{n-p+j}, \tag{8}$$

where the μ_j are independent of n, can be accelerative for all sequences. (The foregoing is a band Toeplitz process with constant diagonals.) A minor modification of his proof permits the generalization that *no band Toeplitz process can be accelerative for all* $\mathscr{C}_C$. Whether *any* Toeplitz method can be accelerative for all $\mathscr{C}_C$ is an important open question.

There are many triangles that sum divergent bounded sequences, but it is a consequence of the Banach–Steinhaus theorem that no regular triangle can sum all bounded sequences (Schur, 1921).

The polynomial

$$P_n(\lambda) = \sum_{k=0}^{n} \mu_{nk} \lambda^k = \prod_{k=1}^{n} \left(\frac{\lambda - \lambda_{nk}}{1 - \lambda_{nk}}\right), \qquad n \geq 0, \tag{9}$$

is called the nth *characteristic polynomial* of the triangle U. The regularity and accelerative properties of U are intimately connected with the location of the complex zeros λ_{nk} of $P_n(\lambda)$.

A useful function, called the *measure of U*, is

$$\sigma(\lambda) = \varlimsup_{n\to\infty} |P_n(\lambda)|^{1/n}. \tag{10}$$

For all $\lambda \in N$,

$$\sigma(\lambda) \leq \varlimsup_{n\to\infty} |M_n|^{1/n} = \kappa(U), \tag{11}$$

and κ is called the *modulus of numerical stability* of U. When U is regular, $\kappa = 1$.

Let $\mathscr{C}_{E^m}(\Gamma) \subset \mathscr{C}_S$ denote the space of all exponential sequences of the form

$$s_n = s + c_1\gamma_1^n + c_2\gamma_2^n + \cdots + c_m\gamma_m^n, \tag{12}$$

where $c_j \neq 0$ is complex and $\gamma_j \in \Gamma$, a nonempty compact subset of the complex plane not containing 0. We assume the γ_j are distinct.

As the following theorem shows, the properties of the measure of U determine whether or not U is regular and accelerative for this important class of sequences.

Theorem 2. Let U be a triangle with measure $\sigma(\lambda)$, $\mathscr{A} = \mathscr{C}_{E^m}(\Gamma)$.

(i) Let $\sigma(\lambda) \neq 1$, $\lambda \in \Gamma$. Then U sums $\mathscr{A}$ with $\bar{s} = s$ iff $\sigma(\lambda) < 1$, $\lambda \in \Gamma$.
(ii) Let $\sigma(\lambda) \neq \lambda$, $\lambda \in \Gamma \subset N$. Then U is accelerative for $\mathscr{A}$ iff $\sigma(\lambda) < |\lambda|$, $\lambda \in \Gamma$.

Proof. The basic inequality from which these statements follow is

$$(\sigma(\lambda) - \varepsilon)^n <: |P_n(\lambda)| <. [\sigma(\lambda) + \varepsilon]^n, \qquad \lambda \in \Gamma, \tag{13}$$

for any $\varepsilon > 0$. To show sufficiency in (ii), for instance, choose an ε so that the above holds for $\lambda = \gamma_1, \gamma_2, \ldots, \gamma_m$, and let $\gamma^* = \sup|\gamma_r|$. Next pick $1 > \delta > 0$ so that

$$\sigma(\gamma_r) \leq |\gamma_r|(1 - \delta). \tag{14}$$

One has

$$\left|\frac{\bar{s}_n - s}{s_n - s}\right| \leq . C\frac{|\bar{s}_n - s|}{\gamma^{*n}} <. C\sum_{r=1}^{m} \frac{|c_r|[|\gamma_r|(1 - \delta) + \varepsilon]^n}{\gamma^{*n}}$$
$$<. C'(1 - \delta + \varepsilon/\gamma^*)^n \qquad \text{for every} \quad \varepsilon > 0. \tag{15}$$

But this must hold for every $\varepsilon > 0$, δ, γ^* being fixed. Thus

$$|(\bar{s}_n - s)/(s_n - s)| = O(\tau^n) \qquad \text{for some} \quad \tau, \quad 0 < \tau < 1, \tag{16}$$

which shows not only that U accelerates convergence of each sequence in $\mathscr{A}$ but does so exponentially.

Remaining details of the proof are left to the reader. ■

If something is known about the zeros of $P_n(\lambda)$, one can say something about the kinds of exponential sequences U sums. In what follows, define the distance functions d, D for sets $\subset \mathscr{C}$ by

$$d(A, B) = \inf_{\substack{z \in A \\ w \in B}} |z - w|, \qquad D(A, B) = \sup_{\substack{z \in A \\ w \in B}} |z - w|, \tag{17}$$

and let $\Lambda = \{\lambda_{nk}\}$. Then clearly,

$$d(\Gamma, \Lambda)/D(\{1\}, \Lambda) \leq |\sigma(\lambda)| \leq D(\Gamma, \Lambda)/d(\{1\}, \Lambda), \qquad \lambda \in \Gamma, \tag{18}$$

so if the *maximum* distance of Γ to $\{\lambda_{nk}\}$ is less than the *minimum* distance of $\{1\}$ to $\{\lambda_{nk}\}$, U sums $\mathscr{C}_{E^m}(\Gamma)$. Also $\sigma(\lambda^*) = 0$ only if λ^* is a limit point of $\{\lambda_{nk}\}$, provided the latter is bounded.

The regularity of U is particularly easy to characterize when λ_{nk} does not depend on n. These are the so-called $(f, -\lambda_k)$ means (Jakimovski, 1959).

Theorem 3. Let the triangle U be defined by $P_0 = 1$ and

$$P_n(\lambda) = \prod_{k=1}^{n} \frac{\lambda - \lambda_k}{1 - \lambda_k}, \qquad n \geq 1, \tag{19}$$

where $\lambda_k \neq 1$.

(i) Let $0 <. \lambda_k < M$; then U is regular iff $\sum \lambda_k$ converges.
(ii) Let $\lambda_k <. 0$; then U is regular iff $\sum \lambda_k^{-1}$ diverges.

Proof. We prove statement (i) first.

Let the last $\lambda_k \leq 0$ be λ_m. Clearly, the regularity of the triangle associated with $P_n(\lambda)$ is unaffected if each factor $(\lambda - \lambda_k)/(1 - \lambda_k)$ in $P_n(\lambda)$, $1 \leq k \leq m$, is replaced by, say, $2\lambda - 1$. Thus we can assume, without loss of generality, that $\lambda_k > 0$. Note also that already

$$\sum_{k=0}^{n} \mu_{nk} = P_n(1) = 1. \tag{20}$$

$\Rightarrow$: Since $\lambda_k > 0$, the coefficients in $P_n(\lambda)$ alternate in sign. Thus

$$|P_n(-1)| = \prod_{k=1}^{n} \left| \frac{1 + \lambda_k}{1 - \lambda_k} \right| = \sum_{k=0}^{n} |\mu_{nk}| < M. \tag{21}$$

$|P_n(-1)|$ is obviously monotone increasing and thus convergent. This means that $\lambda_k \to 0$ (Knopp, 1947, p. 219), which means the infinite product $\prod (1 + \lambda_k)/(1 - \lambda_k)$ converges absolutely. Thus $\sum 2\lambda_k/(1 - \lambda_k)$ is convergent (Knopp, 1947, p. 224), and so $\sum \lambda_k$ is convergent.

$\Leftarrow$:

$$|\mu_{nk}| = \frac{1}{2\pi} \left| \int_{|t| = R > 1} t^{k-n-1} \prod_{k=1}^{n} \left(\frac{1 - t\lambda_k}{1 - \lambda_k} \right) dt \right| \tag{22}$$

$$\leq R^{k-n} \prod_{k=1}^{n} \left[1 + \frac{(R + 1)\lambda_k}{|1 - \lambda_k|} \right]. \tag{23}$$

But the convergence of $\sum \lambda_k$ guarantees the convergence of the above product, since $0 < \lambda_k < M$ (Knopp, 1974, p. 274). Thus

$$|\mu_{nk}| \leq AR^{k-n} \tag{24}$$

and

$$\sum_{k=0}^{n} |\mu_{nk}| \leq \frac{A(R^{n+1} - 1)}{R^n(R-1)} < \frac{AR}{R-1}, \tag{25}$$

and letting $n \to \infty$ in (24) finishes the first part of the proof.

For (ii), let $-\lambda_k = \tau_k > 0$, without loss of generality. Note $\sum |\mu_{nk}| = \sum \mu_{nk} = 1$.

$\Rightarrow$:

$$\mu_{n0} = \prod_{k=1}^{n} \left(1 + \frac{1}{\tau_k}\right)^{-1}. \tag{26}$$

If $\mu_{n0} \to 0$, then $\sum \tau_k^{-1}$ must be divergent, since the product in (26) must diverge (to zero).

$\Leftarrow$: As before

$$|\mu_{nk}| = \frac{1}{2\pi}\left| \int_{|t|=\varepsilon} \frac{1}{t^{k+1}} \prod_{k=1}^{n} \left(\frac{t + \tau_k}{1 + \tau_k}\right) dt \right|$$

$$\leq \frac{1}{\varepsilon^k} \prod_{k=1}^{n} \left(1 + \frac{\varepsilon - 1}{1 + \tau_k}\right), \qquad 0 < \varepsilon < 1. \tag{27}$$

The product diverges as $n \to \infty$ if $\sum (1 + \tau_k)^{-1}$ is divergent. But, as is easily seen, this is true for $\sum \tau_k^{-1}$ divergent. Thus the product diverges and obviously to 0 since its terms are < 1. This means $\mu_{nk} \to 0$, so U is regular. ■

For this very simple class of triangles, the measure can be computed explicitly.

Theorem 4. Let the triangle U be defined by $P_0 = 1$ and

$$P_n(\lambda) = \prod_{k=1}^{n} \left(\frac{\lambda - \lambda_k}{1 - \lambda_k}\right), \qquad n \geq 1, \tag{28}$$

where $\lambda_k \neq 1$, $k \geq 1$, and $\lambda_k = o(1)$. Then

$$\sigma(\lambda) = |\lambda|. \tag{29}$$

Proof

$$|P_n(\lambda)|^{1/n} = \exp\left(\frac{1}{n} \sum_{k=1}^{n} \ln \left|\frac{\lambda - \lambda_k}{1 - \lambda_k}\right|\right), \tag{30}$$

or the left-hand side is the exponential of the Cesaro means of the sequence $s_n^* = \ln|(\lambda - \lambda_n)/(1 - \lambda_n)|$. This sequence is convergent—in fact, $s^* = \ln|\lambda|$—and the theorem results. ■

Example 1. Let $\lambda_k = \sigma^{-k}$, $\sigma > 1$, U is then the triangle corresponding to the Romberg integration procedure (see Section 3.1).

In the next result, the zeros of P_n are allowed to depend on n.

Theorem 5. Let $P_n(\lambda)$ be as in (28) but with $\lambda = \lambda_{nk}$. Furthermore, let all but a finite number of the $\lambda_{nk} \in \mathscr{D}$, which is a compact subset of $\{\mathrm{Re}\, z \leq 0\}$, and let

$$M_n = \sum_{k=0}^{n} |\mu_{nk}| = O(1). \tag{31}$$

Then U is regular. Furthermore, if $\mathscr{D} = [-a, 0]$, $a > 0$, (31) may be omitted.

Proof. Using the previous contour integral, we find

$$\begin{aligned} |\mu_{nk}| &\leq. R^{k-n} \sup_{|t|=R} \prod_{j=1}^{n} \left| \frac{1 - t\lambda_{nj}}{1 - \lambda_{nj}} \right| \\ &= R^k \prod_{j=1}^{n} \left[\frac{(1/R - \mathrm{Re}\, \lambda_{nj})^2 + (\mathrm{Im}\, \lambda_{nj})^2}{(1 - \mathrm{Re}\, \lambda_{nj})^2 + (\mathrm{Im}\, \lambda_{nj})^2} \right]^{1/2}. \end{aligned} \tag{32}$$

But since $\mathrm{Re}\, \lambda_{nj}$, $\mathrm{Im}\, \lambda_{nj}$ are bounded and $\mathrm{Re}\, \lambda_{nj} \leq 0$, each term in the product is less than or equal to $\eta < 1$, so

$$|\mu_{nk}| \leq. R^k \eta^n = o(1) \quad \text{in} \quad n. \tag{33}$$

That (31) may be omitted for $\mathscr{D}$ real follows from

$$M_n = P_n(1) = 1. \quad ■ \tag{34}$$

Example 2. The case where $\{P_n\}$ is a system orthonormal with respect to a distribution function $\psi \in \Psi$ with support (that is, the set of points of increase) in $[-a, 0]$ is very important (Section 2.3.6).

Of course, it is also important to know when a method is *not* regular.

Theorem 6. Let $\lambda_{nk} \in. [0, \infty]$, $k \geq 1$, and let $m(n)$ of these be bounded and bounded away from zero, $m(n) \to \infty$ as $n \to \infty$. Then U is not regular.

Proof

$$M_n = \sum_{k=0}^{n} |\mu_{nk}| =. |P_n(-1)| =. \prod_{k=1}^{n} \left| \frac{1 + \lambda_{nk}}{1 - \lambda_{nk}} \right|. \tag{35}$$

For the $m(n)$ zeros,

$$\left|\frac{1+\lambda_{nk}}{1-\lambda_{nk}}\right| \geq . \ 1+\delta, \qquad \delta > 0, \tag{36}$$

so

$$M_n \geq . \ (1+\delta)^{m(n)} \to \infty. \quad \blacksquare \tag{37}$$

Theorem 7. Any one of the $n+1$ conditions below is necessary for U to be regular:

$$\prod_{k=1}^{n} |1-\lambda_{nk}|^{-1} = O(1);$$
$$S_p(\lambda_{n1}, \lambda_{n2}, \ldots, \lambda_{nn}) \prod_{k=1}^{n} |1-\lambda_{nk}|^{-1} = o(1), \qquad 1 \leq p \leq n, \tag{38}$$

where S_p is the pth symmetric function of the roots of $P_n(\lambda)$.

Proof. Obvious. ■

Definition. Let U be a triangle. U is said to be *equivalent to convergence* if $\lim_{n\to\infty} s_n = s$ iff $\lim_{n\to\infty} \bar{s}_n = s$. [Note that this definition requires s_n (or $\bar{s}_n$) to exist only for n sufficiently large.]

Triangles equivalent to convergence are, generally speaking, pretty weak computationally—they, as it were, try to do too much. The triangle U tends to be heavily weighted toward the diagonal $[\mu_{ii}]$ and so gives excessive weight to the latest member used in the sequence $\{s_n\}$. But the latest member of the sequence carries very little information. The following criterion is due to Agnew (1952).

Theorem 8. Let U be a regular triangle and

$$\underline{\lim}_{n\to\infty} (2\mu_{nn} - M_n) > 0. \tag{39}$$

Then U is equivalent to convergence.

Proof. See Agnew (1952). ■

2.3. Important Triangles

2.3.1. *Weighted Means*

$$\mu_{nk} = p_k/P_n, \tag{1}$$

$$P_n = p_0 + p_1 + \cdots + p_n \neq 0, \qquad n \geq 0. \tag{2}$$

U is regular if and only if $\sum_{k=0}^{n} |p_k| = O(P_n)$ and $P_n \to \infty$. When $p_n > 0$, U is regular if and only if $P_n \to \infty$.

2.3.2. *Euler Means*

$$\mu_{nk} = \binom{n}{k} p^{n-k}/(p+1)^n, \qquad p > 0. \tag{1}$$

U is regular, because

$$\binom{n}{k} = O(n^k), \quad k \geq 0; \qquad P_n(\lambda) = \left(\frac{\lambda + p}{1 + p}\right)^n; \qquad \sigma(\lambda) = \left|\frac{\lambda + p}{1 + p}\right|. \tag{2}$$

When $p = 1$, U is called the binomial method. For further properties of U, see Section 12.2.

2.3.3. *Hausdorff Transformations*

Let ϕ be of bounded variation in $[0, 1]$, $\phi(0) = 0$, $\phi(1) = 1$, $\int_0^1 d\phi = 1$, and define

$$\mu_{nk} = \binom{n}{k} \int_0^1 x^k (1 - x)^{n-k} \, d\phi. \tag{1}$$

Obviously, U is a triangle.

Theorem 1. U is regular iff ϕ is continuous at 0.

Proof

$$\begin{aligned} \sum_{k=0}^{n} |\mu_{nk}| &\leq \int_0^1 \sum_{k=0}^{n} \binom{n}{k} x^k (1 - x)^{n-k} |d\phi| \\ &= \int_0^1 |d\phi| < \infty. \end{aligned} \tag{2}$$

Thus U is regular iff $\lim_{n \to \infty} \mu_{nk} = 0$. If ϕ is continuous at 0,

$$|\mu_{nk}| \leq \int_0^1 \binom{n}{k} x^k (1 - x)^{n-k} |d\phi| \tag{3}$$

and for every ε, $0 < \varepsilon < 1$,

$$
\begin{aligned}
|\mu_{nk}| &\leq \int_0^\varepsilon |d\phi| + \int_\varepsilon^1 \binom{n}{k} x^k (1-x)^{n-k} |d\phi| \\
&\leq \int_0^\varepsilon |d\phi| + \binom{n}{k} (1-\varepsilon)^{n-k} \int_0^1 |d\phi| \\
&= \int_0^\varepsilon |d\phi| + o(1), \qquad n \to \infty, \qquad (4)
\end{aligned}
$$

and so it follows that $\lim_{n\to\infty} \mu_{nk} = 0, k \geq 0$. Conversely, for every $\varepsilon, 0 < \varepsilon < 1$,

$$
\begin{aligned}
|\mu_{nk}| &\geq \left| \int_0^\varepsilon d\phi \right| - \left| \int_\varepsilon^1 \binom{n}{k} x^k (1-x)^{n-k}\, d\phi \right| \\
&\geq \left| \int_0^\varepsilon d\phi \right| - \binom{n}{k} (1-\varepsilon)^{n-k} \int_0^1 |d\phi| \\
&= \left| \int_0^\varepsilon d\phi \right| + o(1), \qquad n \to \infty. \qquad (5)
\end{aligned}
$$

If $\lim_{n\to\infty} \mu_{nk} = 0$ and $\phi(0) = 0$, it follows that ϕ must be continuous at 0. ■

Hausdorff weights yield interesting quadrature formulas for Stieltjes integrals.

Theorem 2. Let $f \in C[0, 1]$. Then

$$\lim_{n\to\infty} \sum_{k=0}^{n} f\left(\frac{k}{n}\right) \mu_{nk} = \int_0^1 f\, d\phi. \qquad (6)$$

Proof. The proof is elementary. By the uniform approximating properties of the Bernstein polynomials (Davis, 1963),

$$B_n(f; x) = \sum_{k=0}^{n} \binom{n}{k} f\left(\frac{k}{n}\right) x^k (1-x)^{n-k}. \qquad (7)$$

Note that $\phi \equiv x$ yields the trapezoidal formula. ■

For additional properties of Hausdorff transformations, see Petersen (1966) and Peyerimhoff (1969).

2.3.4. Salzer Means

Salzer means are given by

$$\mu_{nk} = (-1)^{n+k} \frac{(\gamma + k)^n}{n!} \binom{n}{k}, \qquad \gamma > 0 \qquad (1)$$

(Salzer, 1955, 1956; Wynn, 1956a; Salzer and Kimbro, 1961; Wimp, 1972, 1975). U is not regular since

$$M_n = \sum_{k=0}^{n} |\mu_{nk}| > (n + \gamma)^n/n! \to \infty. \tag{2}$$

In fact the following result holds.

Theorem. Let $\lambda \neq 0$ be arbitrary complex. Then for U defined by (1),

$$\sigma(\lambda) = \omega^{-1}, \tag{3}$$

where $\omega(\lambda)$ is the modulus of the smallest (in magnitude) root(s) of

$$e^{-z} + \lambda ez = 0. \tag{4}$$

In particular,

$$\kappa(U) = \sigma(-1) = 3.5911. \tag{5}$$

Proof. Using

$$\frac{(\gamma + k)^n}{n!} = \frac{1}{2\pi i} \int_{c-i\infty}^{c+i\infty} \frac{e^{pt}}{p^{n+1}}\, dp, \qquad t = \gamma + k, \quad c > 0, \tag{6}$$

we have

$$P_n(\lambda) = \frac{(-1)^n}{2\pi i} \int_{c-i\infty}^{c+i\infty} \frac{e^{p\gamma}}{p} \left(\frac{1 - \lambda e^p}{p}\right)^n dp, \tag{7}$$

and the theorem follows by a straightforward application of the method of steepest descents.

Rouché's theorem shows that Eq. (4) for $\lambda = -1$ has exactly one root in $\operatorname{Re} z > -1$. This root is real,

$$z_0 = -0.278464543, \tag{8}$$

and since the μ_{nk} alternate, Eq. (5) follows immediately. ∎

Equation (4) is rather interesting, and has received much attention. If $0 < \lambda < 1$, it has no real roots. For all $\lambda \neq 0$, it has a string of roots lying asymptotically within an arbitrarily narrow sector enclosing the imaginary axis. An asymptotic formula for these roots is known; see Bellman and Cooke (1963) and Wright (1955).

Equation (5) indicates the method obtained from the Salzer weights is numerically very unstable. What happens, of course, is that the weights grow large and alternate in sign.

These considerations would seem reason enough to dismiss U as a summation method suitable for any practical applications. The reader will therefore be surprised to learn that U is one of the most important summation methods. It is regular for a large and important class of sequences, and performs better in summing these sequences than even the most powerful nonlinear methods. Furthermore, those sequences, which have the property that they approach their limits algebraically (and are thus logarithmically convergent), are the sequences which pose the greatest challenge to any summation method.

To explore this idea, observe that

$$\sum_{k=0}^{n} \mu_{nk}(\gamma + k)^{-r} = \frac{1}{n!} \Delta^n \gamma^{n-r} = \delta_{0r}, \qquad 0 \le r \le n. \tag{9}$$

This means that, ultimately, U is *exact* when applied to sequences that are in $\text{Lin}(1, (\gamma + n)^{-1}, (\gamma + n)^{-2}, \ldots, (\gamma + n)^{-m})$; i.e., $\bar{s}_n =. s$. [In fact, this is a consequence of the manner of derivation of the method; see Wimp (1975) for details.]

Actually U is effective—but not necessarily exact—on a much larger class of sequences.

Let $\mathscr{A}$ be the class of sequences $\mathbf{s}$ with

$$s_n = s + \sum_{r=1}^{\infty} \frac{c_r}{(n + \beta)^r}, \qquad n \ge 0, \tag{10}$$

where $\mathbf{c} \in \mathscr{C}_S$ and the series converges absolutely for $n = 0$. Assume $0 < \beta < \gamma$. Rearranging (10) gives (the also absolutely convergent series)

$$s_n = s + \sum_{r=1}^{\infty} \frac{c_r^*}{(n + \gamma)^r}, \qquad n \ge 0. \tag{11}$$

With the representation

$$(k + \gamma)^{-r} = \frac{1}{\Gamma(r)} \int_0^\infty e^{-(k+\gamma)t} t^{r-1}\, dt, \tag{12}$$

we have

$$\begin{aligned} \bar{r}_n = \bar{s}_n - s &= \sum_{r=1}^{\infty} c_{r+n}^* \sum_{k=0}^{n} \frac{\mu_{nk}}{(k + \gamma)^r}, \\ &= \frac{(-1)^n}{n!} \sum_{r=1}^{\infty} \frac{c_{r+n}^*}{\Gamma(r)} \int_0^\infty e^{-\gamma t}(1 - e^{-t})^n t^{r-1}\, dt, \end{aligned} \tag{13}$$

so

$$|\bar{r}_n| \le \frac{1}{n!} \sum_{r=1}^{\infty} \frac{|c_{r+n}^*|}{\Gamma(r)} \int_0^\infty e^{-\gamma t} t^{r-1}\, dt = \frac{\gamma^n}{n!} \sum_{r=1}^{\infty} \frac{|c_{r+n}^*|}{\gamma^{r+n}} \tag{14}$$

or

$$|\bar{r}_n| \le \frac{\gamma^n}{n!} M(\gamma), \qquad M(\gamma) = \sum_{r=1}^{\infty} \frac{|c_r^*|}{\gamma^r}. \tag{15}$$

Table I

n	$\bar{s}_n$
0	1
1	1.5
2	1.625
3	1.643518
4	1.644965
5	1.644951
6	1.644935185
7	1.644933943
8	1.644934041

Thus U is regular for $\mathscr{A}$. In fact, we can find a constant C and an integer m such that

$$|(\bar{s}_n - s)/(s_n - s)| \leq C\gamma^n n^m/n!, \tag{16}$$

m being the index of the first $c_m^* \neq 0$. Thus U is dramatically accelerative for $\mathscr{A}$.

U also seems to do well on sequences which have Poincaré asymptotic expansions given by the right-hand side of (10). However, without assuming something more about the character of **s**, there seems no way to establish regularity. Nevertheless, as an example, take

$$s_n = (PI^2)_n = \sum_{k=0}^{n} \frac{1}{(k+1)^2} \sim \frac{\pi^2}{6} + \frac{c_1}{n} + \frac{c_2}{n^2} + \cdots, \tag{17}$$

as the work of Section 1.6 shows. Taking $\gamma = 1$ gives the values shown in Table I. The error in the last entry is 2.6×10^{-8}.

Can U sum divergent sequences? If $|\lambda| > 1$, (4) has one zero in N, so for these values of λ, $\sigma(\lambda) > 1$. Thus U sums no divergent exponential sequences. If s_n is a real convergent alternating sequence, $s_n = (-\lambda)^n$, $0 < \lambda < 1$, a direct argument shows Salzer's method is regular when $0 < \lambda < 1/e$ and produces a divergent $\bar{s}_n$ when $1/e < \lambda < 1$.

The Salzer weights are best applied using a lozenge procedure; see Section 3.3, Example 3.

2.3.5. Other Nonregular Methods

There is a class of nonregular methods that work on the same kind of sequences as the Salzer methods but that are easier to analyze theoretically. These are triangles given by

$$P_n(\lambda) = (-1)^n P_n^{(\tau-1,0)}(1 - 2\lambda), \qquad \tau > 0, \tag{1}$$

so

$$\mu_{nk} = \frac{(\tau + k)_n (-1)^{n+k}}{n!} \binom{n}{k}. \tag{2}$$

All the zeros of $P_n(\lambda)$ lie in (0, 1) and, in fact, are equidistributed there. Thus U is not regular [Theorem 2.2(6)]. However, let $\mathscr{A}$ be the class of all sequences of the form

$$s_n = s + \sum_{r=1}^{\infty} \frac{c_r}{(n+\tau)_r}, \qquad n \geq 0. \tag{3}$$

U is very effective for such sequences.

Theorem 1. Let

$$c_n = \rho^n v_n, \qquad v_n = O(1). \tag{4}$$

Then

$$|\bar{r}_n| \leq |\rho|^{n+1} \sup_{r>n} |v_r| \frac{(n-1)!}{(2n)!} e^{|\rho|}, \qquad n \geq 1. \tag{5}$$

Proof.

$$\bar{r}_n = \frac{(-1)^n \Gamma(n+\tau)}{n!} \sum_{r=1}^{\infty} \frac{c_r \Gamma(r)}{\Gamma(r+\tau+n)\Gamma(r-n)} \tag{6}$$

results by using the known formula for ${}_2F_1(1)$. Thus

$$\begin{aligned} |\bar{r}_n| &\leq \frac{1}{n!} \sum_{r=0}^{\infty} \frac{|c_{r+n+1}|(n+r)!}{(n+r)_{n+r+1} r!} \\ &\leq |\rho|^{n+1} \sup_{r>n} |v_r| \frac{\Gamma(n)}{\Gamma(2n+1)} \Phi(n+1, 2n+1; |\rho|), \end{aligned} \tag{7}$$

Φ being Tricomi's Φ-function, since $(n+\tau)_{n+r+1}$ is increasing in τ. However, each term in the Taylor series for Φ is decreasing in n. Letting $n = 0$ gives an upper bound, and the theorem results. ■

The foregoing also shows that U is, ultimately, exact ($\bar{r}_n \doteq 0$) for sequences of the form

$$s_n = s + \sum_{r=1}^{m} \frac{c_r}{(n+\tau)_r}. \tag{8}$$

Since the weights μ_{nk} alternate in sign, the numerical stability of this method is

$$\overline{\lim_{n\to\infty}} |P_n(-1)|^{1/n} = \overline{\lim_{n\to\infty}} |P_n^{(\tau-1,0)}(3)| \tag{9}$$

or

$$\kappa(U) = 3 + \sqrt{8} = 5.828, \tag{10}$$

even worse than the Salzer method.

The method is regular for another important class of sequences.

Theorem 2. Let $\mathscr{A}$ be the class of $\mathbf{s} \in \mathscr{C}_S$ whose remainder sequences $\{(s_n - s)\} = \mathbf{r} \in \mathscr{R}_{TM}$. U is regular for $\mathscr{A}$.

Proof. For some $\psi \in \Psi$,

$$r_n = \int_0^1 t^n \, d\psi, \tag{11}$$

and thus

$$\begin{aligned} \bar{r}_n &= (-1)^n \int_0^1 P_n^{(\tau-1,0)}(1-2t)\, d\psi, \\ &= (-1)^{n+1} \int_{-1}^1 P_n^{(\tau-1,0)}(t)\, d\phi, \qquad \phi(t) = \psi\left(\frac{1-t}{2}\right). \end{aligned} \tag{12}$$

The following facts are in Erdélyi (1953, vol. 2, 10.14, 10.18):

$$|P_n^{(\tau-1,0)}(t)| \le 1, \qquad t \in [-1, 1]; \tag{13}$$

$$P_n^{(\tau-1,0)}(\cos\theta) \approx K(\theta) n^{-1/2} \cos[(n + \tau/2)\theta + C], \tag{14}$$

(14) holding uniformly on compact subsets of $(0, \pi)$, K (>0) being integrable on such subsets. Pick δ, $0 < \delta < 1$, and write

$$|\bar{r}_n| \le \int_{-1}^{\delta-1} d\phi + C' n^{-1/2} \int_{\delta-1}^{1-\delta} K(\theta)\, d\phi + \int_{1-\delta}^1 d\phi, \qquad \theta = \arccos t. \tag{15}$$

Now pick δ to make the first and last integrals $<\varepsilon/3$; the second will be $<.\ \varepsilon/3$. Thus $|\bar{r}_n| <.\ \varepsilon$ or $\bar{s}_n = s + o(1)$. ■

2.3.6. *Orthogonal Methods*

Let $a > 0$ and $\{p_n(t)\}$ be a system of polynomials orthonormal with respect to some $\psi \in \Psi$ with support in $[-1, 1]$. Further, let

$$\int_{-1}^1 \frac{\ln \psi'(t)}{\sqrt{1-t^2}}\, dt \tag{1}$$

be bounded. If we define

$$P_n(\lambda) = p_n(2\lambda/a + 1)/\sigma_n(a), \qquad \sigma_n(a) = p_n(2/a + 1), \tag{2}$$

$P_n(\lambda)$ has its zeros in $(-a, 0)$, so a regular triangle is obtained, by virtue of

Theorem 2.2(5). Further, by the well-known asymptotic properties of p_n (Freud, 1966, p. 245ff.),

$$\sigma(\lambda) = [(\lambda^{1/2} + \sqrt{\lambda + a})/\rho]^2, \qquad \lambda \notin [-a, 0], \quad \rho = 1 + \sqrt{1 + a} \geq 2, \tag{3}$$

branch cuts being taken between $-a$ and 0.

The Toeplitz methods obtained by taking the polynomial P_n in (2) as the characteristic polynomial of U are perhaps the most powerful of all linear summation methods. Furthermore, they lend themselves to an elegant computational scheme that requires no explicit knowledge of the zeros of P_n or the weights μ_{nk}. This algorithm is derived in Chapter 3.

Here, let us explore the regularity of methods based on (2). We first need some geometric concepts. The values of λ for which $\lambda(\sigma) = |\lambda|$ are given by the roots of

$$\frac{\lambda^{1/2} + \sqrt{\lambda + a}}{\rho} = z\lambda^{1/2}, \qquad z = e^{i\theta}, \tag{4}$$

or by

$$\frac{\lambda}{a} = \frac{(\cos 2\theta - (2/\rho)\cos\theta) - i(\sin 2\theta - (2/\rho)\sin\theta)}{\rho^2 - 4\rho\cos\theta + 4}. \tag{5}$$

As θ varies between $-\arccos 1/\rho$ and $\arccos 1/\rho$, the latter having its principal value, the λ-values in Eq. (5) trace out the outer loop of a limaçon-type figure intersecting the real axis at the points $-a/\rho^2$ and 1. This figure is shown in Fig. 1 for $a = 8$.

Let $\Gamma_1(a)$ denote the region of the cut λ-plane exterior to this curve. There (and only there) $\sigma(\lambda) < |\lambda|$.

Let $\Gamma_2(a)$ denote the region interior to an ellipse, center $(-a/2, 0)$, major semiaxis on the real axis of length $a/2 + 1$, minor semiaxis parallel to the imaginary axis of length $\sqrt{a + 1}$, and exterior to $[-a, 0]$. For $\lambda \in \Gamma_2(a)$, $\sigma(\lambda) < 1$ (and only there). These observations and Theorem 2.2(2) lead to the following result.

Theorem 1. Let U be a triangle determined from Eq. (2). Then U sums $\mathscr{C}_{E^m}(\Gamma)$, $\Gamma \subset \Gamma_2$, and accelerates convergence of $\mathscr{C}_{E^m}(\Gamma)$ for $\Gamma \subset \Gamma_1 \cap E_1$.

The asymptotic theory for $p_n((2\lambda/a) + 1)$ on the cut is rather complicated and to get an idea of what can happen, it is best to specialize, taking the case of the Jacobi polynomials, since these generate the most useful triangles. If

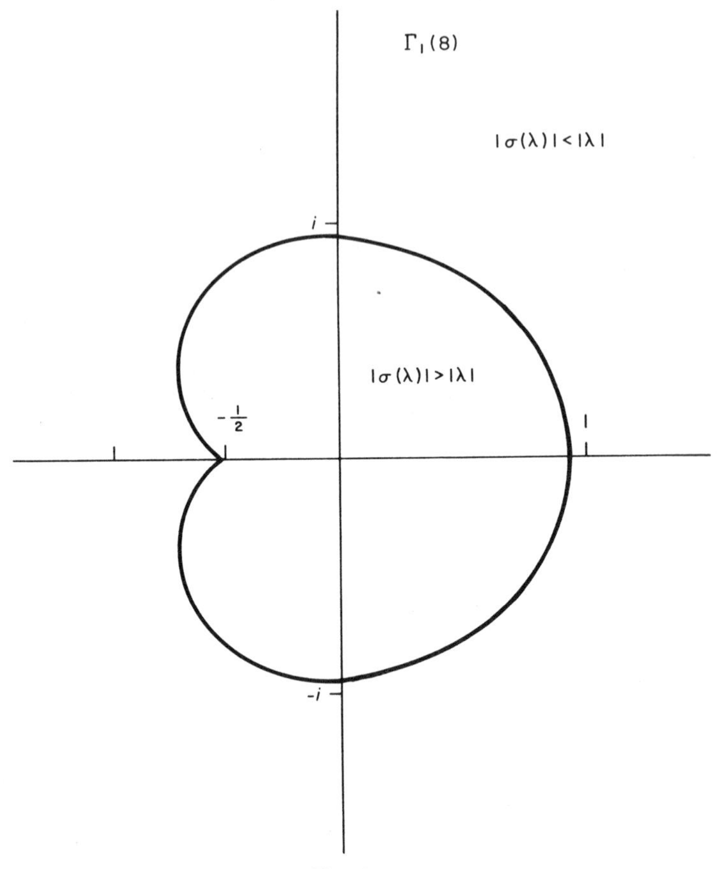

Fig. 1.

$p_n(x) = h_n^{-1/2} P_n^{(\alpha, \beta)}(x)$, h_n denoting the usual orthonormalization constant, then

$$\mu_{nk} = \binom{n}{k} \frac{(n+v)_k}{(\alpha+1)_k} \left(\frac{\lambda}{a}\right)^k \bigg/ \tau_n(a), \qquad v = \alpha + \beta + 1, \tag{6}$$

$$\tau_n(a) = \sum_{k=0}^{n} \binom{n}{k} \frac{(n+v)_k}{(\alpha+1)_k} \left(\frac{1}{a}\right)^k.$$

From Erdélyi *et al.* (1953, vol. II, 10.14 (10)),

$$P_n(\lambda) \approx \frac{C \cos\{[n + \frac{1}{2}(\alpha + \beta + 1)]\theta - (\frac{1}{2}\alpha + \frac{1}{4})\pi\}}{\rho^{2n}},$$

$$\theta = \arccos\left(\frac{2\lambda}{a} + 1\right), \quad \lambda \in (-a, 0). \tag{7}$$

Since $\overline{\lim}|\cos(an + b)|^{1/n} = 1$, $a \neq 0$, this shows

$$\sigma(\lambda) = \rho^{-2}. \tag{8}$$

If $\max|\gamma_r|$ in the sequence s_n [see Eq. 2.2(12)] is larger than this, U will accelerate s_n. For $a = 1$, this value is 0.17157, and then U is accelerative if all $\gamma_r \in (-1, -0.17157)$. However, if some $\gamma_r \in (-0.17157, 1)$, then U is *not* accelerative and, in fact, an application of U will harm rather than help convergence, at least for such exponential sequences. This is consistent with the general philosophy that regular methods are ineffective for summing sequences that approach their limits monotonically, for instance, sequences like $\{1 + (\frac{1}{2})^n\}$. In such cases, one's recourse must be to nonregular methods.

2.3.7. *The Chebyshev Weights*

What is probably the best of all the positive triangles results when $p_n(x)$ of the previous section is chosen to be the Chebyshev polynomial $T_n(x)$ of degree n. The efficiency of U in this case is a consequence of the extraordinary interpolatory properties of $T_n(x)$ that cause that system of polynomials to play such an important role in numerical analysis and approximation theory. The definition is

$$T_n(x) = \cos n\theta, \qquad \theta = \arccos x, \quad n \geq 0, \tag{1}$$

so $T_0 = 1$, $T_1 = x$, $T_2 = 2x^2 - 1, \ldots$. Obviously,

$$T_n(x) = \tfrac{1}{2}[(x + i\sqrt{1 - x^2})^n + (x - i\sqrt{1 - x^2})^n]. \tag{2}$$

Another useful representation is

$$T_n(x) = {}_2F_1\left(\begin{matrix} -n, n \\ \frac{1}{2} \end{matrix}\middle| \frac{1 - x}{2}\right). \tag{3}$$

Letting $x = 2\lambda/a + 1$ gives a positive triangle with entries

$$\mu_{nk} = \frac{(n)_k}{(\frac{1}{2})_k a^k}\binom{n}{k}\bigg/\sigma_n(a), \tag{4}$$

$$\sigma_n(a) = [(\sqrt{a + 1} + 1)^{2n} + (\sqrt{a + 1} - 1)^{2n}]/2a^n, \qquad k, n \geq 0.$$

Usually one takes $a = 1$. As an example of the power of U take $a = 1$ in Eq. (4) and

$$s_n = (\text{LN } 2)_n = \sum_{k=0}^{n} \frac{(-1)^k}{k + 1} \to \ln 2 = 0.69314718. \tag{5}$$

The result is given in Table II.

Table II

n	$\bar{s}_n$
0	1
1	0.66666667
2	0.68627451
3	0.69360269
4	0.69312536
5	0.69315096
6	0.69314685
7	0.69314723
8	0.69314717
9	0.69314718

Of course, U is a positive triangle, and thus works poorly on positive monotone sequences.

The application of U is best accomplished by a lozenge method rather than using $\sum \mu_{nk} s_k$ (see Chapter 3).

2.3.8. Lotockiĭ's Method

For Lotockiĭ's method,

$$P_n(\lambda) = \frac{(\lambda + \alpha)_n}{(1 + \alpha)_n} = \frac{(\lambda + \alpha)(\lambda + \alpha + 1)\cdots(\lambda + \alpha + n - 1)}{(1 + \alpha)(2 + \alpha)\cdots(n + \alpha)}, \qquad \alpha > 0. \quad (1)$$

[See Lotockiĭ (1953), Vučkovič (1958), and for applications and other references Cowling and King (1962/1963) and Agnew (1957).] Theorem 2.2(3) as it stands provides no information about the regularity of U, but, starting with Eq. 2.2(24) the proof is easily modified:

$$\begin{aligned} |\mu_{nk}| &\le R^k \prod_{j=1}^{n} \frac{(\alpha + j - 1 + 1/R)}{(\alpha + j)}, \qquad R > 1 \\ &= CR^k \frac{\Gamma(n + \alpha + 1/R)}{\Gamma(n + \alpha + 1)} \\ &= R^k O(n^{(1/R)-1}) = o(1) \quad \text{in } n. \end{aligned} \quad (2)$$

Since U is a positive triangle, this is all that is required to show that U is regular.

2.3.9. Romberg Weights

The Romberg weights are a triangle, not a positive one, that bears a close relationship to an extrapolation procedure attributed to Richardson and also to a method attributed to Romberg for improving the accuracy of integration by the trapezoidal rule. Both procedures are treated at length in most books on numerical analysis; see, for instance, Isaacson and Keller (1966), Bauer *et al.* (1963), or the articles of Bulirsch and Stoer (1966, 1967).

Take $\sigma > 1$, $P_0 = 1$, and

$$P_n(\lambda) = \prod_{k=1}^{n} \frac{\lambda - \sigma^{-k}}{1 - \sigma^{-k}}, \qquad n \geq 1. \tag{1}$$

μ_{nk} then is the kth symmetric function of $(\sigma^{-1}, \sigma^{-2}, \ldots, \sigma^{-n})$; but it is not necessary, nor even desirable, to compute μ_{nk} to use U. It is much more convenient to use a lozenge algorithm (see Chapter 3).

We have shown that U is regular. Furthermore

$$\frac{P_n(\lambda)}{\lambda^n} = \prod_{k=1}^{n} \left(\frac{1 - 1/\lambda\sigma^k}{1 - 1/\sigma^k} \right), \qquad \lambda \neq 0, \tag{2}$$

and the product on the right is convergent as $n \to \infty$. It cannot converge to zero unless one of the factors is zero. Thus U accelerates an $\mathbf{s}$ in the space of exponential sequences $\mathscr{C}_{E^m}(\Gamma)$ iff the γ_r of largest modulus [see 2.1(12)] is equal to σ^{-k} for some $k \geq 1$.

Since

$$(\sigma^{n+1} - 1)P_{n+1}(\lambda) = (\lambda\sigma^{n+1} - 1)P_n(\lambda), \tag{3}$$

one gets, by equating powers of λ, the recursion formula

$$\begin{gathered}(\sigma^{n+1} - 1)\mu_{n+1,k} = \sigma^{n+1}\mu_{n,k-1} - \mu_{n,k}, \qquad n, k \geq 0 \\ (\mu_{nk} = 0, \quad k < 0, \quad k > n).\end{gathered} \tag{4}$$

2.3.10. Higgins Weights

Higgins weights are designed for sequences having the following behavior.

$$s_n \sim s + (-1)^n \sum_{r=1}^{\infty} \frac{c_r}{(n + \beta)^r}; \tag{1}$$

cf. Eq. 2.3.3(10). In contrast to the summation formulas of Section 2.3.3, the method to be derived here is regular.

Let

$$\mu_{nk} = \binom{n}{k}(k + \gamma)^n \Big/ W_n, \qquad W_n = \sum_{k=0}^{n} \binom{n}{k}(k + \gamma)^n, \qquad \gamma \in \mathscr{C}. \tag{2}$$

By the same steepest descent argument used in Section 2.3.4,

$$W_n/n! \approx \sqrt{2\pi/n}(z_0 + 1)e^{\gamma(z_0+1)}(-z_0)^{-n}, \qquad z_0 = -0.278464543. \tag{3}$$

The Toeplitz limit theorem shows U is regular.

We now demonstrate two theorems about this process. According to Theorem 1.6(6), (1) may be rewritten

$$s_n \sim s + (-1)^n \sum_{r=1}^{\infty} \frac{c_r'}{(n+\gamma)^r}, \tag{4}$$

or

$$s_n =. s + (-1)^n \sum_{r=1}^{m} \frac{c_r'}{(n+\gamma)^r} + \frac{\xi_n^{(m)}}{(n+\gamma)^{m+1}}, \qquad m \geq 1, \tag{5}$$

$\xi^{(m)}$ a bounded sequence. [(5) holds also when (4) is convergent.]

Note that

$$\sum_{k=0}^{n} \frac{(-1)^{n+k}\mu_{nk}}{(k+\gamma)^j} = 0, \qquad 1 \leq j \leq n, \tag{6}$$

and this provides the following theorem.

Theorem 1. Let (1) hold with $\gamma > 0$. Then

$$|\bar{s}_n - s| \leq \sup_n |\xi_n^{(m)}|/\gamma^{m+1}, \qquad m < n. \tag{7}$$

Proof. Left to the reader. ■

When the series on the right-hand side of (4) is absolutely convergent for $n = 0$, much more can be said.

Theorem 2. Let the series on the right-hand side of (4) converge absolutely, $\gamma > 0$. Then

$$|\bar{r}_n|/|r_n| \leq . M(\gamma)\gamma^n n^k/W_n \tag{8}$$

where M is independent of n and k is the smallest integer such that $c_k' \neq 0$.

Proof. The integral representation 2.3.3(12) gives

$$\bar{r}_n = \frac{1}{W_n} \sum_{r=1}^{\infty} \frac{c_{r+n}'}{\Gamma(r)} \int_0^{\infty} t^{r-1} e^{-\gamma t}(1 - e^{-t})^n \, dt. \tag{9}$$

Therefore,

$$|\bar{r}_n| \leq \frac{1}{W_n} \sum_{r=1}^{\infty} \frac{|c_{r+n}'|}{\gamma^r} \leq \frac{1}{W_n} c(\gamma)\gamma^n, \qquad c(\gamma) = \sum_{r=1}^{\infty} \frac{|c_r'|}{\gamma^r}. \tag{10}$$

Choosing k to be the smallest integer with $c_k' \neq 0$, we can describe the error sequence asymptotically by

$$|r_n| \approx |c_k'|/n^k \tag{11}$$

and combining this with (10) proves the theorem. ■

The computational gain on using the transformation (2) on series of the kind (1) is spectacular, the convergence being accelerated more than exponentially, actually, by a factor λ^n/n^n. It is rare indeed that a linear method performs this well.

Of course since the method is a positive triangle, it is numerically stable ($\kappa = 1$). The analysis of this procedure illustrates very well the gulf that exists between the acceleration of sequences that oscillate about their limits and those that do not (logarithmically convergent sequences being cases of the latter). For the former, there exists a profusion of highly efficient summation procedures, while for the latter, the suitable methods are much less efficient and invariably nonregular. For sequences that neither oscillate about their limits nor approach their limits monotonically, almost all known methods fail. (Recent numerical evidence, however, indicates that the implicit procedures of Chapter 9 hold some promise for such sequences.)

2.3.11. Inverse Methods

Some interest attaches to the *inverse* of a Toeplitz method U, i.e., the triangle $U^* = [\mu_{nk}^*]$ defined by

$$\bar{s}_n = \sum_{k=0}^{n} \mu_{nk} s_k, \qquad s_n = \sum_{k=0}^{n} \mu_{nk}^* \bar{s}_k \tag{1}$$

for all sequences $\mathbf{s}$, $\bar{\mathbf{s}}$.

The characteristic polynomials $P_n^*(\lambda)$ of U^* usually cannot be found explicitly. In one case, however, this can be accomplished, that is, for the nonregular methods discussed in Section 2.3.5.

If

$$P_n(\lambda) = (-1)^n P_n^{(\tau-1,0)}(1 - 2\lambda), \tag{2}$$

and application of the formula in Erdélyi *et al.* (1953, vol. 2, 10.20(3)) gives

$$P_n^*(\lambda) = \frac{(\tau + 2\lambda\, d/d\lambda)}{n + \tau} V_n(\lambda) \tag{3}$$

where

$$V_n(\lambda) = F\left(\begin{matrix} -n, 1 \\ n + \tau + 1 \end{matrix} \middle| -\lambda\right). \tag{4}$$

Furthermore,

$$\sigma^*(\lambda) = \overline{\lim_{n\to\infty}} |P_n^*(\lambda)|^{1/n} = \begin{cases} 1, & |\lambda| < 1 \\ \max(1, |\lambda + 1|^2/4|\lambda|), & |\lambda| > 1. \end{cases} \tag{5}$$

For $|\lambda| < 1$, (5) follows from (4) by dominated convergence and taking a termwise limit. For $|\lambda| > 1$, consider the integral

$$\begin{aligned} I(\alpha, \beta, \lambda) &= \int_0^1 (1-t)^{n+\alpha}(1+\lambda t)^{n+\beta}\, dt \\ &= (1+\lambda)^{n+\beta} \int_0^1 z^{n+\alpha}(1-\gamma z)^{n+\beta}\, dz, \qquad \gamma = \lambda/(1+\lambda) \\ &= (1+\lambda)^{n+\beta}\left[\int_0^{1/\gamma} + \int_{1/\gamma}^1\right] \\ &= (1+\lambda)^{2n+\alpha+\beta+1}\lambda^{-n-\alpha-1}B(n+\alpha+1, n+\beta+1) \\ &\quad - (1/\lambda)I(\beta, \alpha; 1/\lambda) \end{aligned} \tag{6}$$

The use of Stirling's formula and the relation

$$(n+\alpha+1)I(\alpha, \beta; \lambda) = F\left(\begin{matrix} -n-\beta, 1 \\ n+\alpha+2 \end{matrix} \middle| -\lambda\right) \tag{7}$$

shows (5) for $|\lambda| > 1$.

A quick computation shows U^* is regular, but it is a poor method to use on exponential sequences since $\bar{\mathbf{s}}$ will converge more slowly than λ^n for all $|\lambda| < 1$.

A considerable amount of research has been done on inverse methods. The paper by Wilansky and Zeller (1957) contains some important results and a number of useful references.

2.4. Toeplitz Methods Applied to Series of Variable Terms; Fourier Series and Lebesgue Constants

Often it is important to discern the effect of U on a series of variable terms:

$$\begin{aligned} f(z) = s(z) &= \sum_{k=0}^{\infty} f_k(z), \\ s_n(z) &= \sum_{k=0}^{n} f_k(z), \\ \bar{s}_n(z) &= \sum_{k=0}^{n} \mu_{nk} s_k(z) = \sum_{k=0}^{n} \nu_{nk} f_k(z), \\ \nu_{nk} &= \sum_{r=k}^{n} \mu_{nr}. \end{aligned} \tag{1}$$

A straightforward application of Cauchy's integral formula shows that for U to sum a Taylor series about the origin anywhere within its circle of convergence, it is sufficient that $P_n(\lambda) \to 0$ uniformly for all $|\lambda| \leq 1 - \delta$, for every $0 < \delta < 1$. Obviously this is a weaker condition than regularity. Necessary *and* sufficient conditions are presented later [Theorem 4.3.1(1)].

Applications to Fourier series present somewhat different problems. Let $f \in L(-\pi, \pi)$ and let a_k, b_k be the Fourier coefficients generated by f,

$$a_k = \frac{1}{\pi}\int_{-\pi}^{\pi} f(x)\cos kx\,dx, \qquad b_k = \frac{1}{\pi}\int_{-\pi}^{\pi} f(x)\sin kx\,dx. \tag{2}$$

Let

$$s_n(x) = \tfrac{1}{2}a_0 + \sum_{k=1}^{n} (a_k \cos kx + b_k \sin kx). \tag{3}$$

Assume that U is a real triangle and that $\bar{s}_n(x)$ is the result of applying the summability method U to $s_n(x)$. The convergence of $\bar{s}_n$ can be related to the constants

$$L_n(U) = \frac{2}{\pi}\int_0^{\pi/2} \frac{|\operatorname{Im} e^{i\theta}P_n(e^{2i\theta})|}{\sin\theta}\,d\theta, \tag{4}$$

called the *Lebesgue constants* for U. The standard theorem establishing the connection is due to Hardy and Rogosinki (1956).

Theorem 1. Let U be regular with $L_n(U)$ bounded. Then $\bar{s}_n(x)$ converges to

$$\tfrac{1}{2}[f(x^+) + f(x^-)] \tag{5}$$

wherever this exists. If f is continuous on a compact set $K \subset [-\pi, \pi]$, then $\bar{s}_n(x)$ converges uniformly to $f(x)$ on K.

Conversely, if $L_n(U)$ is unbounded there is an $f \in C[-\pi, \pi]$ for which $\bar{s}_n(x) \not\to f(x)$ at some point $x \in [-\pi, \pi]$.

Proof. See Hardy and Rogosinski (1956, pp. 58ff.). ■

An important related result is due to Nikolskii (1948).

Theorem 2. $\lim_{n\to\infty} \bar{s}_n(x) = f(x)$ at every Lebesgue point of f iff

(i) $\lim_{n\to\infty} \mu_{nk} = 0$ and
(ii) $L_n(U)$ is bounded.

Proof. The proof is established by an appeal to results of Banach on weak convergence in Banach spaces [see Nikolskii (1948)]. ■

As a philosophical consequence of such theorems, much research has centered on describing the asymptotic properties of $L_n(U)$ for various summation methods.

Concerning the Hausdorff transformation

$$\mu_{nk} = \binom{n}{k} \int_0^1 x^k(1-x)^{n-k}\, d\phi \tag{6}$$

(see Section 2.3.3), Lorch and Newman (1961), improving the earlier work of Livingston (1954), have found the following result.

Theorem 3. Let U for (6) be regular. Then

$$L_n(U) = C_\phi \ln n + o(\ln n), \tag{7}$$

where

$$C_\phi = \frac{2}{\pi^2}|\phi(1) - \phi(1^-)| + \frac{1}{\pi}\mathscr{M}\left|\sum_k [\phi(\xi_k^+) - \phi(\xi_k^-)] \sin \xi_k x\right|, \tag{8}$$

the sum extending over all the discontinuities ξ_k (at most countable) of ϕ, and $\mathscr{M}(f(x))$ represents the mean value of the almost periodic function $f(x)$.

Furthermore,

$$0 \le C_\phi \le \frac{4}{\pi^2}\int_0^1 |d\phi| \tag{9}$$

and $C_\phi = 0$ iff ϕ is continuous.

Theorem 4. Let $\varepsilon \in \mathscr{R}_N^+$ be monotone. Then there exists an increasing absolutely continuous ϕ for which

$$L_n(U) \ne o(\varepsilon_n \ln n). \tag{10}$$

This result establishes that the error term $o(\ln n)$ in (7) is the best possible and cannot be improved even for an increasing absolutely continuous ϕ.

For the Cesaro method

$$\begin{aligned} L_n(U) &= \frac{2}{\pi(n+1)}\int_0^{\pi/2} \frac{\sin^2(n+1)\theta}{\sin^2\theta}\, d\theta \\ &< \frac{M}{n+1}\int_0^{\pi/2} \frac{\sin^2(n+1)\theta}{\theta^2}\, d\theta = M\int_0^{(n+1)\pi/2} \frac{\sin^2 u}{u^2}\, du \end{aligned} \tag{11}$$

and the latter is a convergent integral. This yields the well-known fact that the Fourier series of a continuous function is Cesaro summable. On the other hand, the constants $L_n(U)$ for the binomial method are not bounded.

The Lebesgue constants for the methods displayed in Theorem 2.2(3),

$$P_n(\lambda) = \prod_{k=1}^{n} \frac{\lambda - \lambda_k}{1 - \lambda_k}, \tag{12}$$

have been discussed by Lorch and Newman (1962). (Note this includes Lotockiĭ's method of Section 2.3.8.)

In what follows it is assumed that $\lambda_k < 0$. Let

$$u_n = -2\sum_{k=1}^{n} \frac{\lambda_k}{(1 - \lambda_k)^2}, \qquad v_n = 1 + 2\sum_{k=1}^{n} \frac{1}{1 - \lambda_k}. \tag{13}$$

Then, if u_n is bounded,

$$L_n(U) = (4/\pi^2) \ln v_n + O(1), \tag{14}$$

while if u_n is unbounded,

$$L_n(U) = (2/\pi^2)\ln(v_n^2/u_n) + \alpha + o(1),$$

$$\alpha = -\frac{2}{\pi^2}\gamma + \frac{2}{\pi}\int_0^1 \frac{\sin t}{t}\,dt - \frac{2}{\pi}\int_1^\infty \frac{1}{t}\left(\frac{2}{\pi} - |\sin t|\right)dt, \tag{15}$$

$$\gamma = 0.57721 \quad \text{(Euler's constant)}.$$

Statement (14), coupled with Theorem 2.2(3ii), shows that for regular $(f, -\gamma_k)$ methods with u_n bounded and $\gamma_k < 0$, the Lebesgue constants are unbounded.

For generalizations to methods (not triangles) defined by

$$\sum_{k=1}^{n} \frac{f(\lambda) - \lambda_k}{f(1) - \lambda_k} = \sum_{k=0}^{\infty} \mu_{nk} \lambda^k, \tag{16}$$

f analytic at 0, see Shoop (1979).

Concerning the behavior of the Lebesgue constants for the powerful methods based on orthogonal polynomials (Section 2.3.6) nothing at all is known. Undoubtedly, they will prove to be unbounded, another consequence of the rule of thumb that only weak methods (Cesaro summability, for instance) are regular for large classes of sequences, in this case, the partial sums of the Fourier series for continuous functions.

How much improvement can be expected on applying a Toeplitz triangle to a Fourier series involves the concept of *saturation*. Let X denote either $C[-\pi, \pi]$ or $L_p[-\pi, \pi]$, $1 \le p < \infty$ with the norm defined in the usual manner. Let U be a triangle and

$$\bar{s}_n(x) = \frac{a_0}{2} + \sum_{k=1}^{n} v_{nk}(a_k \cos kx + b_k \sin kx). \tag{17}$$

Theorem 5. Let $f \in X$. If there is an $\alpha > 0$ such that for each $k > 0$

$$\underline{\lim}_{n\to\infty} n^{\alpha} |v_{nk} - 1| > 0, \tag{18}$$

then

$$\underline{\lim}_{n\to\infty} n^{\alpha} \|\bar{s}_n(x) - f(x)\| = 0 \tag{19}$$

implies f is constant almost everywhere (a.e.).

Proof. Let

$$f_k = \frac{1}{2\pi} \int_{-\pi}^{\pi} f(x) e^{-ikx}\, dx.$$

Then

$$\frac{1}{2\pi} \int_{-\pi}^{\pi} [\bar{s}_n(x) - f(x)] e^{-ikx}\, dx = (v_{nk} - 1) f_k, \tag{20}$$

and so

$$|f_k||v_{nk} - 1| \le \frac{1}{2\pi} \int_{-\pi}^{\pi} |\bar{s}_n(x) - f(x)|\, dx. \tag{21}$$

There exists a subsequence $\{n_j\}$, $n_j \to \infty$, such that both $\lim_{j\to\infty} n_j^{\alpha} |v_{n_j,k} - 1| = c_k > 0$ and also, by Hölder's inequality, $\lim_{j\to\infty} n_j^{\alpha} \|\bar{s}_{n_j}(x) - f(x)\|_1 = 0$. But this implies $c_k |f_k| = 0$ for each $k > 0$. Since a function in X is uniquely characterized by its Fourier coefficients, f must be a constant. ■

This shows that the approximation in norm of $\bar{s}_n(x)$ to $f(x)$ by Toeplitz methods satisfying (18) cannot be improved beyond the critical order $n^{-\alpha}$ no matter how smooth f is. Saturation theory deals with the optimal order of approximation to functions $\in E \subset X$ by a triangle U. For instance, consider the Cesaro means, $v_{nk} = (n + 1 - k)/(n + 1)$. One cannot have $\|\bar{s}_n(x) - f(x)\| = o(n^{-1})$ for $f \in C[-\pi, \pi]$ no matter how smooth f is, since $|v_{nk} - 1| = k/(n + 1)$ and $\alpha = 1$ in the previous theorem. For all nonconstant functions in $C[-\pi, \pi]$, $\bar{s}_n(x)$ approximates f with an order at most $O(n^{-1})$. In fact, this order is actually attained since for $f = e^{ix}$, $\|\bar{s}_n(x) - f(x)\| = 1/(n + 1)$. One says the Cesaro triangle *is saturated in* $C[-\pi, \pi]$ *with order* $O(n^{-1})$.

One problem is to characterize those elements in X for which the optimal order is attained. In some cases this can be done. Define

$$\tilde{f}(x) = \frac{1}{2\pi} \int_{-\pi}^{\pi} f(x - t) \cot \frac{t}{2}\, dt, \tag{22}$$

the integral being a principal value integral.

Theorem 6. Let $\bar{s}_n(x)$ be the Cesaro means of the Fourier series for $f(x)$, $X = C[-\pi, \pi]$. Then

$$\|\bar{s}_n(x) - f(x)\| = O(n^{-1}) \tag{23}$$

iff $\tilde{f}(x) \in C[-\pi, \pi]$, ess sup$|\tilde{f}'(x)| < \infty$.

Proof. See Butzer and Nessel (1971, Chap. 12). ■

It can be shown that the typical means defined by $v_{nk} = 1 - [k/(n+1)]^\kappa$, $\kappa > 0$, are saturated in C or L^p with order $n^{-\kappa}$.

Zemansky (1949) has studied the case

$$v_{nk} = g(k/n), \tag{24}$$

where g is a polynomial with $g(0) = 1$, $g(1) = 0$, $g^{(j)}(0) = 0$, $1 \le j \le p-1$, $g^{(p)}(0) \ne 0$. U is saturated in $C[-\pi, \pi]$ with order n^{-p}. The saturation class of U is the subclass of functions such that $\tilde{f}$ (resp. f) is $p-1$ times differentiable and satisfies a Lipschitz condition of order 1 for p even (resp. odd).

Many of the previous ideas can be generalized to abstract spaces. See Butzer and Nessel (1971).

Much work has been done on the summability of expansions in general orthogonal functions, for instance, by Olevskiĭ (1975 and the references given there). A discussion of these results is outside the scope of this book. However, one result is particularly intriguing: *if* $\mathbf{a} \in l^2$ *and* $\boldsymbol{\phi}(t)$ *is an orthonormal system on* $[0, 1]$ *and* $s_n(t) = \sum_{k=1}^{n} a_k \phi_k(t)$ *is summable a.e. by a real regular triangle* U, *then some subsequence of* $\mathbf{s}(t)$ *converges a.e.* [see Cooke (1955, p. 90)].

2.5. Toeplitz Methods and Rational Approximations; The Padé Table

Let $s_n(z)$ denote the partial sums of the power series of a function analytic at 0,

$$s(z) = \sum_{k=0}^{\infty} a_k z^k, \qquad s_n(z) = \sum_{k=0}^{n} a_k z^k. \tag{1}$$

Let $\gamma \in \mathscr{C}$ and σ_{nk} be an infinite lower triangular array of numbers. Define

$$A_n(z, \gamma) = \sum_{k=0}^{n} \gamma^{-k} \sigma_{nk} s_k(z), \qquad B_n(\gamma) = \sum_{k=0}^{n} \gamma^{-k} \sigma_{nk}, \qquad R_n(z, \gamma) = \sum_{k=0}^{n} \gamma^{-k} r_k(z), \tag{2}$$

so that

$$s(z) = \bar{s}_n(z, \gamma) - \bar{r}_n(z, \gamma), \quad \bar{s}_n(z, \gamma) = A_n(z, \gamma)/B_n(\gamma), \quad \bar{r}_n(z, \gamma) = R_n(z, \gamma)/B_n(\gamma). \tag{3}$$

Explicitly,

$$A_n(z, \gamma) = \sum_{k=0}^{n} a_k \left(\frac{z}{\gamma}\right)^k \sum_{r=0}^{n} \sigma_{n, r+k} \gamma^{-r} = \sum_{k=0}^{n} \gamma^{-k} \sum_{r=0}^{n-k} a_r \sigma_{n, r+k} \left(\frac{z}{\gamma}\right)^r. \tag{4}$$

For γ fixed, $\bar{s}_n$ is, of course, just the Toeplitz means of $s_0, \ldots, s_n$ with weights

$$\mu_{nk} = \gamma^{-k} \sigma_{nk} \Big/ \sum_{k=0}^{n} \gamma^{-k} \sigma_{nk}. \tag{5}$$

However, if we put $\gamma = z$ then, defining $\bar{s}_n(z, z) = \bar{s}_n(z)$, we have

$$A_n(z, z) = A_n(z), \qquad \bar{r}_n(z, z) = \bar{r}_n(z), \qquad \bar{s}_n(z) = A_n(z)/B_n(z) = z^n A_n(z)/z^n B_n(z) \tag{6}$$

and the latter is the ratio of two polynomials in z and thus a rational approximation. Clearly

$$[z^n B_n(z)]s(z) - [z^n A_n(z)] = O(z^{n+1}), \qquad z \to 0; \tag{7}$$

i.e., the rational approximation agrees with the power series through $n + 1$ terms. As will be shown, for certain functions s and certain choices of weights, considerably greater agreement is possible.

What is a "reasonable" choice for σ_{nk}? Certainly, some of the most powerful Toeplitz methods are those based on orthogonal polynomials (Section 2.3.6). Thus one could take

$$\sigma_{nk} = \binom{n}{k} \frac{(n + \nu)_k (-1)^k}{(\beta + 1)_k}, \qquad \nu = \alpha + \beta + 1, \quad \alpha, \beta > -1. \tag{8}$$

The characteristic polynomial for the method defined by (2) and (3) is then

$$P_n(\lambda) = R_n^{(\alpha, \beta)}(\lambda/\gamma)/R_n^{(\alpha, \beta)}(1/\gamma), \tag{9}$$

so P_n has its zeros on the ray connecting 0 and γ. An argument based on Eq. 2.3.6(3) and Theorem 2.2(5) shows that U is regular iff γ is real and $\gamma < 0$. In this case $\bar{s}_n(z, \gamma) \to s(z)$ for all z interior to the circle of convergence of (1). Also, the rational approximation $\bar{s}_n(z)$ will converge for all z real, negative, and interior to the circle of convergence of (1).

For many important functions, however, this appraisal of convergence is far too weak. These are the functions that have a representation as Stieltjes integrals

$$s(z) = \int_0^\infty \frac{d\psi}{1 - zt}, \qquad \psi \in \Psi^*, \quad z \notin \operatorname{Supp} \psi. \tag{10}$$

Theorem 1. Let the representation (5) hold and ψ have compact support. Define

$$a = \sup\{t \mid t \in \operatorname{Supp} \psi\}. \tag{11}$$

Let $\gamma \in \mathscr{C}$, $1/\gamma \notin [0, 1]$, $za/\gamma \in R$, $0 \le za/\gamma \le 1$, $\alpha > -1$, $\beta > -1$.
Then

$$|\bar{r}_n(z, \gamma)| \le K_n(z), \tag{12}$$

$$K_n(z) \approx 2 \sup_{0 \le t \le a} \left| \frac{zt}{1 - zt} \right| \frac{\sqrt{\pi a} |1/\gamma - 1|^{\alpha/2 + 1/4} |\gamma|^{-\beta/2 - 1/4} n^{q + 1/2}}{q! |\gamma^{-1/2} + \sqrt{1/\gamma - 1}|^{2n + \nu}},$$

$$q = \max(\alpha, \beta, -\tfrac{1}{2}), \quad n \to \infty. \tag{13}$$

Proof.

$$\bar{r}_n(z, \gamma) = -R_n^{(\alpha, \beta)}(1/\gamma)^{-1} \int_0^\infty [zt/(1 - zt)] R_n^{(\alpha, \beta)}(zt/\gamma)\, d\psi. \tag{14}$$

The proof will require the following well-known estimate (Szegö, 1959, p. 194). For $w \notin (0, 1)$,

$$R_n^{(\alpha, \beta)}(w) \approx \frac{1}{2\sqrt{\pi n}} (w - 1)^{-\alpha/2 - 1/4} w^{-\beta/2 - 1/4} (w^{1/2} + \sqrt{w - 1})^{2n + \nu}, \tag{15}$$

branch cuts for $(w - 1)^\mu$ and w^ν being taken along $(-\infty, 1]$ and $(-\infty, 0]$, respectively. This result holds uniformly on compact subsets of $\mathscr{C} - [0, 1]$. Using the fact that $R_n^{(\alpha, \beta)}(x)$ can be bounded algebraically (Erdélyi *et al.*, 1953, vol. II, 10.18(12)) completes the proof. ■

Corollary. Under the conditions stated above, $\bar{s}_n(z)$ converges exponentially to $s(z)$. Further, the rational approximations $\bar{s}_n(z, az)$ converge uniformly to $s(z)$ on every compact subset S of $\mathscr{C} - [1/a, \infty)$, also exponentially; i.e.,

$$|s(z) - \bar{s}_n(z, az)| \le L_n, \qquad z \in S,$$

$$L_n \approx M n^\theta \tau^n, \tag{16}$$

$$\tau = \sup_{z \in S} |(az)^{-1/2} + \sqrt{1/az - 1}|^{-2},$$

for some M and θ. Note $\tau < 1$.

Example. Let

$$s(z) = F(1, \beta; \nu; z), \qquad \nu = \alpha + \beta + 1. \tag{17}$$

Then $a = 1$,

$$d\psi = \frac{t^{\beta-1}(1-t)^{\alpha}\Gamma(v)}{\Gamma(\beta)\Gamma(\alpha+1)}\,dt \tag{18}$$

and

$$\bar{r}_n(z) = \frac{-z^{n+1}\Gamma(v)}{z^n R_n^{(\alpha,\beta)}(1/z)\Gamma(\beta)\Gamma(\alpha+1)} \int_0^1 \frac{t^{\beta}(1-t)^{\alpha}}{(1-zt)} R_n^{(\alpha,\beta)}(t)\,dt. \tag{19}$$

On expanding $(1-zt)^{-1}$ in powers of t one finds the first n terms, that is, the coefficients of $1, z, \ldots, z^{n-1}$ vanish by virtue of the orthogonality properties of $R_n^{(\alpha,\beta)}(t)$. Thus

$$[z^n B_n(z)]s(z) - [z^n A_n(z)] = O(z^{2n+1}); \tag{20}$$

i.e., in this case the rational approximation yields the $[n/n]$ entry in the Padé table for $F(1, \beta; \alpha+\beta+1; z)$. These rational approximations, by virtue of the theorem, converge uniformly on compact subsets of $\mathscr{C} - [1, \infty)$. For an extensive discussion of the construction and properties of Padé approximants, see Chapter 6.

Using

$$R_n^{(\alpha,\beta)}(t) = \frac{(-1)^n}{n!}(1-t)^{-\alpha}t^{-\beta}\frac{d^n}{dt^n}[(1-t)^{\alpha+n}t^{\beta+n}], \tag{21}$$

a useful formula for $\bar{r}_n$ can be derived by integration by parts:

$$\bar{r}_n(z) = \bar{s}_n(z) - s(z) = C_n z^{n+1} F(n+1, n+\beta+1; 2n+v+1; z)/R_n^{(\alpha,\beta)}(1/z),$$
$$C_n = \frac{-\Gamma(v)\Gamma(n+\alpha+1)\Gamma(n+\beta+1)}{\Gamma(\beta)\Gamma(\alpha+1)\Gamma(2n+v+1)}. \tag{22}$$

The F in (22) is easily estimated by Watson's formula (Luke, 1969, vol. I, p. 237):

$$\bar{r}_n(z) = \frac{-2\pi\Gamma(v)(1-z)^{\alpha}z^{2n+1}}{\Gamma(\beta)\Gamma(\alpha+1)[1+\sqrt{1-z}]^{4n+2v}}\left[1 + O\left(\frac{1}{n}\right)\right], \tag{23}$$

the branch cut of $\sqrt{1-z}$ being taken along $[1, \infty)$.

A similar error formula holds when the σ_{nk} are the coefficients of any system of polynomials orthogonal with respect to a general distribution function $\psi \in \Psi^*$ with bounded support.

Theorem 2. Let $\psi \in \Psi^*$ with support in $[a, b]$, $-\infty \le a < b \le \infty$,

$$s(z) = \int_a^b \frac{d\psi}{1-zt}, \qquad z \neq \frac{1}{\tau}, \quad a \le \tau \le b, \tag{24}$$

and let σ_{nk} be defined by

$$p_n(t) = \sum_{k=0}^{n} \sigma_{nk} t^k \tag{25}$$

where $\{p_n\}$ is a system of polynomials orthogonal with respect to $t\,d\psi$. Then the rational approximation $\bar{s}_n(z)$ defined by (6) is the $[n/n]$ element of the Padé table for $s(z)$ [where (1) is interpreted as a formal series if its radius of convergence is zero and where $\bar{s}_n(z)$ is the rational function formally agreeing with that series through $2n + 1$ terms].

Furthermore, if $[a, b]$ is bounded (take $a = 0$, $b = 1$ without loss of generality) and if ψ satisfies the conditions

(i) ψ is absolutely continuous,
(ii) $0 < m \le t^{3/2}(1 - t)^{1/2}\psi' \le M$,
(iii) $\int_0^1 (\ln \psi'\, dt/\sqrt{t(1 - t)} > -\infty$, and
(iv) $\int_0^1 |\psi'(t + h) - \psi'(t)|\, dt = O(h)$,

then $\bar{s}_n(z)$ converges to $s(z)$ uniformly on compact subsets of $\mathscr{C} - [1, \infty)$ with

$$\overline{\lim_{n\to\infty}} \left| \bar{r}_n(z)\left(\frac{1 + \sqrt{1 - z}}{z^{1/2}}\right)^n \right|^{1/n} < \infty. \tag{26}$$

Proof. The proof uses the known asymptotic properties of orthogonal polynomials [see Freud (1966), in particular, p. 245] and is straightforward. ■

The diagonal elements of the Padé table for $s(z)$ (in fact, for any power series) may also be generated with the ε-algorithm; see Chapter 6. That chapter also contains a number of convergence results pertaining to the off-diagonal elements of the Padé table for $s(z)$ a Stieltjes integral (10) and an example where the support of ψ is not compact.

It is easily verified that if p_n is orthogonal with respect to $d\psi$ rather than $t\,d\psi$, the $[n - 1/n]$ entry in the Padé table is obtained.

When the support of ψ is not compact, then the power series for $s(z)$ may be only formal, for example, $\psi = -e^{-t}, 0 \le t \le \infty$. In such cases Luke (1979) suggests taking

$$\sigma_{nk} = \binom{n}{k} \frac{(n + \nu)_k(-1)^k}{(\beta + 1)_k a_k}, \qquad a_k \ne 0. \tag{27}$$

It is very difficult to prove anything about the rational approximations associated with such a choice of weights, but the numerical evidence accumulating so far seems to indicate the rational approximations converge.

As an example consider the logarithm of the gamma function,

$$\ln \Gamma(z) = \left(z - \frac{1}{2}\right) \ln z - z + \frac{1}{2} \ln 2\pi + \frac{1}{12z} h(z),$$
$$h(z) \sim \sum_{k=0}^{\infty} a_k z^{-2k}, \quad |\arg z| < \pi, \tag{28}$$

where

$$a_k = 6B_{2k+2}/(2k+1)(k+1), \tag{29}$$

the B_k being Bernoulli numbers, $B_0 = 1$, $B_2 = \frac{1}{6}$, $B_4 = -\frac{1}{30}, \ldots$. Let $\alpha = \beta = 0$, $v = 1$ (the Legendre polynomial weights). Let $z \to 1/z^2$ and $\gamma = 1/z^2$.

The rational approximations are of the form

$$\bar{s}_n(z) = \sum_{k=0}^{n} v_k z^{2k} \Big/ \sum_{k=0}^{n} u_k z^{2k}, \qquad u_k = \binom{n}{k} \frac{(n+1)_k(-1)^k}{k!a_k}, \quad v_k = \sum_{m=0}^{n-k} u_{m+k} a_k. \tag{30}$$

Table III gives an idea of the kind of accuracy that can be expected with such approximations. The rational approximations compare favorably with the Padé, as Table IV shows. In this case $s(z) = h(iz^{-1/2})$ has the representation as a Stieltjes integral,

$$h(iz^{-1/2}) = \int_0^{\infty} \frac{\rho(t)\,dt}{1 - zt}, \qquad \rho(t) = 6t^{-1/2} \int_t^{\infty} \frac{u^{-1/2}\,du}{(e^{2\pi\sqrt{u}} - 1)} \tag{31}$$

[see Erdélyi *et al.*, 1953, 1.9(9)]. Applying Darboux's method to the power series

$$\frac{z}{e^z - 1} = \sum_{n=0}^{\infty} \frac{B_n}{n!} z^n \tag{32}$$

shows that

$$\frac{B_{2k+2}}{(2k+2)!} = \frac{2(-1)^k}{(2\pi)^{2k+2}} + O\left[\frac{1}{(4\pi)^{2k+2}}\right]. \tag{33}$$

Thus

$$a_k = O((2k+1)!R^k) \tag{34}$$

for any $R > 2\pi$. Appealing to a future result [Theorem 6.5(7) with $n = 0$] we can assert that the $[n/n]$ Padé approximants to $h(z)$ converge to $h(z)$ uniformly on compact subsets of $\{\text{Re } z > 0\}$.

Table III

$\bar{r}_n(z) = \bar{s}_n(z) - h(z)$

$n\backslash z$	1	2	5	10
2	1.4(−3)	4.6(−5)	2.6(−7)	4.4(−9)
4	3.1(−5)	5.3(−7)	2.4(−10)	3.0(−13)
6	1.3(−6)	2.5(−9)	4.1(−13)	5.8(−17)
8	8.7(−8)	1.1(−10)	6.8(−16)	2.1(−20)
10	1.6(−8)	2.4(−12)	4.5(−19)	1.0(−23)
12	1.1(−9)	2.1(−13)	1.9(−20)	4.6(−27)

Table IV

$[n/n]$ Padé Approximant $- h(z)$

$n\backslash z$	1	2	5	10
2	2.4(−4)	1.8(−6)	5.3(−10)	6.3(−13)
4	2.8(−5)	2.8(−8)	7.2(−14)	6.6(−19)
6	6.7(−6)	1.8(−9)	1.3(−16)	1.6(−23)
8	2.2(−6)	2.3(−10)	9.8(−19)	3.0(−27)
10	8.5(−7)	4.3(−11)	1.9(−20)	2.2(−30)
12	3.1(−7)	9.5(−12)	6.7(−22)	1.2(−33)

In both types of rational approximations the approximations gain in accuracy as z increases, reflecting the asymptotic character of the series for $h(z)$.

2.6. Other Orthogonal Methods; Pollaczek Polynomials and Padé Approximants

From the previous discussion, it is clear that the problem of producing weights σ_{nk} that, when used in Eqs. 2.5.1(1)–2.5.1(4) with $\gamma = z$, will produce rational approximations having a maximum degree of precision at $z = 0$ for a given function $s(z)$ representable as a Stieltjes integral reduces to the problem of determining polynomials orthogonal to a weight function $d\psi$ that has the property

$$s(z) = \int \frac{d\psi}{1 - zt}, \qquad z^{-1} \notin \text{Supp}\,\psi. \tag{1}$$

Obviously, the weights σ_{nk} can be used to define summability methods. When $[a, b]$ is finite and ψ satisfies Eq. 2.3.6(1) the method will have the regularity properties described in Theorem 2.3.6(1). Interestingly, members of the class of distribution functions to be discussed in this section do *not* satisfy Eq. 2.3.6(1). The most far-reaching results to date on the construction of distribution functions and associated polynomials are due to Pollaczek (1956). [See also Erdélyi *et al.* (1953, Chap. 10).]

Pollaczek's approach is as follows. It is known that any system of polynomials orthogonal with respect to a distribution function $\psi \in \Psi^*$ must satisfy a three-term recursion relationship:

$$P_n(x) = (A_n x + B_n)P_{n-1}(x) - C_n P_{n-2}(x), \qquad n \geq 1,$$
$$A_n \neq 0, \quad C_n/A_{n-1}A_n > 0. \tag{2}$$

(Without loss of generality we may assume $P_{-1} = 0$, $P_1 = 1$.) Suppose a generating function for the set $P_n(x)$ exists:

$$g(z, x) = \sum_{n=0}^{\infty} z^n P_n(x). \tag{3}$$

Let δ denote the operator $\delta = z\, d/dz$. Then

$$\delta g(z, x) = \sum_{n=0}^{\infty} n z^n P_n(x). \tag{4}$$

If the coefficients A_n, B_n, C_n are rational in n the substitution of (3) into (2), multiplying by the lowest common multiple of A_n, B_n, C_n and using the properties of the δ operator, produces an ordinary linear differential equation for g (in the variable z). If a fundamental set of solutions for the related homogeneous equation can be determined, then the equation for g can, in principle, be solved. Once g is found, Pollaczek shows that $\chi(z) = z^{-1}s(z^{-1})$ can be found, where

$$\chi(z) = \int \frac{d\psi}{z - t}, \tag{5}$$

and then, by using the inversion formula for the Hilbert transform (Shohat and Tamarkin, 1943, p. xiv) one can determine ψ:

$$\tfrac{1}{2}[\psi(t^+) + \psi(t^-)] - \tfrac{1}{2}[\psi(0^+) + \psi(0^-)]$$
$$= \lim_{\substack{\varepsilon \to 0 \\ \varepsilon > 0}} -\frac{1}{2\pi i} \int_0^t [\chi(t + i\varepsilon) - \chi(t - i\varepsilon)]\, dt. \tag{6}$$

The appropriate weights for computing the $[n - 1/n]$ Padé approximants to $s(z) = z^{-1}\chi(z^{-1})$ cannot, in general, be given in closed form but can be computed conveniently from (2):

$$\sigma_{nk} = A_n\sigma_{n-1,k-1} + B_n\sigma_{n-1,k} - C_n\sigma_{n-2,k}, \qquad 0 \le k \le n, \quad n \ge 1, \quad k \ge 0;$$
$$\sigma_{ij} = 0, \qquad i < 0, \quad j < 0, \quad \text{or} \quad j > i; \tag{7}$$
$$\sigma_{00} = 1.$$

For the class of transcendental functions to be discussed, this is tantamount to having a closed form expression for the $[n-1/n]$ Padé approximant. The necessity for solving linear equations is avoided (cf. Section 6.5) and, in fact, the ε-algorithm for generating Padé approximants is tedious.

The most general case considered by Pollaczek was for A_n, B_n, C_n bilinear functions of n having a common denominator. Through a suitable normalization the recursion relationship can be written

$$\begin{aligned}(n+c)P_n - 2[(n-1+\lambda+a+c)z+b]P_{n-1}&\\ + (n+2\lambda+c-2)P_{n-2} = 0,&\end{aligned} \tag{8}$$

so $A_n = 2(n-1+\lambda+a+c)/(n+c)$, etc. We shall assume a, b, c, and λ are real and $a > |b|$, $2\lambda + c > 0$, $c \ge 0$, although often an appeal to continuity will enable some of these conditions to be relaxed.

In what follows $\sqrt{z^2-1}$, $z \notin (-1, 1)$, will denote that branch of the function that is positive for z positive and >1.

Let

$$\begin{aligned}
B(z) &= \frac{az+b}{\sqrt{z^2-1}}, & \omega(z) &= z - \sqrt{z^2-1}, & &\\
B_+(t) &= \frac{-i(at+b)}{\sqrt{1-t^2}}, & \omega_+(t) &= t - i\sqrt{1-t^2}, & t &\in (-1, 1),\\
B_-(t) &= \frac{i(at+b)}{\sqrt{1-t^2}}, & \omega_-(t) &= t + i\sqrt{1-t^2}, & t &\in (-1, 1).
\end{aligned} \tag{9}$$

We wish to solve the integral equation

$$\chi(z) = \int_{-\infty}^{\infty} d\psi/(z-t), \qquad z \notin \operatorname{Supp} \psi \tag{10}$$

where

$$\chi(z) = \frac{\pi 2^{2-2\lambda}\Gamma(c+1)\Gamma(c+2\lambda)\omega(z)F\left(\begin{matrix}1-\lambda+B(z), c+1\\ c+\lambda+1+B(z)\end{matrix}\middle|\,\omega(z)^2\right)}{(\lambda+c+B(z))F\left(\begin{matrix}1-\lambda+B(z), c\\ c+\lambda+B(z)\end{matrix}\middle|\,\omega(z)^2\right)} \tag{11}$$

Pollaczek's work guarantees that a solution $\psi \in \Psi^*$ exists; it will be given by the inversion formula (6). It turns out that ψ is differentiable. Writing $d\psi = \rho\, dt$, we have

$$\rho(t) = i2^{1-2\lambda}\Gamma(c+1)\Gamma(c+2\lambda) \times \left[\frac{\omega_+(t)H_+(c+1)}{(\lambda+c+B_+(t))H_+(c)} - \frac{\omega_-(t)H_-(c+1)}{(\lambda+c+B_-(t))H_-(c)}\right], \tag{12}$$

$$H_+(c) = F\left(\begin{matrix}1-\lambda+B_+(t),\, c\\ \lambda+c+B_+(t)\end{matrix}\,\middle|\, \omega_+(t)^2\right),$$

etc. The computations are rather complicated, but straightforward. First, use the fact that $\omega_- = \omega_+^{-1}$, and then Eq. 2.10(2) of Erdélyi *et al.* (1953) on $H_-(c+1)$ and $H_-(c)$. Next use Eqs. 2.8(25) and 2.8(26). The result is

$$\begin{aligned}\rho(t) &= i2^{1-2\lambda}\omega_+^{2c+2\lambda+2B_+}e^{\pi i(1-\lambda-B_+)} \\ &\quad\times (1-\omega_+^2)\Gamma(\lambda+c-B_+)\Gamma(\lambda+c-1+B_+)(u_1u_5' - u_1'u_5), \\ u_1 &= F\left(\begin{matrix}1-\lambda+B_+,\, c\\ \lambda+c+B_+\end{matrix}\,\middle|\, \omega_+^2\right), \\ u_5 &= \omega_+^{2-2c-2\lambda-2B_+}F\left(\begin{matrix}2-2\lambda-c,\, 1-\lambda-B_+\\ 2-c-\lambda-B_+\end{matrix}\,\middle|\, \omega_+^2\right).\end{aligned} \tag{13}$$

The Wronskian of this pair of solutions of the hypergeometric equation is easily determined by the standard techniques, so that finally

$$\rho(t) = \begin{cases} \dfrac{\exp[(2\arccos t - \pi)(at+b)/\sqrt{1-t^2}] \times (1-t^2)^{\lambda-1/2}|\Gamma(\lambda+c+B_+(t))|^2}{\left|F\left(\begin{matrix}1-\lambda+B_+(t),\, c\\ \lambda+c+B_+(t)\end{matrix}\,\middle|\, \omega_+(t)^2\right)\right|^2}, & t\in(-1,1) \\ 0, & |t|>1. \end{cases} \tag{14}$$

Pollaczek has shown that the polynomials defined by (8) are orthogonal with respect to this weight function.

From this basic result a number of other Hilbert transforms and the recursion relationship for the corresponding orthogonal polynomials may be obtained. If $z \to \varepsilon z + \cos\phi, t \to \varepsilon t + \cos\phi, a \to \sin\phi/\varepsilon, b \to -\sin\phi\cos\phi/\varepsilon$, and $\varepsilon \to 0$, then

$$\frac{\pi 2^{2-2\lambda}\Gamma(c+1)\Gamma(c+2\lambda)e^{-i\phi}F\left(\begin{matrix}1-\lambda-iz, c+1\\ \lambda+c+1-iz\end{matrix}\middle| e^{-2i\phi}\right)}{(\sin\phi)^{2\lambda-1}(\lambda+c-iz)F\left(\begin{matrix}1-\lambda-iz, c\\ \lambda+c-iz\end{matrix}\middle| e^{-2i\phi}\right)}$$
$$= \int_{-\infty}^{\infty} \frac{e^{(2\phi-\pi)t}|\Gamma(\lambda+c-it)|^2\,dt}{(z-t)\left|F\left(\begin{matrix}1-\lambda-it, c\\ \lambda+c-it\end{matrix}\middle| e^{-2i\phi}\right)\right|^2}, \tag{15}$$

and the corresponding orthogonal polynomials satisfy

$$(n+c)P_n - 2[(n-1+\lambda+c)\cos\phi + x\sin\phi]P_{n-1} + (n+2\lambda+c-2)P_{n-2} = 0. \tag{16}$$

If we let $z \to z/\phi, t \to t/\phi$, in this result and $\phi \to 0$, the support of ψ collapses in a surprising way to $(-\infty, 0)$. To evaluate the limits, use Erdélyi *et al.* (1953, 2.10(1), 6.5(7)). Redefining things in an obvious way gives

$$\Gamma(a)\Gamma(a+2-c)\frac{\Psi(a+1, c; z)}{\Psi(a, c; z)} = \int_0^{\infty} \frac{e^{-t}t^{1-c}\,dt}{(z+t)|\Psi(a, c; -t)|^2}, \tag{17}$$

and the corresponding polynomials satisfy

$$(n+a)P_n - [2n+2a-c+x]P_{n-1} + (n+a-c)P_{n-2} = 0. \tag{18}$$

Letting $a \to 0$ gives the distribution for the Laguerre polynomials $L_n^{(1-c)}(x)$.

2.7. Other Methods for Generating Toeplitz Transformations

In this section U will denote *any* doubly infinite complex matrix $[\mu_{nk}]$ satisfying

(i) $\sum_{k=0}^{\infty} |\mu_{nk}| \le M$;
(ii) $\sum_{k=0}^{\infty} \mu_{nk} = 1 + o(1)$; and
(iii) $\mu_{nk} = o(1)$, k fixed.

By Theorem 2.1(2), U defines a regular summation process. Such a matrix will be called a *T-matrix*.

The methods previously used in this book to discover T-matrices, or at least those T-matrices that are also triangles, were generally based on a consideration of the zeros λ_{nk} of the characteristic polynomial for the method.

In a way, such an approach is unsatisfactory because no necessary and sufficient conditions on $\Lambda = \{\lambda_{nk}\}$ could be specified guaranteeing the regularity of U.

Garreau (1952), using a totally different approach, has found a way of systematically generating T-matrices that yields *all* T-matrices.

Theorem 1. Let $\boldsymbol{\sigma} \in \mathscr{R}_S^+$, $\sigma_n \to \infty$. Let $\mathbf{f}$ be a sequence of functions satisfying

(i) $\int_0^\infty |f_n(x)|\, dx < M$;
(ii) $\sigma_n^{-1} f_n(x) = o(1)$ uniformly for $x \in [0, h]$, for some $h > 0$; and
(iii) $\int_0^\infty f_n(x)\, dx = 1 + o(1)$.

Let

$$\mu_{nk} = \frac{1}{\sigma_n} \int_k^{k+1} f_n\!\left(\frac{u}{\sigma_n}\right) du, \qquad n, k \geq 0. \tag{1}$$

Then U is a T-matrix.

Proof

$$\begin{aligned} \sum_{k=0}^\infty |\mu_{nk}| &= \sigma_n^{-1} \sum_{k=0}^\infty \left| \int_k^{k+1} f_n\!\left(\frac{u}{\sigma_n}\right) \right| \\ &= \sigma_n^{-1} \sum_{k=0}^\infty \int_k^{k+1} \left| f_n\!\left(\frac{u}{\sigma_n}\right) \right| du \\ &= \int_0^\infty |f_n(x)|\, dx < M. \end{aligned} \tag{2}$$

Also for fixed k,

$$\begin{aligned} |\mu_{nk}| &= \sigma_n^{-1} \left| \int_k^{k+1} f_n\!\left(\frac{u}{\sigma_n}\right) du \right| \\ &\leq \sigma_n^{-1} \max_{k<u<k+1} \left| f_n\!\left(\frac{u}{\sigma_n}\right) \right|. \end{aligned} \tag{3}$$

Application of (ii) shows that $\mu_{nk} = o(1)$. Finally,

$$\sum_{k=0}^\infty \mu_{nk} = \sigma_n^{-1} \sum_{k=0}^\infty \int_k^{k+1} f_n\!\left(\frac{u}{\sigma_n}\right) = \int_0^\infty f_n(x)\, dx = 1 + o(1), \tag{4}$$

and the proof is complete. ■

Theorem 2. Given any T matrix U a sequence $\mathbf{f}$ of complex-valued locally integrable functions can be found that, using Eq. (1), yield U.

Proof. Let

$$\phi_n(k) = \sum_{r=k}^{\infty} \mu_{nr}, \qquad n, k \geq 0. \tag{5}$$

Extend this function to a function $\phi_n(x) \in C'[0, \infty)$. Now let $\boldsymbol{\sigma}$ satisfy the hypotheses of Theorem 1 and define $f_n(x)$ by

$$f_n(x) = -\sigma_n \phi_n'(x\sigma_n). \tag{6}$$

Then

$$\frac{1}{\sigma_n} \int_k^{k+1} f_n\left(\frac{u}{\sigma_n}\right) du = -\int_k^{k+1} \phi_n'(u)\, du = \phi_n(k) - \phi_n(k+1) = \mu_{nk}. \tag{7}$$

[Note that (i) and (ii) of Theorem 1 need not be satisfied.] ■

Example 1. Let $f_n(x) = 0$, $x > (n+1)/\sigma_n$, $k = n + p$, $p \geq 1$. Then

$$\begin{aligned} \mu_{n,n+p} &= \frac{1}{\sigma_n} \int_{n+p}^{n+p+1} f_n\left(\frac{u}{\sigma_n}\right) du \\ &= \int_{(n+p)/\sigma_n}^{(n+p+1)/\sigma_n} f_n(x)\, dx = 0, \end{aligned} \tag{8}$$

so U is lower triangular. Let $\sigma_n = n + 1$. Then $f_n(x) = 0$, $x > 1$, and

$$\mu_{nk} = \int_{k/(n+1)}^{(k+1)/(n+1)} f_n(x)\, dx, \qquad 0 \leq k \leq n. \tag{9}$$

Condition (iii) of Theorem 1 becomes

$$\int_0^1 f_n(x)\, dx = 1 + o(1).$$

Taking $f_n(x) = r(1-x)^{r-1}$, $\operatorname{Re} r > 0$, gives

$$\mu_{nk} = [1 - k/(n+1)]^r - [1 - (k+1)/(n+1)]^r, \qquad 0 \leq k \leq n, \tag{10}$$

which is a generalization of the Cesaro means ($r = 1$) called the *Riesz means.*

Example 2. Take the finite case of Example 1 with

$$f_n(x) = \begin{cases} rx^{r-1}, & 0 \leq x \leq 1, \quad \operatorname{Re} r > 0, \\ 0, & x > 1. \end{cases} \tag{11}$$

This gives

$$\mu_{nk} = [(k + 1)^r - k^r]/(n + 1)^r, \qquad 0 \le k \le n, \tag{12}$$

another generalization of the Cesaro means.

Example 3. Going the other way, let

$$\mu_{nk} = n^k/(n + 1)^{k+1}. \tag{13}$$

(This is called the Abel T-matrix.) Then

$$\phi_n(k) = \sum_{i=k}^{\infty} n^i(n + 1)^{-i-1} = n^k/(1 + n)^k, \tag{14}$$

and we may take $\phi_n(x) = [n/(n + 1)]^x$, $\sigma_n = n$. In this case conditions (i)–(iii) of Theorem 1 are satisfied.

Chapter 3 Linear Lozenge Methods

3.1. Background: Richardson Extrapolation and Romberg Integration

Let us see how linear lozenge techniques can arise in numerical analysis. Often a sequence $\mathbf{t} \in \mathscr{C}_S$ is known to converge to its limit t as

$$e_n = t_n - t \sim n^{\theta} \sum_{r=0}^{\infty} c_r n^{-\omega r}, \qquad \omega > 0, \quad \operatorname{Re} \theta < 0. \tag{1}$$

(This is slightly more general than a B–P series with $Q \equiv 0$ since ω need not be rational.) It is easily seen that for $p > 1$

$$e_{n \cdot p} \sim p^{\theta} n^{\theta} \sum_{r=0}^{\infty} c_r p^{-\omega r} n^{-\omega r}. \tag{2}$$

Multiplying (1) by p^{θ} and subtracting (2) gives

$$\frac{p^{\theta} t_n - t_{n \cdot p}}{p^{\theta} - 1} = t_n^{(1)} = t + e_n^{(1)} \tag{3}$$

and

$$e_n^{(1)} = n^{\theta - \omega} \sum_{r=0}^{\infty} c_r^{(1)} n^{-\omega r}. \tag{4}$$

Obviously, $t_n^{(1)}$ converges to t more rapidly than t_n.

Replacing n by np in (4) and repeating serves to define yet another sequence that converges more rapidly than $t_n^{(1)}$:

$$t_n^{(2)} = \frac{p^{\theta - \omega} t_n^{(1)} - t_{n \cdot p}^{(1)}}{p^{\theta - \omega} - 1} = t + e_n^{(2)}, \tag{5}$$

and so on. There is thus a double sequence defined recursively by

$$t_n^{(k+1)} = \frac{p^{\theta - k\omega} t_n^{(k)} - t_{n \cdot p}^{(k)}}{p^{\theta - k\omega} - 1}, \qquad n, k \geq 0. \tag{6}$$

The transformed sequence may have great advantages provided $t_{n \cdot p}$ may be computed easily once t_n is known. This is sometimes the case, particularly for certain numerical integration formulas.

Now replace n by p^n in (6) and define a new sequence

$$s_n^{(k)} = t_{p^n}^{(k)}. \tag{7}$$

The formula becomes

$$s_n^{(k+1)} = \frac{p^{\theta - k\omega} s_n^{(k)} - s_{n+1}^{(k)}}{p^{\theta - k\omega} - 1}, \qquad s_n^{(0)} = s_n = t_{p^n}, \qquad n, k \geq 0, \tag{8}$$

and this is the form the algorithm usually takes in practice. Obviously it can be applied to any sequence **s**. Furthermore, induction shows that, for some constants μ_{km},

$$s_n^{(k)} = \sum_{m=0}^{k} \mu_{km} s_{n+m}. \tag{9}$$

Putting $s_n \equiv 1$ shows $\sum \mu_{km} = 1$, so (8) is in reality just a triangle applied to the sequence **s** starting with $\{s_n, s_{n+1}, \ldots\}$ rather than $\{s_0, s_1, \ldots\}$. In fact, as is easily shown, the kth characteristic polynomial for the method is

$$P_k(\lambda) = \prod_{r=1}^{k} \left(\frac{\lambda - p^{\theta + (1-r)\omega}}{1 - p^{\theta + (1-r)\omega}} \right) \tag{10}$$

and thus, by Theorem 2.2(3), the U taking s_n into $s_n^{(1)}$ defines a regular method, which means

$$\begin{gathered} \mu_{km} = o(1), \qquad k \text{ in } J^+, \quad m \geq 0, \\ \sum_{m=0}^{k} |\mu_{km}| < M, \qquad k \geq 0. \end{gathered} \tag{11}$$

In fact an application of a result yet to be shown (Theorem 3.2(1)) gives the following theorem.

Theorem 1. The transformation defined by (8) is regular for any path **P**.

The computation scheme for the algorithm is as follows:

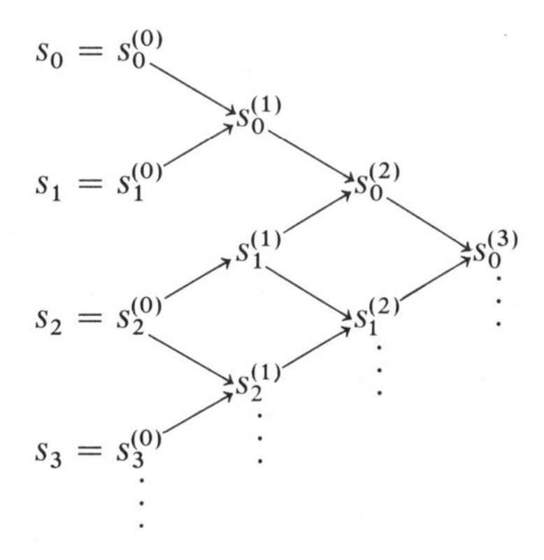

When $\theta = -1, p = \sigma, \omega = 1$, the triangle $[\mu_{km}]$ is the triangle of Romberg weights given in Section 2.3.9.

Example. As an example, consider the problem of evaluating

$$I = \int_0^1 f(x)\,dx \tag{12}$$

by the trapezoidal rule. Define

$$t_n = T_n(f) = \frac{1}{n} \sum_{k=0}^{n}{}'' f\left(\frac{k}{n}\right) \qquad (T_0 = 0), \tag{13}$$

where the double-prime notation means that the first and last terms in the sum are to be halved. Then

$$I = t_n - e_n, \tag{14}$$

and if f is sufficiently smooth [see Krylov (1962, Chap. 11)],

$$e_n \sim \frac{c_1}{n^2} + \frac{c_2}{n^4} + \frac{c_3}{n^6} + \cdots, \qquad c_1 = \tfrac{1}{12}[f'(1) - f'(0)],$$
$$c_2 = -\tfrac{1}{720}[f'''(1) - f'''(0)], \quad \ldots. \tag{15}$$

With $p = 2$, $\theta = -2$, $\omega = 2$ in (8), this means that to start the method one must compute T_0, T_2, T_4, $T_8, \ldots$; that is, each time the number of intervals in (13) is doubled. However, to calculate $T_{2^{n+1}}$ from T_{2^n} means computing the function f at only an additional 2^n points. Thus

$$s_n^{(0)} = T_{2^n} = \frac{1}{2^n} \sum_{k=0}^{2^n}{}'' f\left(\frac{k}{2^n}\right), \qquad n \geq 0, \tag{16}$$

and

$$s_n^{(k+1)} = \frac{4^{k+1} s_{n+1}^{(k)} - s_n^{(k)}}{4^{k+1} - 1}, \qquad s_n^{(0)} = T_{2^n}. \tag{17}$$

The following convergence theorem is useful.

Theorem 2. Let f be bounded and Riemann integrable on $[0, 1]$. Then along any path **P**

$$s_n^{(k)} \to \int_0^1 f(x)\,dx. \tag{18}$$

Further, if $f \in C^{(2\nu)}[0, 1]$, $\nu \geq 1$, then

$$\left| s_n^{(k)} - \int_0^1 f(x)\,dx \right| \leq \frac{K|B_{2\nu}|\,\|f^{(2\nu)}\| \prod_{j=1}^{\nu} (1 + 1/4^j)}{(2\nu)!\,4^{\nu(n+k)}}, \qquad k \geq \nu - 1,$$

$$K = \left(\frac{\theta_2}{\theta_1'}\right)_{q=1/2}^{1/2} = 1.969255. \tag{19}$$

Proof. The first statement is immediate. For the second, we use the following facts:

$$P_k(\lambda) = \prod_{m=1}^{k} \left(\frac{\lambda - 1/4^m}{1 - 1/4^m}\right), \tag{20}$$

$$r_n = \sum_{j=1}^{\nu=1} \alpha_j 2^{-2nj} + \frac{2^{-2\nu n}}{(2\nu)!} B_{2\nu} f^{(2\nu)}(\xi_n) \tag{21}$$

[see Krylov (1962, p. 216)].

Replacing n by $n + m$, multiplying by μ_{km}, and summing from $m = 0$ to k produces

$$\begin{aligned} |r_n^{(k)}| &\leq \frac{2^{-2\nu n}}{(2\nu)!} |B_{2\nu}|\,\|f^{(2\nu)}\| \left| P_k\left(-\frac{1}{4^\nu}\right)\right| \\ &\leq \frac{|B_{2\nu}|\,\|f^{(2\nu)}\|}{(2\nu)!\,4^{\nu(n+k)}} \frac{\prod_{j=1}^{k} (1 + 1/4^{j-\nu})}{\prod_{j=1}^{\infty} (1 - 1/4^j)} \\ &\leq \frac{|B_{2\nu}|\,\|f^{(2\nu)}\|}{(2\nu)!\,4^{\nu(n+k)}} \prod_{j=1}^{\nu} \left(1 + \frac{1}{4^j}\right) \prod_{j=1}^{\infty} \left(\frac{1 + 1/4^j}{1 - 1/4^j}\right), \end{aligned} \tag{22}$$

Table I

Romberg Algorithm, $s = \int_0^1 (x + 0.05)^{-1}\,dx = \ln 21 = 3.044522437723$

$n\backslash k$	0	1	2	3	4	5	6	7
0	10.476							
		4.704						
1	6.147		3.502					
		3.576		3.146				
2	4.219		3.152		3.059			
		3.178		3.060		3.046		
3	3.438		3.061		3.046		3.044574	
		3.068		3.046		3.044575		3.044523
4	3.161		3.046		3.044576		3.044523	
		3.047		3.044581		3.044523		—
5	3.076		3.04460		3.044523		—	
		3.0447		3.044524		—		
6	3.053		3.044525		—			
		3.044541		3.044522436				
7	3.047		3.044522475					
		3.044573						
8	3.045							

and identifying the infinite product in terms of θ-functions is straightforward (Rainville, 1960, Section 173). ■

This theorem seems to indicate that for functions in $C^{(2\nu)}$, diagonal convergence ($k \to \infty$) is not much better than vertical convergence. Numerical evidence seems to bear this out. As an example take $f = (x + 0.05)^{-1}$. The results of applying the algorithm are given in Table I. Note $s_6^{(2)}$ is nearly as accurate as any other entry in the table yet easier to compute and less subject to roundoff error than entries on its right. It is a curious contrast that in most other deltoid algorithms diagonal convergence is more rapid than vertical convergence.

3.2. General Deltoids

The Romberg integration scheme leads one to analyze the more general deltoid scheme defined by

$$s_n^{(k+1)} = a_k s_{n+1}^{(k)} + b_k s_n^{(k)}, \qquad n, k \geq 0, \quad a_k, b_k \text{ real}, \quad s_n^{(0)} = s_n. \tag{1}$$

Let $\mathbf{s} \in \mathscr{C}_C$. The idea is to determine when the above transformation is regular for any $\mathbf{P}$. If $s_n^{(k)} \to s$ is to hold, an induction argument shows

$$b_k = 1 - a_k. \tag{2}$$

From here on assume this holds. As before, write

$$s_n^{(k)} = \sum_{m=0}^{k} \mu_{km} s_{n+m} \tag{3}$$

for some constants μ_{km}. Substituting (3) in (1) shows

$$\sum_{m=0}^{k+1} \mu_{k+1,m} s_{n+m} = a_k \sum_{m=0}^{k} \mu_{km} s_{n+m+1} + b_k \sum_{m=0}^{k} \mu_{km} s_{n+m}, \tag{4}$$

and this will hold for all possible sequences iff

$$\mu_{k+1,m} = a_k \mu_{k,m-1} + b_k \mu_{km} \qquad (\mu_{km} = 0, \quad m < 0, \quad m > k). \tag{5}$$

Multiplying by λ^m and summing from $m = 0$ to $k + 1$ gives

$$P_{k+1}(\lambda) = (a_k \lambda + b_k) P_k(\lambda) \tag{6}$$

or, since $P_k(1) = 1$,

$$P_k(\lambda) = \prod_{m=1}^{k} \left(\frac{\lambda - \lambda_m}{1 - \lambda_m} \right), \qquad \lambda_m = \frac{a_{m-1} - 1}{a_{m-1}}. \tag{7}$$

Furthermore, in the notation of Chrystal (1959, vol. I, p. 431),

$$\mu_{km} = (-1)^{k+m} \mathscr{P}_{k-m}(\boldsymbol{\lambda}), \tag{8}$$

where $\mathscr{P}_r(\boldsymbol{\lambda})$ is the product of $\lambda_1, \ldots, \lambda_k$ taken r at a time.

An application of Theorem 2.2(4) furnishes the next result.

Theorem 1. Let $a_k \notin [0, 1]$. Then the transformation defined by (1) is regular for all paths **P** iff $\sum (1 - a_k^{-1})$ converges.

It often happens, of course, that $s_n^{(k)}$ goes to s along some **P** much more rapidly than s_n goes to s as $n \to \infty$. Then the scheme defined by (1) is computationally desirable, as is the case for Romberg integration.

This algorithm can be derived heuristically on the assumption that the given sequence s_n behaves as

$$s_n = s + \sum_{r=1}^{k} c_r \lambda_r^n, \qquad \lambda_k = \frac{a_{k-1} - 1}{a_{k-1}} \tag{9}$$

(see Chapter 10). Consequently, one would expect the algorithm to be exact ($s_n^{(k)} \equiv s$), when **s** has such a representation. This turns out to be true and does not even require convergence.

Theorem 2. Let **s** have the above representation with $a_j \neq 0$ for some complex constants c_r. Then $s_n^{(k)} \equiv s, n \geq 0$.

Proof. Trivial. ■

3.3. Deltoids Obtained by Extrapolation

Other deltoid formulas are generated by the Neville–Aitken formalization of the Lagrangian interpolation polynomial [see Householder (1953, pp. 202ff.)]. Let **x** be an arbitrary sequence of distinct numbers, and suppose there is a function $f(x)$ such that $f(x_j) = s_j, j \geq 0$, **s** being the sequence to be transformed.

Consider the algorithm given by the following computational scheme:

$$s_n^{(k+1)} = \frac{x_n s_{n+1}^{(k)} - x_{n+k+1} s_n^{(k)}}{x_n - x_{n+k+1}}, \qquad n, k \geq 0, \quad s_n^{(0)} = s_n, \quad n \geq 0. \tag{1}$$

One sees immediately, by referring to the appendix, that $s_n^{(k)}$ is the value at $z = 0$ of the Lagrangian polynomial of degree k in z that interpolates to s_j at $x_j, n \leq j \leq n + k$. Now $f(x_n) = s_n$, so if x_n, for instance, is decreasing and f is reasonably behaved, $f(0) = s$, and for n and/or k large $s_n^{(k)}$ will closely approximate s.

Laurent (1964) has shown necessary and sufficient conditions for diagonal ($k \to \infty$) regularity of the algorithm. A minor additional effort enables us to show the same conditions are equivalent to regularity for any path.

Theorem 1. Let **x** be monotone decreasing to zero. Then the algorithm (1) is regular for all paths iff $x_n/x_{n+1} \geq \alpha > 1, n \geq 0$.

Proof. By the formulas in the appendix, we can write

$$s_n^{(k)} = \sum_{m=0}^{k} \mu_{km}(n) s_{n+m}, \qquad \mu_{km}(n) = \prod_{i=0, i \neq m}^{k} \frac{x_{n+i}}{x_{n+i} - x_{n+m}}. \tag{2}$$

Note $\sum \mu_{km} = 1$.

$\Rightarrow$: Assume that on the contrary

$$x_n/x_{n+1} = 1 + \varepsilon_n, \tag{3}$$

where $\boldsymbol{\varepsilon} \in \mathscr{R}_S^+$ and contains a subsequence converging to zero. Then

$$\mu_{nk} = \prod_{i=0}^{k-1} \frac{x_{n+i}}{x_{n+i} - x_{n+k}} > \frac{x_{n+k-1}}{x_{n+k-1} - x_{n+k}}, \tag{4}$$

since each term in the product is greater than 1. Thus

$$\mu_{nk} > (1 + \varepsilon_{n+k-1})/\varepsilon_{n+k-1} \tag{5}$$

and, taking n fixed, we can pick k values $\to \infty$ such that ε_{n+k} is a member of the aforementioned subsequence of $\boldsymbol{\varepsilon}$. Thus μ_{kn} contains a positively divergent subsequence. Therefore, by the Toeplitz limit theorem, the transformation defined by (1) is not regular (in k). In fact, this shows that (1) is not regular for any diagonal path.

$\Leftarrow$: Taking products of $x_n/x_{n+1} \geq \alpha$ from $n = j$ to $k - 1$ and reciprocating gives

$$x_k/x_j \leq \alpha^{j-k}, \qquad k > j. \tag{6}$$

Write

$$|\mu_{km}| = A \cdot B, \qquad A = \left|\prod_0^{m-1}\right|, \quad B = \left|\prod_{m+1}^{k}\right|. \tag{7}$$

Then

$$A = \prod_{i=0}^{m-1}\left(1 - \frac{x_{n+m}}{x_{n+m-i-1}}\right)^{-1}. \tag{8}$$

But $x_{n+m}/x_{n+m-1-1} \leq \alpha^{-i-1}$, so $A \leq d$, where d is the limiting value of the convergent monotone increasing sequence $\prod^m (1 - \alpha^{-i-1})^{-1}$ [see Knopp (1947, p. 219, Theorem 3)].

Also,

$$\begin{aligned} B &= \prod_{i=0}^{k-m-1} \frac{x_{n+m+i+1}}{x_{n+m} - x_{n+m+i+1}} = \prod_{i=0}^{k-m-1} \frac{x_{n+m+i+1}/x_{n+m}}{1 - x_{n+m+i+1}/x_{n+m}} \\ &\leq \prod_{i=0}^{k-m-1} \frac{\alpha^{-i-1}}{1 - \alpha^{-i-1}} < d\alpha^{-(k-m)(k-m+1)/2}. \end{aligned} \tag{9}$$

Thus

$$\begin{aligned} |\mu_{km}| &\leq d^2\alpha^{-(k-m)(k-m+1)/2}, \\ M_k &= \sum_{m=0}^{k} |\mu_{km}| \leq d^2 \sum_{m=0}^{k} \alpha^{-m(m+1)/2} < C. \end{aligned} \tag{10}$$

Since both the bounds are independent of n and $\mu_{km} \to 0$ as $k \to \infty$, $m \geq 0$ fixed, an application of a result soon to be given, Theorem 5.2(1), completes the proof. ■

Example 1. Let $x_i = \sigma^i$, $0 < \sigma < 1$. This choice yields a special case of the Romberg–Richardson algorithm 3.1(8). The algorithm is regular for all paths.

Example 2. Let $x_i = 1/(i + 1)$. Then

$$\mu_{km} = \frac{(-1)^{m+k}}{k!}(n + m + 1)^k\binom{k}{m}, \tag{11}$$

and

$$s_n^{(k+1)} = \frac{(n+k+2)s_{n+1}^{(k)} - (n+1)s_n^{(k)}}{k+1}, \qquad n, k \geq 0, \quad s_n^{(0)} = s_n, \quad n \geq 0, \tag{12}$$

is a deltoid formalization of Salzer's weight scheme for $\gamma = 1$, Eq. 2.3.4(1). Clearly, the hypotheses of the theorem are violated. In fact, it is easy to show the algorithm (12) is regular on no vertical or diagonal path. Nevertheless, used on appropriate sequences, the technique is very valuable.

Let $s_n = (\mathrm{GAM})_n$ so that

$$a_n = 1/(n+1) + \ln[n/(n+1)], \quad n \geq 1, \qquad a_0 = 1. \tag{13}$$

Then $s_n \to \gamma = 0.5772156649$, Euler's constant. Suppose only ten terms of **s** are available. How well can one do in computing γ? Table II lists the tenth ascending diagonal of $s_n^{(k)}$, each term of which requires $s_0, \ldots, s_9$.

Example 3. Let $x_i = 1/(1+i)^2$. Then

$$\mu_{km} = \frac{2(-1)^{m+k}(n+m+1)^{2k+1}}{k!(2n+m+2)_{k+1}} \binom{k}{m} \tag{14}$$

and

$$s_n^{(k+1)} = \frac{(n+k+2)^2 s_{n+1}^{(k)} - (n+1)^2 s_n^{(k)}}{(k+1)(2n+k+3)}. \tag{15}$$

This algorithm is appropriate for sequences behaving as

$$c_0/n^2 + c_1/n^4 + \cdots;$$

Table II

k	$s_{9-k}^{(k)}$
0	0.626
1	0.578
2	0.577219
3	0.577214
4	0.577215590
5	0.577215682
6	0.577215669
7	0.577215665
8	0.5772156643
9	0.5772156644

Table III

k	$s_{25-k}^{(k)}$
0	0.6932
5	0.6931469
10	0.69314705
15	0.693147089
20	0.693147099
25	0.693147100

e.g., the sequence of iterates T_n in the trapezoidal rule 3.1(13). Table III gives some entries on the 26th ascending diagonal for $f(x) = (x + 1)^{-1}$ (ln 2 = 0.6931471805.)

It is easy to show that for the Toeplitz array U corresponding to $\mu_{km}(0)$ (these weights yield the diagonal entries $s_0^{(k)}$) one has

$$\kappa(u) \geq e^2 = 7.389, \tag{16}$$

worse than that for the methods given in Sections 2.3.4 or 2.3.5. Such numerical instability dictates great caution in the use of (15).

There are several ways of looking at the acceleration properties of lozenge algorithms. One is to compare rapidity of convergence along different paths. Very little work has been done in this area. However, an interesting condition for horizontal acceleration in the previous algorithm is due to Brezinski (1972).

Theorem 2. Let $\mathbf{x}$ be monotone decreasing to zero and $x_n/x_{n+1} \geq \alpha > 1$. Then for $\mathbf{s} \in \mathscr{C}_S$, $s_n^{(k+1)}$ converges more rapidly than $s_n^{(k)}$, $n \to \infty$, k fixed, iff

$$\lim_{n \to \infty} (s_{n+1}^{(k)} - s)/(s_n^{(k)} - s) = \lim_{n \to \infty} (x_{n+k+1}/x_n). \tag{17}$$

Proof. Left to the reader. ■

The algorithm of this section can be derived formally from the assumption that $\mathbf{s}$ behaves as

$$s_n = s + \sum_{r=1}^{k} c_r x_n^r. \tag{18}$$

Not surprisingly, the algorithm is exact for such sequences, even when $\mathbf{x}$ depends on $\mathbf{s}$.

Theorem 3. Let $\mathbf{s}$ have the foregoing representation with $x_i \neq x_j$, $i \neq j$, for some complex constants c_m. Then $s_n^{(k)} \equiv s$, $n \geq 0$.

Proof. Trivial. ■

3.4. Example: Quadrature Based on Cardinal Interpolation

A class of quadrature formulas derived from a general Hermite cardinal interpolation formula provides an excellent example of the summation process of Section 3.3.

It has long been known that the approximation of a doubly infinite integral by a trapezoidal sum

$$I = \int_{-\infty}^{\infty} f(x)\,dx \approx h \sum_{-\infty}^{\infty} f(mh) \tag{1}$$

gives surprisingly good results in many cases; i.e., the series on the right approaches rapidly the value of the integral as $h \to 0$. For instance, if $f = e^{-x^2}$ and $h = 1$, the sum has the value 1.77264, to be compared to $\sqrt{\pi} = 1.77245$. This agreement is nothing short of phenomenal, considering how few values of f are required to define the sum, and indicates that something profound is going on.

In many instances, however, there are knotty computational problems associated with (1). It may happen that the right-hand side is indeed a good approximation to the integral, but converges very slowly; in fact, those small values of h that give a good approximation produce a slowly convergent series. An example is

$$1 = \frac{1}{\pi} \int_{-\infty}^{\infty} \frac{dx}{1 + x^2} \approx \frac{h}{\pi} \sum_{-\infty}^{\infty} \frac{1}{1 + (mh)^2}. \tag{2}$$

One would like a procedure to calculate I based on as few evaluations of the sum as possible. One approach is to truncate the sum at N, which depends on h, and to try to find, given h and a suitable class of functions f, the values of N that produce optimal accuracy. This approach is the basis of the so-called *tanh rule*. However the iterates in that rule are not suitable for the application of the present summation procedure. A more general procedure, the BL protocol, is required; the subject is discussed in Section 11.3.

Note that any procedure to compute I is adaptable to the computation of finite integrals; for instance, the substitution $x = \tanh t$ gives an integral over $(-1, 1)$. (Some writers have conjectured that this change of variable is, in some sense, the best choice; again see the discussion in Section 11.3.)

The quadrature formulas to which Section 3.3 is to be applied are generalizations of (1). Let $f: \mathscr{R} \to \mathscr{C}$ and $h > 0$. The series

$$T_h(f)(z) = \sum_{-\infty}^{\infty} f(mh) \frac{\sin \omega_m}{\omega_m}, \qquad z \in \mathscr{C}, \qquad \omega_m = \frac{\pi}{h}(z - mh), \tag{3}$$

is called the *cardinal interpolation series* of the function f with respect to h. Obviously,

$$T_h(f)(mh) = f(mh), \qquad m \in J. \tag{4}$$

This formula and its remainder have been thoroughly investigated [e.g., Kress (1971); McNamee *et al.* (1971)]. Here a more general interpolation series is required.

Define the $p + 1$ entire functions $t_q(z)$, $0 \le q \le p$, by

$$t_q(z) = \frac{z^q}{q!}\left[\frac{\sin(\pi z/h)}{\pi z/h}\right]^{p+1} \sum_{\substack{r=0 \\ r\,\text{even}}}^{2\langle(p-q)/2\rangle} a_r \left[\frac{\pi z}{h}\right]^r, \tag{5}$$

$\langle\alpha\rangle$ indicating largest integer contained in α. $a_r \equiv a_r(p)$ are the coefficients of the Taylor series,

$$\left[\frac{z}{\sin z}\right]^{p+1} = \sum_{\substack{r=0 \\ r\,\text{even}}}^{\infty} a_r z^r, \qquad |z| < \pi. \tag{6}$$

Lemma. Let $0 \le r, q \le p$. Then

$$t_q^{(r)}(0) = \delta_{qr}, \qquad t_q^{(r)}(mh) = 0, \quad |m| \ge 1. \tag{7}$$

Proof. Since

$$t_q(z) = z^q/q! + z^{p+1}u_q(z), \qquad 0 \le q \le p, \tag{8}$$

where u_q is entire, (7) is immediate. ∎

Now let $f: \mathscr{R} \to \mathscr{C}$, $f \in C^p(\mathscr{R})$, $h > 0$. The series

$$T_{p,h}(f)(z) = \sum_{-\infty}^{\infty} \sum_{q=0}^{p} f^{(q)}(mh)t_q(z - mh) \tag{9}$$

is called the pth *cardinal interpolation series* of f with respect to h. Clearly

$$\frac{d^q}{dz^q} T_{p,h}(f)(z)|_{z=mh} = f^{(q)}(mh), \qquad 0 \le q \le p, \quad m \in J. \tag{10}$$

For functions f analytic and bounded in a strip $[-ia, ia] \times \mathscr{R}$, a remainder formula is easy to derive [see Kress (1972)]. Its exact form is not important for our purposes. It suffices to say that for all x the remainder is $O(e^{-\pi(p+1)a/h})$. It is the integration of (9) that provides the desired quadrature formula:

$$I_{p,h}(f) = h \sum_{-\infty}^{\infty} \sum_{\substack{q=0 \\ q\ \text{even}}}^{p} b_q f^{(q)}(mh), \qquad b_q = \frac{1}{h}\int_{-\infty}^{\infty} t_q(x)\,dx. \tag{11}$$

A simple recursion formula exists for the computation of b_q (Kress, 1972). If the first several such formulas are recorded, it turns out that I_{0h} is given by (1), $I_{2p+1,h} = I_{2p,h}$, and

$$\begin{aligned} I_{2h} &= I_{0h} + \frac{h^3}{4\pi^2} \sum_{-\infty}^{\infty} f''(mh), \\ I_{4h} &= I_{0h} + \frac{5h^3}{16\pi^2} \sum_{-\infty}^{\infty} f''(mh) + \frac{h^5}{64\pi^4} \sum_{-\infty}^{\infty} f''''(mh). \end{aligned} \tag{12}$$

A remainder formula can be computed directly from that for (9).

Theorem. Let f be analytic in $[-ia, ia] \times \mathscr{R}$, $f(x + iy) \to 0$ as $x \to \pm\infty$ uniformly for $-a \le y \le a$, and

$$A_{\pm} = \int_{-\infty \pm ia}^{-\infty \pm a} |f(z)||dz| < \infty. \tag{13}$$

Then $I = \int_{-\infty}^{\infty} f(x)\,dx$ exists and

$$|I - I_{p,h}| \le \frac{e^{-\pi a/h}}{2(\sinh \pi a/h)^{p+1}} (A_+ + A_-). \tag{14}$$

This is a generalization of a result ($p = 1$) first given, apparently, by Luke (1969, vol. 2, p. 217).

Now for $p, f, N > 1$ fixed (and thus a, which may be taken as the distance from $\mathscr{R}$ to the nearest singularity of f), let $h = N/(n + 1)$ and define

$$s_n = I_{p,N/(n+1)}, \qquad 0 \le n \le N - 1. \tag{15}$$

Equations (14) and 3.3(18) suggest taking

$$x_n = e^{-\pi a(p+2)n/N} \tag{16}$$

in the summation formula 3.3(1). Thus one can compute the $s_n^{(k)}$ array for $0 \le n + k \le N - 1$. Equation (16) turns out to be a very happy choice. Take as an example $f(x) = 1/\pi(x^2 + 1)$, $p = 0$, and consider formula (2). Then $a = 1$. The lozenge formula is

$$s_n^{(k+1)} = \frac{s_{n+1}^{(k)} - e^{-2\pi(k+1)/N} s_n^{(k)}}{1 - e^{-2\pi(k+1)/N}}, \qquad 0 \le n + k \le N - 1. \tag{17}$$

Table IV gives the results for $N = 4$. Thus we have I to almost eight significant figures with only four evaluations of the sum in (2).

Table IV

n	s_n	$s_n^{(1)}$	$s_n^{(2)}$	$s_n^{(3)}$
0	1.524868619			
		0.976293939		
1	1.090331411		1.000214887	
		0.999181169		0.999999598
2	1.018129443		1.000001532	
		0.999966081		
3	1.003741873			

3.5. General Rhombus Lozenges

This section shows how a lozenge algorithm can be developed for the orthogonal triangles discussed in Section 2.3.6.

Theorem 1. Let $\{p_k(x)\}$ be a system of polynomials orthogonal on $[-1, 1]$ with respect to $\psi \in \Psi$ with

$$p_{k+1}(x) = (A_k x + B_k)p_k(x) - C_k p_{k-1}(x), \qquad k \geq 0, \quad p_{-1} \equiv 0. \tag{1}$$

Then the sequence transformation defined by

$$s_n^{(k+1)} = a_k s_n^{(k)} + b_k s_{n+1}^{(k)} + c_k s_n^{(k-1)}, \qquad n, k \geq 0, \quad s_n^{(-1)} = 0, \quad s_n^{(0)} = s_n, \tag{2}$$

where

$$\begin{aligned} a_k &= (B_k + A_k)\sigma_k/\sigma_{k+1}, \\ b_k &= 2A_k\sigma_k/a\sigma_{k+1}, \\ c_k &= -C_k\sigma_{k-1}/\sigma_{k+1}, \\ \sigma_k &= \sigma_k(a) = p_k(2/a + 1), \qquad a > 0, \end{aligned} \tag{3}$$

is regular for any path **P**.

Proof. First, note that σ_k satisfies

$$\sigma_{k+1} = [(2/a + 1)A_k + B_k]\sigma_k - C_k\sigma_{k-1}. \tag{4}$$

This shows that $a_k + b_k + c_k = 1$. Thus for some constants μ_{km},

$$s_n^{(k)} = \sum_{m=0}^{k} \mu_{km} s_{n+m}, \tag{5}$$

and putting $s_n \equiv 1$ in (2) shows that $[\mu_{km}]$ is a triangle. Proceeding as before, we find that $P_k(\lambda)$ satisfies

$$P_{k+1}(\lambda) = (b_k\lambda + a_k)P_k(\lambda) + c_k P_{k-1}(\lambda), \qquad P_{-1} \equiv 0, \tag{6}$$

and this is precisely the recursion relationship satisfied by $p_k(2\lambda/a + 1)/p_k(2/a + 1)$, and since the two agree when $k = 0$ and $k = 1$, identically

$$P_k(\lambda) = \frac{p_k(2\lambda/a + 1)}{p_k(2/a + 1)}. \tag{7}$$

Theorem 2.2(5) then asserts that U is regular. The rest of the proof is as in Theorem 3.2(1). ■

The computational scheme for the algorithm is as follows:

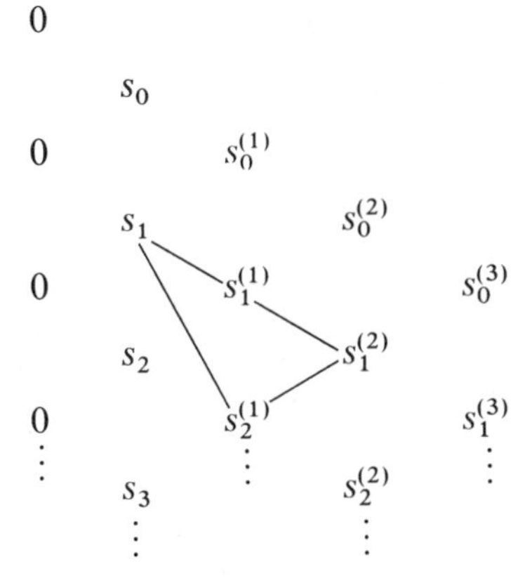

In the algorithm it is much more efficient to compute

$$t_n^{(k)} = \sigma_k s_n^{(k)} \tag{8}$$

and then divide $t_n^{(k)}$ by σ_k (which itself satisfies a simple recursion relation) to get $s_n^{(k)}$. The algorithm becomes

$$\begin{aligned} t_n^{(k+1)} &= (B_k + A_k)t_n^{(k)} + \frac{2A_k}{a} t_{n+1}^{(k)} - C_k t_n^{(k-1)}, \qquad n, k \geq 0, \\ t_n^{(-1)} &\equiv 0, \quad t_n^{(0)} = s_n; \\ s_n^{(k)} &= t_n^{(k)}/\sigma_k; \\ \sigma_{k+1} &= [(2/a + 1)A_k + B_k]\sigma_k - C_k\sigma_{k-1}, \qquad k \geq 1. \end{aligned} \tag{9}$$

As an example consider the Chebyshev polynomials $T_k(x)$ with $a = 1$. These satisfy

$$\begin{aligned} T_{k+1}(x) &= \varepsilon_k x T_k(x) - T_{k-1}(x), \qquad k \geq 0, \\ T_{-1} &\equiv 0, \qquad \varepsilon_k = \begin{cases} 1, & k = 0 \\ 2, & k > 0, \end{cases} \end{aligned} \tag{10}$$

Thus

$$t_n^{(k+1)} = \varepsilon_k t_n^{(k)} + 2\varepsilon_k t_{n+1}^{(k)} - t_n^{(k-1)}, \quad t_n^{(-1)} = 0; \qquad t_n^{(0)} = s_n. \tag{11}$$

Even simpler is the algorithm given by the Chebyshev polynomials of the second kind $U_k(x)$:

$$U_{k+1}(x) = 2xU_k(x) - U_{k-1}(x), \quad k \geq 0, \quad U_{-1} \equiv 0; \qquad U_0 = 1, \tag{12}$$

and so

$$t_n^{(k+1)} = 2t_n^{(k)} + 4t_{n+1}^{(k)} - t_n^{(k-1)}. \tag{13}$$

For T_k

$$\sigma_k(1) = \{1, 3, 17, 99, 577, 3363, 19601, 114243, \ldots\}, \tag{14}$$

and for U_k

$$\sigma_k(1) = \{1, 6, 35, 204, 1189, 6930, 40391, 235416, \ldots\}. \tag{15}$$

Both satisfy

$$\sigma_{k+1} = 6\sigma_k - \sigma_{k-1}, \qquad k \geq 1. \tag{16}$$

Applying the T_k algorithm to the sequence $s_n = \sum_{k=0}^{n} (-1)^k$ with the computational scheme yields Table V for $t_n^{(k)}$. This sequence is divergent, and thus $s_n^{(k)}$ cannot sum **s** along all paths **P** (**s** must converge if that is to happen). However, it is easy to show that $s_n^{(k)} \to \frac{1}{2}$ as $k \to \infty$. This, in fact, is the traditional "sum" assigned to the sequence [see Knopp (1947, Chap. XIII) for a historical discussion]. The sequence $\{s_0^{(k)}\}$ obtained from the preceding table is one of dramatic precision:

$$\{1, 0.333, 0.529, 0.495, 0.501, 0.49985, 0.500025, 0.499996, \ldots\}. \tag{17}$$

Table V

$t_n^{(k)}$

0								
	1							
0		1						
	0		9					
0		2		49				
	1		8		289			
0		1		50		1681		
	0		9		288		9801	
0		2		49		1682		57121
	1		8		289		9800	
0		1		50		1681		
	0		9		288			
0		2		49				
	1		8					
0		1						
	0							
0								

It is likely that no other linear method is more efficient than T_k for sequences that alternate around their limits.

Two-dimensional algorithms can also be developed for the nonregular class of triangles of Section 2.3.5. The case $\tau = 1$ (Legendre polynomials) is particularly simple:

$$s_n^{(k+1)} = \frac{(2k+1)[2s_{n+1}^{(k)} - s_n^{(k)}] - ks_n^{(k-1)}}{(k+1)}, \qquad n, k \geq 0, \tag{18}$$

$$s_n^{(-1)} = 0, \quad s_n^{(0)} = s_n.$$

In general, this algorithm will be regular only along vertical paths. However for an important class of sequences it is regular along any path.

Theorem 2. Let $\mathbf{r} \in \mathscr{R}_{\mathrm{TM}}$. Then the transformation defined by (18) is regular along any $\mathbf{P}$.

Proof

$$r_n^{(k)} = s_n^{(k)} - s = \sum_{m=0}^{k} \mu_{km} r_{n+m} = \int_0^1 t^n P_k(2t-1)\, d\psi, \tag{19}$$

so for any δ, $0 < \delta < 1$,

$$\begin{aligned} |r_n^{(k)}| &\leq \int_0^\delta d\psi + \int_{1-\delta}^1 d\psi + \left| \int_\delta^{1-\delta} t^n P_k(2t-1)\, d\psi \right| \\ &\leq \int_0^\delta d\psi + \int_{1-\delta}^1 d\psi + \frac{C(\delta)}{k}(1-\delta)^n \end{aligned} \tag{20}$$

by Erdélyi (1953, 10.18, (1), (5)). Now pick δ so that the first two terms are less than ε and take $\overline{\lim}$ as $n + k \to \infty$ along $\mathbf{P}$. This shows $|r_n^{(k)}| \to 0$ or $s_n^{(k)} \to s$ along $\mathbf{P}$. ■

The exactness problem for the algorithms of this section is easily disposed of.

Theorem 3. Let $s_n^{(k)}$ be defined by the algorithm of Theorem 1 except that now a in (3) may be any nonzero complex number.

If $\mathbf{s}$ has the representation

$$s_n = s + \sum_{r=1}^{k} c_r \lambda_{kr}^n, \tag{21}$$

λ_{kr} a zero of $p_k(2\lambda/a + 1)$, then $s_n^{(k)} \equiv s$, $n \geq 0$.

Obviously, if the interval of orthogonality of $p_k(2\lambda/a + 1)$ lies partly outside N, the algorithm will sum some divergent sequences.

For other such algorithms, see Wimp (1974c).

Chapter 4 Optimal Methods and Methods Based on Power Series

4.1. Best Methods for Laplace Moment Sequences

In this section we assume that $\mathbf{s} \in \mathscr{C}_C$ may be represented as its limit s plus a certain moment sequence as follows:

$$s_n = s + \bar{f}(\alpha + n), \qquad \operatorname{Re} \alpha > 0, \quad n \geq 0, \tag{1}$$

$$\bar{f}(\xi) = \int_0^\infty e^{-\xi t} f(t)\, dt, \qquad f \in L_1(0, \infty). \tag{2}$$

(The motivation for representing $\mathbf{s}$ this way will be discussed later.)

Theorem 1. Let $f \in L_1(0, \infty)$, let $f^{(r)}$ be locally integrable for some $r \geq 1$, and suppose that

$$f^{(r)} e^{-\alpha c u} \in L_p(0, \infty) \qquad \text{for some} \quad p > 1, \quad 0 < c < 1. \tag{3}$$

Further, let the nth characteristic polynomial of a triangle U satisfy the r conditions

$$\int_0^\infty e^{-\alpha t} P_n(e^{-t}) t^j \, dt = 0, \qquad 0 \leq j \leq r - 1, \quad n \geq r. \tag{4}$$

Then

$$|\bar{r}_n| \leq \| f^{(r)} e^{-\alpha c u} \|_p E_n^{(q)}, \qquad n \geq r, \tag{5}$$

where

$$E_n^{(q)} = \left\| \sum_{k=0}^n \frac{\mu_{nk} e^{-\alpha d - k u}}{(\alpha + k)^r} \right\|_q, \qquad c + d = 1, \quad \frac{1}{p} + \frac{1}{q} = 1. \tag{6}$$

Equality in (5) is attained for some f in the class of functions (3).

Proof. f may be represented

$$f(t) = \sum_{i=0}^{r-1} \frac{f^{(i)}(0)t^i}{i!} + \frac{1}{\Gamma(r)} \int_0^\infty f^{(r)}(u)(t-u)^{r-1}E(t-u)\,du, \tag{7}$$

where

$$E(t) = \begin{cases} 1, & t > 0 \\ 0, & t < 0. \end{cases} \tag{8}$$

Substituting this in the representation (1) and (2), summing, and using (4) gives after an easily justified change in the order of integration

$$\begin{aligned} \bar{r}_n &= \int_0^\infty f^{(r)}(u)K_n(u)\,du, \\ K_n(u) &= \frac{1}{\Gamma(r)} \int_u^\infty e^{-\alpha t}P_n(e^{-t})(t-u)^{r-1}\,dt. \end{aligned} \tag{9}$$

If a change of variable is made ($t - u = x$), this may be written

$$K_n(u) = e^{-\alpha u} \sum_{k=0}^{n} \frac{\mu_{nk}e^{-ku}}{(\alpha + k)^r} = O(e^{-\alpha u}), \qquad k \to \infty. \tag{10}$$

Because of the integrability condition (3), Hölder's inequality may be invoked, proving the theorem. [That an f may be found yielding equality in (5) is a standard argument: see Krylov (1962, pp. 134ff.).] ∎

It is desirable to have a more workable representation of $K_n(u)$, and so of $E_n^{(q)}$. To do this we define the polynomial

$$P_{nr}(\lambda) = \sum_{k=0}^{n} \frac{\mu_{nk}\lambda^k}{(\alpha + k)^r}, \tag{11}$$

so that

$$K_n(u) = e^{-\alpha u}P_{nr}(e^{-u}). \tag{12}$$

Lemma. $P_{nr}(\lambda)$ has the representation

$$P_{nr}(\lambda) = (1 - \lambda)^r \tau_{nr}(\lambda), \tag{13}$$

where $\tau_{nr}(\lambda)$ is a polynomial of degree $n - r$ satisfying

$$\tau_{nr}(1) = (-1)^r/r! \tag{14}$$

Proof. Condition (4) becomes

$$\int_0^\infty e^{-\alpha t} \sum_{k=0}^n e^{-kt} \mu_{nk} t^j \, dt = \sum_{k=0}^n \frac{\mu_{nk}}{(\alpha + k)^{j+1}} = 0, \qquad 0 \le j \le r-1, \tag{15}$$

or

$$\left. \frac{\partial^j}{\partial t^j} \sum_{k=0}^n \frac{\mu_{nk} e^{-t(\alpha+k)}}{(\alpha+k)^r} \right|_{t=0} = 0, \qquad 0 \le j \le r-1. \tag{16}$$

But this means

$$\frac{\partial^j}{\partial t^j} e^{-\alpha t} P_{nr}(e^{-t})|_{t=0} = 0, \tag{17}$$

so

$$P_{nr}(\lambda) = (1-\lambda)^r \tau_{nr}(\lambda). \tag{18}$$

For $j = r$, one must have

$$\frac{\partial^r}{\partial t^r} e^{-\alpha t} P_{nr}(e^{-t}) = (-1)^r e^{-\alpha t} \sum_{k=0}^n \mu_{nk} e^{-kt} = (-1)^r e^{-\alpha t} P_n(e^{-t}). \tag{19}$$

Next, $\tau_{nr}(1)$ can be calculated by

$$\begin{aligned} \frac{\partial^r}{\partial t^r} e^{-\alpha t} P_{nr}(e^{-t})|_{t=0} = (-1)^r &= \frac{\partial^r}{\partial t^r} [e^{-t}(1-e^{-t})^r \tau_{nr}(e^{-t})]|_{t=0} \\ &= \sum_{m=0}^r \binom{r}{m} \frac{\partial^{r-m}}{\partial t^{r-m}} (1-e^{-t})^r \frac{\partial^m}{\partial t^m} [e^{-\alpha t} \tau_{nr}(e^{-t})]|_{t=0}, \end{aligned} \tag{20}$$

or

$$r! e^{-\alpha t} \tau_{nr}(e^{-t})|_{t=0} = (-1)^r, \tag{21}$$

and (14) follows, as claimed. ■

The formula for $E_n^{(q)}$ is now much more manageable:

$$\begin{aligned} E_n^{(q)} &= \|e^{\alpha c u} K_n(u)\|_q = \|e^{-\alpha d u} P_{nr}(e^{-u})\|_q \\ &= \left[\int_0^1 x^{\alpha d q - 1} (1-x)^{rq} |\tau_{nr}(x)|^q \, dx \right]^{1/q}. \end{aligned} \tag{22}$$

There are two cases in which the minimization of $E_n^{(q)}$ can be easily effected. Performing the minimization will yield the weights $[\mu_{nk}]$.

4.1.1. The Case $p = q = 2$

Now we want to find

$$\min_{\tau_{nr}} \int_0^1 x^{2\alpha d - 1} (1-x)^{2r} \tau_{nr}^2(x) \, dx = M_n \tag{1}$$

subject to condition $\tau_{nr}(1) = (-1)^r/r!$. This minimum problem is associated with the Jacobi polynomials. The solution, fully worked out in Freud (1966), is

$$\tau_{nr}(x) = \frac{(-1)^r(n-r)!(2r+1)!}{r!(n+r+1)!} R_{n-r}(x), \qquad R_k(x) \equiv R_k^{(2r+1,\, 2\alpha d-1)}(x),$$

$$M_n = \frac{\Gamma(n-r+2\alpha d)(n-r)!(2r)!(2r+1)!}{\Gamma(n+r+2\alpha d+1)(n+r+1)!r!^2} = \frac{(2r!)^2(2r+1)n^{-4r-2}}{r!^2}[1+O(n^{-1})]. \tag{2}$$

With the notation

$$\delta_k f = \{\text{coefficient of } t^k \text{ in } f,\ f \text{ a polynomial}\} \tag{3}$$

an explicit formula for the weights in U is

$$\mu_{nk} = \frac{(\alpha+k)^r(-1)^r(n-r)!(2r+1)!}{r!(n+r+1)!}\delta_k\{(1-t)^r R_{n-r}(t)\}. \tag{4}$$

Equation 4.1(5) gives

$$|\bar{r}_n| \le \|f^{(r)}e^{-\alpha cu}\|_2 \frac{(2r)!}{r!}\sqrt{2r+1}\, n^{-2r-1}\left(1+\frac{C}{n}\right), \tag{5}$$

$n \ge r$ for some C independent of f, equality being attained for some f in the class 4.1(3).

It is useful to have an explicit representation of P_n. Let

$$\nu_{nk} = \frac{(-1)^r(n-r)!(2r+1)!}{r!(n+r+1)!}\delta_k[(1-t)^r R_{n-r}(t)]. \tag{6}$$

Then

$$\sum_{k=0}^{n} \nu_{nk}\lambda^k = Q_n(\lambda) = \frac{(-1)^r(n-r)!(2r+1)!}{r!(n+r+1)!}(1-\lambda)^r R_{n-r}(\lambda), \quad P_n(\lambda) = \lambda^{-\alpha}(\lambda D)^r[\lambda^\alpha Q_n(\lambda)], \tag{7}$$

or

$$P_n(\lambda) = \frac{\lambda^{-\alpha}(-1)^r(n-r)!(2r+1)!}{r!(n+r+1)!}(\lambda D)^r[\lambda^\alpha(1-\lambda)^r R_{n-r}(\lambda)]. \tag{8}$$

Since all zeros of R_k are simple and in (0, 1) and dense there, an application of Rolle's theorem shows the zeros of $P_n(\lambda)$ are in (0, 1) and dense there. Thus, by Theorem 2.2(6), U is not regular. Of course U is regular for the class of sequences defined through conditions 4.1(1) and 4.1(2).

Note that 4.1(11) and 4.1(12) yield

$$K_n(u) = e^{-\alpha u}(1 - e^{+u})^r(-1)^r \frac{(n-r)!(2r+1)!}{r!(n+r+1)!} R_{n-r}(e^{-u}). \tag{9}$$

Theorem 1. Let $p = q = 2$ in Theorem 4.1 and let the set $\{m \mid f^{(m)}(0) \neq 0, 0 \leq m \leq r-1\}$ be nonempty, j being its smallest member.

Then U accelerates **s**. In fact

$$\bar{r}_n = n^{j-2r-1/2} O(r_n). \tag{10}$$

Proof

$$\bar{r}_n = \frac{(2r+1)!(-1)^r n^{-2r-1}}{r!} (1 + O(n^{-1})) \int_0^\infty e^{-\alpha u}(1 - e^{-u})^r R_{n-r}(e^{-u}) f^{(r)}(u)\, du$$

$$= O(n^{-2r-3/2}), \tag{11}$$

by the use of the asymptotic estimate for the Jacobi polynomial. Now

$$r_n = \sum_{i=0}^{r-1} \frac{f^{(i)}(0)}{(\alpha+n)^{i+1}} + \frac{1}{(\alpha+n)^r} \int_0^\infty e^{-(\alpha+n)t} f^{(r)}(t)\, dt$$

$$= \frac{f^{(j)}(0)}{(\alpha+n)^{j+1}} + \cdots + \frac{f^{(r-1)}(0)}{(\alpha+n)^r} + \frac{o(1)}{(\alpha+n)^r}, \tag{12}$$

or, $r_n \approx Cn^{-j-1}$. Combining this with (11) gives the theorem. ■

In Wimp (1977) it is shown that μ_{nk} satisfies the recursion relation

$$a_n \mu_{n+1,k} = b_n \mu_{nk} + c_n \mu_{n,k-1} + d_n \mu_{n-1,k}, \qquad 0 \leq k \leq n+1, \quad n > r, \tag{13}$$

with the interpretation $\mu_{nk} = 0$, $k < 0$, $k > n$, where c_n depends on k,

$$a_n = (n + r + 2)(n + r + 2\alpha d + 1)(n + \alpha d)$$

$$b_n = (2n + 2\alpha d + 1)[(r + \alpha d)(r + 1 - \alpha d) - (n + \alpha d)(n + \alpha d + 1)],$$

$$c_n = 2(2n + 2\alpha d + 1)(n + \alpha d)(n + \alpha d + 1)(\alpha + k)^r(\alpha + k - 1)^{-r},$$

$$d_n = -(n - r)(n - r + 2\alpha d - 1)(n + \alpha d + 1). \tag{14}$$

To start the computations one needs

$$\begin{aligned} \mu_{rk} &= (\alpha + k)^r(-1)^{r+k}/k!(r-k)!, && 0 \le k \le r, \\ \mu_{r+1,k} &= (\alpha + k)^r(-1)^{r+k+1}(k + \alpha d)/k!(r-k+1)!, && 0 \le k \le r+1. \end{aligned} \tag{15}$$

Example. Let

$$\begin{aligned} s_n &= (\mathrm{GAM})_n = \sum_{k=1}^{n+1} \frac{1}{k} - \ln(n+1) = \gamma + \int_0^\infty e^{-nt} f(t)\,dt, \\ f(t) &= e^{-t}[t^{-1} - (e^t - 1)^{-1}] \in C^\infty(0, \infty), \\ \gamma &= 0.5772156649 \qquad \text{(Euler's constant)}. \end{aligned} \tag{16}$$

Let, e.g., $r = 3$, $\alpha = 1$, $n = 5$, $c = d = \frac{1}{2}$. Then

$$\bar{s}_5 = 0.577209, \qquad \bar{r}_5 = 6.1\,(-6). \tag{17}$$

The improvement in convergence is dramatic; $s_6 = 0.658$ is not accurate to even one significant figure. Very often monotone sequences that come up in applications can be put in the form of Laplace moment sequences, the above being an example.

It can be shown that μ_{nk} is alternating in k. Therefore, for $\lambda > 0$,

$$\begin{aligned} V_n(\lambda) &= \sum_{k=0}^{n} |\mu_{nk}| \lambda^k \\ &= \frac{(n-r)!(2r+1)!}{r!(n+r+1)!} \sum_{k=0}^{n} (\alpha + k)^r \lambda^k |\delta_k[(1+t)^r R_{n-r}(-t)]|. \end{aligned} \tag{18}$$

Putting $\alpha + k = \alpha$ provides a lower bound, $\alpha + k = \alpha + n$ an upper bound. For some C, C',

$$Cn^{-2r}(1+\lambda)^r |R_{n-r}(-\lambda)| < V_n(\lambda) < C'n^{-2r}(1+\lambda)^r(\alpha+n)^r |R_{n-r}(-\lambda)| \tag{19}$$

and using the asymptotic properties of R_{n-r} (Freud, p. 245) shows

$$C'' < V_n(\lambda) n^{2r+1/2}[(2\lambda + 1) + 2\sqrt{\lambda(\lambda+1)}]^n < C'''. \tag{20}$$

Setting $\lambda = 1$, taking nth roots and lim sup shows $K = 5.828$, the same coefficient of numerical instability, in fact, as the triangle discussed in Section 2.3.5. For the present methods, however, the presence of the factor $(\alpha + k)^r(\alpha + k - 1)^{-r}$ in the recursion formula for μ_{nk} [see (14)] seems to preclude formalizing U as a lozenge algorithm.

4.1.2. The Case $r \to \infty$

To interpret formally the case $r \to \infty$ ($f \in C^{\infty}(0, \infty)$), rewrite condition 4.1(4) as

$$\int_0^{\infty} e^{-\alpha t} P_n(e^{-t}) t^j \, dt = 0, \qquad 0 \le j \le n-1, \tag{1}$$

or

$$\sum_{k=0}^{n} \frac{\mu_{nk}}{(\alpha + k)^{j+1}} = \delta_{-1,j}, \qquad -1 \le j \le n-1, \tag{2}$$

and this gives

$$\mu_{nk} = \binom{n}{k} (-1)^{k+n} \frac{(\alpha + k)^n}{n!}, \tag{3}$$

i.e., the Salzer weights. For the error,

$$\begin{aligned} |\bar{r}_n| &\le \|f^{(n)} e^{-\alpha c u}\|_p \left\| \sum_{k=0}^{n} \frac{\mu_{nk} e^{-\alpha d u - k u}}{(\alpha + k)^n} \right\|_q, \\ &= \|f^{(n)} e^{-\alpha c u}\|_p \left\| \frac{(-1)^n}{n!} e^{-\alpha d u} (1 - e^{-u})^n \right\|_q, \end{aligned} \tag{4}$$

or

$$\begin{aligned} |\bar{r}_n| &\le \|f^{(n)} e^{-\alpha c u}\|_p \frac{1}{n!} \left[\frac{\Gamma(\alpha d q)\Gamma(nq + 1)}{\Gamma(nq + 1 + \alpha d q)} \right]^{1/q}, \\ &= \|f^{(n)} e^{-\alpha c u}\|_p \frac{\Gamma(\alpha d q)^{1/q}}{n!} (nq)^{-\alpha d} \left[1 + \frac{A}{n} \right]. \end{aligned} \tag{5}$$

The same argument as in Krylov (1962, p. 150) shows that the error estimate (5) cannot be decreased. (Note that A is independent of f.)

Higgins, in his thesis (1976), has extended the methods of the present section to sequences whose remainders can be represented as moments of an arbitrary weight function.

4.2. Optimal Approximations in l^1 and $\mathscr{R}_C$

The larger the linear subspace of sequences of $\mathscr{C}_C$ to which a Toeplitz method is to be applied, the less effective the member will be for some members of that class. In short, you cannot get something for nothing.

Some interesting recent research has been devoted to the study of the efficiency of summability methods as applied to large sequence spaces. These results are due mainly to Baranger (1970) and Germain-Bonne (1978).

The efficiency of a Toeplitz method for any linear subspace of l^1 can be characterized very nicely through the concept of a Hilbertian subspace, see Schwartz (1961/1962). In what follows it is assumed that all sequences are real, but this entails no loss of generality. Also, $\mathbf{a} \in l^1$ means, as usual, that a_n is the $(n+1)$th term of an absolutely convergent series.

Definition. $\mathscr{H}$ is a *Hilbertian subspace of* l^1 means $\mathscr{H}$ is a Hilbert space with inner product $(\cdot, \cdot)$ and related norm $\|\cdot\|_H$ for which

$$\|\mathbf{a}\| \leq C\|\mathbf{a}\|_H, \qquad \mathbf{a} \in \mathscr{H}, \tag{1}$$

where

$$\|\mathbf{a}\| = \sum_{k=0}^{\infty} |a_k|. \tag{2}$$

It is known (Schwartz, 1961/1962) that the functionals

$$F: \mathbf{a} \to \sum_{k=0}^{\infty} a_k = s, \qquad F_k: \mathbf{a} \to a_k \tag{3}$$

are then bounded linear functionals on $\mathscr{H}$. By the Riesz representation theorem, there are representers $\zeta, \zeta^{(k)} \in \mathscr{H}$ such that

$$\sum_{k=0}^{\infty} a_k = (\zeta, \mathbf{a}) \qquad \text{for all} \quad \mathbf{a} \in \mathscr{H}; \tag{4}$$

$$a_k = (\zeta^{(k)}, \mathbf{a}) \qquad \text{for all} \quad \mathbf{a} \in \mathscr{H}. \tag{5}$$

Let n be fixed, $v_{nk} \in \mathscr{R}$, and define

$$\mathbf{e} = \zeta - \sum_{k=0}^{n} v_{nk} \zeta^{(m_k)}. \tag{6}$$

Now for any $\mathbf{a} \in \mathscr{H}$, we can consider approximations to s by finite linear combinations of the elements of $\mathbf{s}$ with coefficients v_{nk} as follows;

$$\sum_{k=0}^{\infty} a_k \sim \sum_{k=0}^{n} v_{nk} a_{m_k}. \tag{7}$$

But notice that

$$s - \sum_{k=0}^{n} v_{nk} a_{m_k} = s - \sum_{k=0}^{n} v_{nk} (\zeta^{(m_k)}, \mathbf{a}) = (\mathbf{e}, \mathbf{a}). \tag{8}$$

The problem, then, of finding a "best" summability method for all $\mathscr{H}$ can be formulated in terms of determining v_{nk}, m_k to minimize $\|\mathbf{e}\|_{\mathscr{H}}$.

There are two ways of looking at this optimization problem; one when the m_k are given, the other when they are not known. We shall treat only a special case of the former problem. For additional material, the reader should consult Baranger's paper (1970).

Denote by $\mathscr{A}_\rho$ that linear subspace of $\mathscr{C}_\sigma \cap \mathscr{R}_S \subset l^1$ such that

$$\overline{\lim_{n\to\infty}} |a_n|^{1/n} = \rho, \qquad 0 < \rho < 1. \tag{9}$$

Let $0 < \lambda < 1$. The space

$$\mathscr{H} = \left\{ \mathbf{a} \in l^1 \,\middle|\, \sum_{k=0}^{\infty} \frac{(\Delta a_k)^2}{\lambda^k} < \infty \right\} \tag{10}$$

is easily shown to contain $\mathscr{A}_\rho$ when $\lambda^2 > \rho$. Furthermore, it is a Hilbert space with inner product

$$(\mathbf{a}, \mathbf{b}) = \sum_{k=0}^{\infty} \frac{\Delta a_k \, \Delta b_k}{\lambda^k}. \tag{11}$$

Theorem 1. The optimization problem for $\mathscr{H}$ has a unique solution given by

$$\begin{aligned} v_{n0} &= C_0 + (1 - \lambda)^{-1}; \\ v_{nk} &= C_{k+1} - C_k, \qquad 1 \le k \le n - 1; \\ v_{nn} &= m_n - C_{n-1}, \end{aligned} \tag{12}$$

where

$$C_k = (m_k \lambda^{m_k} - m_{k+1}\lambda^{m_{k+1}})/(\lambda^{m_k} - \lambda^{m_{k+1}}). \tag{13}$$

In particular, when $m_k = k$, $v_{n0} = v_{n1} = \cdots = v_{n,n-1} = 1$, $v_{nn} = (1 - \lambda)^{-1}$, and $\|\mathbf{e}\|_{\mathscr{H}} = \lambda^{n+1}/(1 - \lambda)^3$.

This approximation, which works "best" for all $\mathscr{H}$, is, then, $a_0 + a_1 + \cdots + a_{n-1} + a_n(1 - \lambda)^{-1}$. Thus

$$\bar{s}_n = s_n + a_n \lambda/(1 - \lambda), \tag{14}$$

and the transformation offers little improvement over ordinary convergence.

Things are not much better for $\mathscr{R}_C$, the space of convergent real sequences, as will be shown. Let A be an infinite real matrix $[a_{ij}]$, $i, j \ge 0$, and denote by A_n the $(n + 1) \times (n + 1)$ truncate of A, and by $\mathbf{s}_n \in \mathscr{R}^{n+1}$ the $n + 1$ truncate $(s_0, \ldots, s_n)$ of $\mathbf{s} \in \mathscr{R}_S$.

A is assumed to satisfy the following three hypotheses:

(1) A_n is positive definite;
(2) $\sup_{n,i} |a_{ni}| \le M$;
(3) $\lim_{n\to\infty} a_{nk} = a_{0k}$, $k \ge 0$.

Now define

$$\mathscr{H} \equiv \mathscr{H}(A) = \left\{ \mathbf{s} \in \mathscr{R}_S \,|\, \sup_n \mathbf{s}_n A_n^{-1} \mathbf{s}_n^{\mathrm{T}} < \infty \right\}. \tag{15}$$

Germain-Bonne (1978) has established the following.

Theorem 2. $\mathscr{H}(A)$ is a Hilbert space with inner product

$$(\mathbf{s}, \mathbf{t}) = \lim_{n \to \infty} \mathbf{s}_n A_n^{-1} \mathbf{t}_n^{\mathrm{T}}. \tag{16}$$

Furthermore, $\mathscr{H}(A)$ is a Hilbertian subspace of both l^∞ and $\mathscr{R}_C$. In either case the representer $\boldsymbol{\zeta}^{(k)}$ having the property

$$s_k = (\boldsymbol{\zeta}^{(k)}, \mathbf{s}) \qquad \text{for all} \quad \mathbf{s} \in \mathscr{H}(A) \tag{17}$$

is the $(k + 1)$th row of A. In the latter case the representer $\boldsymbol{\zeta}$ having the property

$$s = (\boldsymbol{\zeta}, \mathbf{s}) \qquad \text{for all} \quad \mathbf{s} \in \mathscr{H}(A) \tag{18}$$

is the first row of A.

Now let

$$\mathbf{e} = \boldsymbol{\zeta} - \sum_{k=0}^{n} \mu_{nk} \boldsymbol{\zeta}^{(k)}. \tag{19}$$

The problem is to determine μ_{nk} to minimize $\|\mathbf{e}\|_{\mathscr{H}}$. It can be shown that the problem has a solution, and the resulting Toeplitz method is regular for $\mathscr{H}(A)$.

As an example, let $d > 1$ and take

$$A = \begin{bmatrix} 1 & 1 & 1 & \cdots \\ 1 & d & 1 & \cdots \\ 1 & 1 & d & \cdots \\ \vdots & \vdots & \vdots & \ddots \end{bmatrix}. \tag{20}$$

Then A_n is positive definite and

$$A_n^{-1} = \frac{1}{d-1} \begin{bmatrix} n+d-1 & -1 & -1 & \cdots & -1 \\ -1 & & & & \\ -1 & & (\mathrm{I}) & & \\ \vdots & & & & \\ -1 & & & & \end{bmatrix}. \tag{21}$$

One has

$$\mathbf{s}_n A_n^{-1} \mathbf{t}_n^{\mathrm{T}} = s_0 t_0 + \frac{1}{d-1} \sum_{k=0}^{n} (s_k - s_0)(t_k - t_0), \tag{22}$$

and the space $\mathscr{H}$ is the space of sequences in $\mathscr{R}_C$ for which $\lim \mathbf{s}_n A_n^{-1} \mathbf{s}_n^{\mathrm{T}} < \infty$, or, those satisfying

$$\sum_{k=1}^{\infty} (s_k - s_0)^2 < \infty, \tag{23}$$

i.e., those converging in l^2 to their first element. The associated inner product is

$$(\mathbf{s}, \mathbf{t}) = s_0 t_0 + \frac{1}{d-1} \sum_{k=1}^{\infty} (s_k - s_0)(t_k - t_0). \tag{24}$$

As far as the minimum problem associated with (19) is concerned, this will entail no loss of generality, for we can put, quite arbitrarily, $\lim s_n = s = s_0$, $\lim t_n = t = t_0$ and require $\mu_{n0} = 0$.

The Hilbertian subspace of $\mathscr{R}_C$ then consists of those sequences converging in l^2 to their limits. The μ_{nk} can be computed explicitly for this case, and the result is

$$\mu_{nk} = 1/(n + d - 1), \quad n \geq 1, \quad 1 \leq k \leq n; \qquad \mu_{n0} = 0. \tag{25}$$

This is, of course, Cesaro's method.

For a number of other examples, consult Germain-Bonne's thesis.

4.3. Methods Based on Power Series

4.3.1. Construction and Properties of Methods

The natural space of sequences to consider for these methods is $\mathscr{C}_\sigma$. Recall $\mathbf{s} \in \mathscr{C}_\sigma$ means

$$\overline{\lim_{n \to \infty}} |a_n|^{1/n} < 1. \tag{1}$$

$\mathscr{C}_\sigma$ contains all linearly convergent sequences. The material in this section is due to Gordon (1975).

First, we have a regularity theorem.

Theorem 1. A triangle U is regular for $\mathscr{C}_\sigma$ iff

(i) given $\varepsilon > 0$,

$$|v_{nk}| <. (1 + \varepsilon)^k, \quad 0 \leq k \leq n, \tag{2}$$

where $v_{nk} = \sum_{j=k}^{n} \mu_{nj}$, and

(ii) $\mu_{nk} = o(1)$, k fixed.

Proof. It clearly suffices to do this for μ_{nk} real.

$\Leftarrow$: We shall show more, namely, if $s(z) = \sum a_k z^k$ is analytic in N_R, $R > 1$, then U sums the partial sum of the series to $s(z)$ uniformly in any smaller circle. Since s is analytic in $\bar{N}$, the sufficiency part of the theorem will follow when $z = 1$.

Let $R'' < R' < R$. A simple computation with Cauchy's formula gives

$$r_n(z) = \frac{1}{2\pi i} \int_{t \in \partial N_{R'}} \left[\sum_{k=0}^{k_0} (v_{nk} - 1)\tau^k - \frac{\tau^{k_0+1}}{1-\tau} + \sum_{k=k_0+1}^{n} v_{nk}\tau^k \right] \frac{s(w)\,dw}{w}, \tag{3}$$

$$\tau = z/w, \quad z \in N_{R''}.$$

Now pick ε so that $(1 + \varepsilon)R''/R' < 1$ and N so that (i) holds for $n > N$. Then choose $k_0 \geq N$ to make the contribution of the second and third terms to the integral $<\varepsilon/2$, and then n large enough so the first term contributes $<\varepsilon/2$. (Note $v_{nk} \to 1$, k fixed.)

$\Rightarrow$: This requires a much more subtle argument. We shall only sketch the proof. The reader unable to fill in details should consult Gordon's paper (1975). (ii) is obvious. To establish the necessity of (i), assume it is false. This means given $\varepsilon > 0$, $|v_{nk}| \geq (1 + \varepsilon)^k$ holds for two n and k sequences with the n sequence unbounded. But, clearly, the k sequence must also be unbounded.

Now we construct two sequences inductively, $\{n_r\}$, $\{k_r\}$, as follows:

(i) Let $n_0 = k_0 = 0$;
(ii) assume n_{r-1}, k_{r-1} have been chosen;
(iii) pick $n'_r > n_{r-1}$ so that

$$|v_{ni}| < 1 + \varepsilon, \qquad 0 \leq i \leq k_{r-1}, \quad n > n'_r \tag{4}$$

(this is possible because U is a triangle);

(iv) choose $n_r > n'_r$, $k'_r > k_{r-1}$ so that

$$\left| \sum_{j=k'_r}^{n_r} \mu_{n_r,j} \right| = |v_{n_r,k'_r}| \geq (1 + \varepsilon)^{k'_r}; \tag{5}$$

(v) choose $k_r > n_r$.

We now construct a sequence $\mathbf{s} \in \mathscr{C}_\sigma$. Choose $\delta \in (0, 1)$ so that

$$\delta(1 + \varepsilon) = \rho > 1. \tag{6}$$

Let

$$a_0 = 1, \qquad a_i = \delta^i \operatorname{sgn} v_{n_r,i}, \quad k_{r-1} < i \leq k_r, \tag{7}$$

with empty sums interpreted as 0 and sgn 0 = 0.

Then

$$|\bar{s}_{n_r}| = \left|\sum_{j=0}^{n_r} s_j \mu_{n_r,j}\right| = \left|\sum_{j=0}^{n_r} a_j v_{n_r,j}\right|$$

$$= \left|\sum_{j=0}^{k_{r-1}} a_j v_{n_r,j} + \sum_{j=k_{r-1}+1}^{n_r} \delta^j |v_{n_r,j}|\right|$$

$$\geq \delta^{k'_r}(1+\varepsilon)^{k'_r} - \sum_{j=0}^{k_{r-1}} |a_j||v_{n_r,j}|$$

$$> \rho^{k'_r} - \sum_{j=0}^{k_{r-1}} \delta^j(1+\varepsilon) \geq \rho^{k'_r} - \frac{1+\varepsilon}{1-\delta} \to \infty. \tag{8}$$

Thus $\bar{\mathbf{s}}$ contains an unbounded sequence. ■

It is easy to show that the regularity of U for $\mathscr{C}_\sigma$ implies $P_n(\lambda) = o(1)$ for each $\lambda \in N$. However, regularity for $\mathscr{C}_\sigma$ does not imply regularity. Take

$$\mu_{nk} = \begin{cases} 0, & 0 \leq k \leq n-2 \\ -n, & k = n-1 \\ n+1, & k = n. \end{cases} \tag{9}$$

Then

$$\bar{r}_n = r_n + na_n \to 0, \tag{10}$$

but U is not regular.

A *Taylor sequence* $\mathbf{s}$ is a sequence such that $|s_n|^{1/n} = O(1)$. Clearly, the space of Taylor sequences contains $\mathscr{C}_\sigma$. Bajšanski and Karamata (1960) have shown that U (not necessarily a triangle) takes a Taylor sequence into a Taylor sequence iff, given $\varepsilon > 0$, there is an M such that $|\mu_{nk}| < \varepsilon^k M^n$, $n, k \geq 0$. For additional material on such sequences, see the thesis by Heller referenced in Bajšanski and Karamata (1960).

The transformation we wish to study here is defined by formal power series. Let $\mathbf{s} \in \mathscr{C}_S$ and

$$G(z) = \sum_{k=1}^{\infty} a_{k-1} z^k. \tag{11}$$

Let

$$z = f(w) = \sum_{k=1}^{\infty} f_{k-1} w^k, \qquad f(1) = 1. \tag{12}$$

Define the sequence $\mathbf{b}$ by

$$\sum_{k=1}^{\infty} b_{k-1} w^k = G(f(w)) = \sum_{k=1}^{\infty} a_{k-1} \left(\sum_{r=1}^{\infty} f_{r-1} w^r\right)^k. \tag{13}$$

Finally define $T(\mathbf{s}) = \bar{\mathbf{s}}$ by

$$\bar{s}_n =. \sum_{k=0}^{n} b_k \Big/ \sum_{k=0}^{n} f_k, \qquad n \geq 0. \tag{14}$$

(If $f_0 + f_1 + \cdots + f_n \neq 0$, we can use this to define $\bar{s}_n$ for all n.)

Further exploring the properties of U requires its matrix representation. Let

$$[f(w)]^m = \sum_{k=m}^{\infty} f_{k-1,m} w^k, \qquad m \geq 1. \tag{15}$$

Define

$$\beta_{nk} = \sum_{j=k}^{n} f_{j,k+1}, \qquad 0 \leq k \leq n. \tag{16}$$

Thus $f_{k,1} = f_k$, $\beta_{n0} = 1 + o(1)$. Substituting the series (15) into (13) and interchanging the order of summation gives

$$b_n = \sum_{k=0}^{n} a_k f_{n,k+1} \tag{17}$$

so

$$\begin{aligned} \bar{s}_n &= \beta_{n0}^{-1} \sum_{j=0}^{n} b_j = \beta_{n0}^{-1} \sum_{k=0}^{n} a_k \beta_{nk} \\ &= \beta_{n0}^{-1} \left[\sum_{k=0}^{n-1} s_k(\beta_{nk} - \beta_{n,k+1}) + s_n \beta_{nn} \right]. \end{aligned} \tag{18}$$

Define

$$\mu_{nk} = \begin{cases} \beta_{n0}^{-1} \beta_{nn}, & k = n \\ \beta_{n0}^{-1}(\beta_{nk} - \beta_{n,k+1}), & k < n. \end{cases} \tag{19}$$

An application of Theorem 1 then furnishes immediately the next result.

Theorem 2. U is regular for $\mathscr{C}_\sigma$ iff

(i) given $\varepsilon > 0$,

$$|\beta_{nk}| <. (1 + \varepsilon)^k, \qquad 0 \leq k \leq n; \tag{20}$$

(ii) $\beta_{nk} = 1 + o(1)$, in n.

Proof. Obvious. ■

Theorem 3. Let f have a radius of convergence greater than 1. Then U is regular for $\mathscr{C}_\sigma$ iff

$$\sup_{|w|=1} |f(w)| = 1. \tag{21}$$

Proof. $\Leftarrow$: First, note that the coefficient of w^k on the right-hand side of (13) involves $a_0, a_1, \ldots, a_{k-1}$, or $\bar{s}_n$ involves $s_0, s_1, \ldots, s_n$. Next, let $a_0 = 1$, $a_k = 0, k > 0$. This means $s_n \equiv 1$. But then $b_k = f_k$ and $\bar{s}_n = \sum_{k=0}^{n} \mu_{nk} = 1$, so the transformation defined by T is a triangle, U.

$f(w)$ is analytic for $|w| < R, R > 1$. Let the condition (21) be satisfied. If $\mathbf{s} \in \mathscr{C}_\sigma$ then $G(z)$ is analytic for $|z| < 1 + \varepsilon$, $\varepsilon > 0$. But then $G(f(w))$ is analytic for $|w| < 1 + \delta$, for some $\delta > 0$. In particular

$$\sum_{k=0}^{n} b_k \bigg/ \sum_{k=0}^{n} f_k \to \sum_{k=0}^{\infty} b_k \bigg/ 1 = G(f(1)) = G(1) = \lim_{n\to\infty} \sum_{k=0}^{n} a_k = s. \tag{22}$$

Thus U is regular for $\mathbf{s}$.

$\Rightarrow$: Assume (21) is not satisfied. Then there is a point w^* such that $f(w^*) = z^*$ where $|w^*| < 1$ and $|z^*| > 1$ since $f(1) = 1$ and by the continuity of $f(w)$. Let

$$G(z) = \frac{z}{1 - z/z^*} = \sum_{k=1}^{\infty} a_{k-1} z^k. \tag{23}$$

The corresponding sequence $\mathbf{s}$ is in $\mathscr{C}_\sigma$. Since $G(f(w))$ has a pole at $w = w^*$ the radius of convergence of this series is $|w^*| < 1$. The series, therefore, cannot converge at $w = 1$, so $\lim_{n\to\infty} \bar{s}_n$ does not exist. ■

Corollary. Let $f(u)$ have a radius of convergence >1. Then the two conditions $\sup_{|w|=1} |f(w)| = 1$ and $|\beta_{nk}| <. (1 + \varepsilon)^k$, $0 \le k \le n$, are equivalent in the sense that either is necessary and sufficient for U to be regular for $\mathscr{C}_\sigma$.

Proof. I need only show Theorem 2(ii). $f(w)^m$ is analytic in the closed unit disk. Since $f(1) = 1, f(w)^m|_{w=1} = 1$, or

$$\sum_{k=m}^{\infty} f_{k-1,m} = 1, \qquad m \ge 1, \tag{24}$$

or $\beta_{n,m-1} \to 1, m \ge 1$. ■

Not yet discussed is the regularity of U. There is a simple positivity condition on the Taylor series coefficients of f that ensures this.

Theorem 4. Let $f(w)$ have radius of convergence greater than 1 and $\mathbf{f} \in \mathscr{R}_S^0$. Then U is regular.

Proof. First

$$\beta_{n0}^{-1}(\beta_{nk} - \beta_{n,k+1}) = \beta_{n0}^{-1}\left(\sum_{j=k}^{n} f_{j,k+1} - \sum_{j=k+1}^{n} f_{j,k+2}\right) \to 0, \qquad n \to \infty. \tag{25}$$

Further, it is easily established by induction that $\mu_{nk} \geq 0$. Thus

$$\sum_{k=0}^{n} \mu_{nk} = \sum_{k=0}^{n} |\mu_{nk}| = 1$$

and the Toeplitz limit theorem can be invoked. ■

4.3.2. Applications

Example 1. Let

$$f(w) = \frac{1}{1+q}\left[\frac{w}{1 - qw/(1+q)}\right], \qquad q \geq 0. \tag{1}$$

This gives, for all practical purposes, the Euler (E, q) method, although this formulation differs slightly from the standard one, the weights here having row sum of 1. Then

$$\begin{aligned} b_k &= \frac{1}{(1+q)^{k+1}} \sum_{r=0}^{k} q^{k-r}\binom{k}{r} a_r, \qquad & k \geq 0 \\ \bar{s}_n &= \sum_{k=0}^{n} b_k \Bigg/ \left[1 - \frac{q}{q+1}\right]^{n+1}, \qquad & n \geq 0. \end{aligned} \tag{2}$$

Theorem 4 provides the well-known result that U is regular.

Example 2. Consider the case where f is a polynomial of degree 1, 2, or 3. Usually condition 4.3.1(21) is easier to check than Eq. 4.3.1(20). If $f(w) = w - w^2$ for example, condition (21) shows that U is not regular for $\mathscr{C}_\sigma$ since $f(-1) = -2$.

For first-degree polynomials, the only $\mathscr{C}_\sigma$-regular methods arise from $f(w) = w$, the identity transformation. For second-degree polynomials, (21) requires that $f(w) = w(w + \alpha)/(1 + \alpha)$ with α real and positive. Again, U is regular. For third-degree polynomials, both regularity and $\mathscr{C}_\sigma$ regularity arise. The results are as follows: first

$$f(w) = w(w^2 + \alpha w + \beta)/(1 + \alpha + \beta). \tag{3}$$

For real α, β, (21) holds iff at least one of the following holds:

(i) $\alpha, \beta \geq 0$;
(ii) $-1 < \beta < 0, \quad \alpha \geq 4|\beta|/(1 - |\beta|)$;
(iii) $\beta < -1, \quad \alpha \leq -4|\beta|/(|\beta| - 1)$.

For case (i) U is regular; for cases (ii) and (iii) only $\mathscr{C}_\sigma$ regular.

Example 3. The heat conduction equation

$$\frac{\partial v}{\partial t} = \frac{\partial^2 v}{\partial x^2} \tag{4}$$

can be approximated by the difference equation

$$v_j^{(n+1)} - v_j^{(n)} = p(v_{j+1}^{(n)} - 2v_j^{(n)} + v_{j-1}^{(n)}) \tag{5}$$

where

$$p = \Delta t/(\Delta x)^2 \geq 0. \tag{6}$$

To study the instability of (5) one can substitute $v_j^{(n)} = \xi^n w^j$, $w = e^{i\theta}$. This gives

$$\xi = (1 - 2p) + p(w + 1/w). \tag{7}$$

For stability it is required that $|\xi| \leq 1$ for all θ [see Richtmeyer (1957)]. This is equivalent to $|f(w)| \leq 1$ for $|w| = 1$, where

$$f(w) = w[p + (1 - 2p)w + pw^2]. \tag{8}$$

Thus the difference scheme (5) is stable iff U is $\mathscr{C}_\sigma$ regular. Example 2, Case (3), then yields the well-known result that (5) is stable only if $1 - 2p \geq 0$.

Chapter 5 | Nonlinear Lozenges; Iteration Sequences

5.1. General Theory of Nonlinear Lozenge Algorithms

One cannot expect general lozenge algorithms 1.3(1)–1.3(3) to satisfy a theory as simple and as elegant as the linear theory of Chapter 3. What made the linear theory possible was, of course, the Toeplitz limit theorem.

Only recently has work been done on developing a general description of regularity and acceleration properties for nonlinear methods, much of the work having been done in France, particularly by Germain-Bonne (1973, 1978). The theory doesn't cover all cases—in fact, it excludes some rather important algorithms and in its present formulation can handle only vertical convergence. Nevertheless it is elegant and seems to point the way to further study.

In what follows let k be fixed and ≥ 0. Let $G: \mathscr{C}^{k+1} \to \mathscr{C}$. For any $\mathbf{s} \in \mathscr{C}_C$ define

$$s_n^{(k)} = G(s_n, s_{n+1}, \ldots, s_{n+k}). \tag{1}$$

During the analyses to follow G will be required to fulfill one or more of the following conditions:

(i) G is *continuous* on $\mathscr{C}^{k+1}$;
(ii) G is *homogeneous*, $G(\lambda x) = \lambda G(x)$, $\lambda \in \mathscr{C}$, $x \in \mathscr{C}^{k+1}$;
(iii) G is *translative*, $G(x + \alpha e) = G(x) + \alpha$, $x \in \mathscr{C}^{k+1}$, and $\alpha \in \mathscr{C}$, $e = (1, 1, 1, \ldots, 1)$.

Note that any function satisfying (ii) and (iii) also satisfies $G(0) = 0$ and $G(\alpha e) = G(0) + \alpha$, $\alpha \in \mathscr{C}$.

Let $\mathscr{D}_{k+1}$ be that subset of $\mathscr{C}^{k+1}$ such that

$$x_i = x_j \qquad \text{iff} \quad i = j, \qquad 1 \le i, j \le k+1. \tag{2}$$

Let $\mathscr{F}_{k-1}$ be that subset of $\mathscr{C}^{k-1}$ such that $x_i \neq 0$, $1 \le i \le k-1$. Let $\mathscr{G}_{k-1}$ be the subset of $\mathscr{F}_{k-1}$ interior to the hypercube $|x_i| < 1$.

Theorem 1. Let G be continuous on $\mathscr{D}_{k+1}$ and satisfy (ii) and (iii). Then G can be written

$$\begin{aligned} G(x) &= G(x_0, x_1, \dots, x_k) \\ &= x_0 + (x_1 - x_0) g\left[\frac{x_2 - x_1}{x_1 - x_0}, \frac{x_3 - x_2}{x_2 - x_1}, \dots, \frac{x_k - x_{k-1}}{x_{k-1} - x_{k-2}}\right], \end{aligned} \tag{3}$$

where g is defined and continuous on $\mathscr{F}_{k-1}$. Conversely, any function defined and continuous on $\mathscr{F}_{k-1}$ can be associated with a G as above that is continuous on $\mathscr{D}_{k+1}$ and satisfies (ii) and (iii) if one defines $G(0) = 0$.

Proof. Let the x_j be distinct. Then

$$G(x_0, \dots, x_k) = x_0 + (x_1 - x_0) G\left(0, 1, \frac{x_2 - x_0}{x_1 - x_0}, \dots, \frac{x_k - x_0}{x_1 - x_0}\right). \tag{4}$$

Now define

$$\begin{aligned} X_j &= \frac{x_{j+1} - x_j}{x_j - x_{j-1}}, \qquad 1 \le j \le k-1; \\ Y_j &= \frac{x_j - x_0}{x_1 - x_0}, \qquad j \ge 0. \end{aligned} \tag{5}$$

Then $Y_0 = 0$, $Y_1 = 1$, and

$$Y_j = 1 + \sum_{i=1}^{j-1} X_1 X_2 \dots X_i, \qquad 2 \le j \le k. \tag{6}$$

Define

$$g(X) \equiv g(X_1, X_2, \dots, X_{k-1}) = G(Y) \equiv G(0, 1, Y_2, \dots, Y_k). \tag{7}$$

As X varies continuously over $\mathscr{F}_{k-1}$, Y varies continuously over $\mathscr{D}_{k+1}$, and vice versa. This proves the theorem. ■

Theorem 2. Let G satisfy (ii) and (iii) and be bounded on the hypercube $\mathscr{N} = \{x \mid \max_{0 \le j \le k} |x_j| \le 1\}$. Then G in (1) is regular as $n \to \infty$.

Proof. Let

$$\delta_n = \sup_{j \ge n} |r_j|. \tag{8}$$

If $\delta_n \equiv .0$, the theorem is clearly true, so assume $\delta_n \neq . 0$. Then

$$r_n^{(k)} = . \delta_n G(x), \qquad x_j = . r_{n+j}/\delta_n, \quad 0 \le j \le k, \tag{9}$$

so $x \in \mathcal{N}$ and

$$|r_n^{(k)}| \le . \delta_n M \to 0. \quad \blacksquare \tag{10}$$

Theorem 3. Let G satisfy (ii) and (iii) and be continuous on $\mathcal{D}_{k+1}$. Let $\mathbf{s} \in \mathscr{C}_C$ and $|a_{n+1}/a_n|$ be bounded and bounded away from 0 for n large. Then G is regular for $\mathbf{s}$.

Proof

$$s_n^{(k)} = s_n + \Delta s_n g\left(\frac{a_{n+2}}{a_{n+1}}, \ldots, \frac{a_{n+k}}{a_{n+k-1}}\right) = s_n + \Delta s_n g(Y). \tag{11}$$

Now for n sufficiently large, Y belongs to the compact set $0 < C_1 \le |Y_j| \le C_2$, and so the theorem follows immediately. $\blacksquare$

Corollary. G in the above theorem is regular for $\mathscr{C}_l$.

Proof. Use Theorem 1.4(1). $\blacksquare$

Remark. Often it happens that $s_n^{(k)}$ is given ab initio in the form (11) rather than (1). Then, of course, conditions (ii) and (iii) are superfluous.

Recall that G is said to accelerate $\mathbf{s} \in \mathscr{C}_C$ if

$$(s_n^{(k)} - s)/(s_n - s) = o(1) \qquad \text{in} \quad n. \tag{12}$$

We employ the usual notations

$$\begin{aligned} \rho_n &= \frac{s_{n+1} - s_n}{s_n - s_{n-1}} = \frac{a_{n+1}}{a_n}, \qquad n \ge 0 \quad (s_{-1} = 0), \\ r_n &= s_n - s, \qquad h_n = r_{n+1}/r_n, \qquad r_n^{(k)} = s_n^{(k)} - s. \end{aligned} \tag{13}$$

and assume in what follows that G satisfies (ii) and (iii). Therefore we can write

$$r_n^{(k)} = r_n + (r_{n+1} - r_n) g(\rho_{n+1}, \ldots, \rho_{n+k-1}). \tag{14}$$

Let $\mathscr{A}_\rho \subset \mathscr{C}_l$ be the class of sequences with

$$\lim_{n\to\infty} a_{n+1}/a_n = \rho, \qquad \rho \quad \text{fixed,} \quad 0 < |\rho| < 1. \tag{15}$$

Without loss of generality we can assume $a_n \neq 0$ for any of these sequences. Further [Theorem 1.4(1)] ρ_n, h_n tend to ρ.

Theorem 4. Let g be defined on $\mathscr{G}_{k-1}$, $k \geq 2$. Then G accelerates $\mathscr{A}_\rho$ iff g is continuous in a neighborhood of ρe and

$$g(\rho e) = 1/(1 - \rho). \tag{16}$$

Proof. We can write

$$r_n^{(k)}/r_n = [1 + (h_n - 1)g(\rho_{n+1}, \rho_{n+2}, \ldots, \rho_{n+k-1})]. \tag{17}$$

Note that (14) shows that G is regular for $\mathscr{A}_\rho$. Clearly g's continuity and value (15) at ρe is necessary and sufficient for G to accelerate $\mathscr{A}_\rho$. ■

Note that if G *fails* to be regular for $\mathscr{A}_\rho$, then g must be unbounded in some neighborhood of ρ.

Example 1. Anticipating a little, we now discuss the Aitken δ^2-process, defined by

$$\bar{s}_n = \frac{s_n s_{n+2} - s_{n+1}^2}{s_n - 2s_{n+1} + s_{n+2}} = s_n + a_{n+1}\left(1 - \frac{a_{n+2}}{a_{n+1}}\right)^{-1}. \tag{18}$$

Note $\bar{s}_n$ is *always* defined for n sufficiently large since $a_n \neq. a_{n+1}$, $\mathbf{s} \in \mathscr{A}_\rho$. Here $k = 2$ and $g(x) = 1/(1 - x)$. The theorem confirms the well-known fact that Aitken's δ^2-process accelerates all sequences in $\mathscr{A}_\rho$ for each ρ, i.e., accelerates $\mathscr{C}_l$.

Smith and Ford (1979) have given a useful one-way generalization of the previous theorem in which the functions g are allowed to depend on n as well as k. We again assume $G \equiv G_n$ satisfies (ii) and (iii).

Theorem 5. Let $k \geq 2$ and the functions g_n be continuous in a common neighborhood K of ρe and converge uniformly in K as $n \to \infty$ to a function g with the property

$$g(\rho e) = 1/(1 - \rho).$$

Then G as defined by (1) is accelerative for $\mathscr{A}_\rho$.

Proof. By uniform convergence

$$g_n(\rho_{n+1}, \ldots, \rho_{n+k-1}) = 1/(1 - \rho) + \eta_n, \tag{19}$$

where $\boldsymbol{\eta}$ is a null sequence. Thus

$$\left|\frac{r_n^{(k)}}{r_n}\right| = 1 + (\rho - 1 + \delta_n)\left(\frac{1}{1 - \rho} + \eta_n\right) = (\rho - 1)\eta_n + \frac{\delta_n}{(1 - \rho)} + \delta_n \eta_n, \tag{20}$$

$\boldsymbol{\delta}$ a null sequence, and this gives the theorem. ■

Example 2. Let $\mathbf{y}$ be a fixed sequence $\in \mathscr{C}_S$.
Define the *generalized* Aitken δ^2-process by

$$\bar{s}_n = s_n + a_{n+1}(1 - (a_{n+2}y_n/a_{n+1}y_{n+1}))^{-1}. \tag{21}$$

(It is assumed that the denominator in Eq. (21) is nonzero—more about this later.) Again $k = 2$, but

$$g(x) \equiv g_n(x) = (1 - x(y_n/y_{n+1}))^{-1}. \tag{22}$$

Thus if $y_n/y_{n+1} = 1 + o(1)$, (21) is accelerative for $\mathscr{C}_l$. [Then, of course (21) is defined for n large enough. If the denominator of the right-hand side vanishes for any n^*, it is customary to put $\bar{s}_{n^*} = s_{n^*}$.] ■

5.2. Path Regularity for Certain Lozenges

When each element $s_n^{(k)}$ in a lozenge can be written as a weighted sum of $s_n, \ldots, s_{n+k}$, then certain simple conditions ensure the path regularity of $s_n^{(k)}$ even though the algorithm may be nonlinear.

To be precise, let

$$s_n^{(k)} = \sum_{m=0}^{k} \mu_{km} s_{n+m}, \qquad n, k \geq 0, \tag{1}$$

where $\mu_{km} \equiv \mu_{km}(n, \mathbf{s})$.

Theorem 1. For $\mathbf{s} \in \mathscr{C}_C$ let

(i) $\sum_{m=0}^{k} |\mu_{km}| \leq M$, M independent of n;
(ii) $\sum_{m=0}^{k} \mu_{km} = 1$; and
(iii) $\mu_{km} = o(1)$, $k \to \infty$, uniformly in n, along $\mathbf{P}$.

Then the algorithm defined by (1) is regular for $\mathbf{s}$ along $\mathbf{P}$.

Proof. By (ii) we can write, for $\mathbf{s} \in \mathscr{C}_C$,

$$\begin{aligned} r_n^{(k)} = s_n^{(k)} - s &= \sum_{m=0}^{k} r_{n+m}\mu_{km} \\ &= \sum_{m=0}^{k_0} + \sum_{m=k_0+1}^{k}. \end{aligned} \tag{2}$$

Applying (i) gives

$$|r_n^{(k)}| \leq \sup_{j \geq n} |r_j| \sum_{m=0}^{k_0} |\mu_{km}| + \sup_{j > n+k_0} |r_j| M. \tag{3}$$

Either $k \to \infty$ or $n \to \infty$ on **P** (or both). If $k \to \infty$ pick k_0 to make the second term $<\varepsilon/2$ for all n and then the first term will be less than $\varepsilon/2$ for k large. If $n \to \infty$ simply take the direct limit. This shows $r_n^{(k)} \to 0$ on **P**, or $s_n^{(k)} \to s$. ■

This theorem means, in effect, that an analog of the Toeplitz limit theorem holds one way for certain nonlinear algorithms. This provides us with a one-way generalization of the linear deltoid obtained by extrapolation, Theorem 3.3(1).

Theorem 2. Let

$$s_n^{(k+1)} = \frac{x_n s_{n+1}^{(k)} - x_{n+k+1} s_n^{(k)}}{x_n - x_{n+k+1}}, \qquad n, k \geq 0, \qquad s_n^{(0)} = s_n, \qquad n \geq 0, \tag{4}$$

where **x** (which may depend on **s**) is monotone decreasing to zero. Then if $x_n/x_{n+1} \geq \alpha > 1$, $n \geq 0$, the above algorithm is regular for all paths.

Proof. Left to the reader. ■

5.3. Iteration Sequences

5.3.1. The GBW Transformation

The algorithm to be studied in this section is a case of algorithm 5.2(4) with $x_n = \Delta s_n = a_{n+1}$.

$$s_n^{(k+1)} = \frac{a_{n+1} s_{n+1}^{(k)} - a_{n+k+2} s_n^{(k)}}{a_{n+1} - a_{n+k+2}}, \qquad n, k \geq 0, \qquad s_n^{(0)} = s_n, \qquad n \geq 0. \tag{1}$$

Equation 3.3(2) gives

$$s_n^{(k)} = \sum_{m=0}^{k} s_{n+m} \prod_{\substack{i=0 \\ i \neq m}}^{k} \left(\frac{a_{n+i+1}}{a_{n+i+1} - a_{n+m+1}} \right). \tag{2}$$

If $a_i \neq a_j$, $i \neq j$, then $s_n^{(k)}$ is defined. This restriction on **s** will be held in force throughout.

The transformation (1) was discovered independently by Wimp (1970) and Germain-Bonne (1973).

$E_k(s_n) \equiv s_n^{(k)}$ is homogeneous, translative, and exact when **s** has the form

$$s_n = s + c_1 \Delta s_n + c_2 (\Delta s_n)^2 + \cdots + c_k (\Delta s_n)^k \tag{3}$$

for some c_j not all zero, Theorem 3.2(3). This rather obscure nonlinear difference equation offers little clue to the behavior of **s** itself, and one would expect the exactness problem for this algorithm to be rather intractable, certainly in comparison with the easy analyses of the preceding chapters. It turns out, however, that E_k is a very natural transformation to use for a class of sequences of fundamental importance in numerical analysis.

Definition. $\mathbf{s} \in \mathscr{C}_S$ is an *iteration sequence at* s if there is a nonconstant function $\phi(z)$ analytic in a region $\mathscr{A}$ with $s \in \mathscr{A}$, $\phi(s) = s$, $|\phi'(s)| < 1$, and

$$s_{n+1} = \phi(s_n), \qquad n \geq 0, \quad s_n \in \mathscr{A}. \tag{4}$$

It is easy to show that if ϕ is analytic in a disk centered at s and $s_0 - s = r_0$ is sufficiently small, then s_n lies in the disk for all n and $s_n \to s$. As the reader may know, these sequences arise in the attempt to find solutions of scalar equations of the form $z = \phi(z)$, or, in abstract spaces, solutions of operator equations $x = Ax$. For the latter, the abstract Brezinski–Håvie process with $f_n^{(j)} = (\Delta s_n)^j$ is the process to study (see Chapter 10). Here we discuss only the scalar case.

It is easy to show the class of convergent iteration sequences is a subset of $\mathscr{C}_l$ provided $0 < |\phi'(\alpha)| < 1$ and $|s - s_0|$ is sufficiently small for then $(s_{n+1} - s)/(s_n - s) \to \phi'(\alpha)$. In practice, however, s_n does not usually converge either because insufficient information is available to enable one to choose s_0 close enough to s or because the function $\phi(z)$ lacks the property $|\phi'(s)| < 1$. For computational aspects of such sequences, see, for instance, Isaacson and Keller (1966) or Householder (1953), and, for a discussion in abstract spaces, Kantorovich and Akilov (1964).

The first result is a representation theorem.

Theorem 1. Let **s** satisfy (3) with $|1 + 1/c_1| < 1$. Then for r_0 sufficiently small, **s** is an iteration sequence at s.

Conversely if **s** is an iteration sequence at s and the function inverse to $\phi(z + s) - (z + s)$ at $z = 0$ is a polynomial of degree $\leq k$ vanishing at 0, then $s_n^{(k)} \equiv s$, $n \geq 0$.

Proof. $\Rightarrow$: Rewriting (3) as

$$r_n = c_1 \Delta r_n + \cdots + c_k(\Delta r_n)^k \tag{5}$$

and reversing this series gives

$$\Delta r_n =. c_1' r_n + c_2' r_n^2 + \cdots, \qquad c_1' = 1/c_1, \tag{6}$$

or

$$r_{n+1} =. (1 + 1/c_1) r_n + c_2' r_n^2 + \cdots. \tag{7}$$

Now let

$$\phi(z) = s + (1 + 1/c_1)(z - s) + c_2'(z - s)^2 + \cdots. \tag{8}$$

Then ϕ has all the required properties.

$\Leftarrow$: If **s** is an iteration sequence, we can write

$$s_{n+1} = \phi(s_n) = s + c_1(s_n - s) + c_2(s_n - s)^2 + \cdots \tag{9}$$

or

$$\Delta r_n = (c_1 - 1)r_n + c_2 r_n^2 + \cdots = \phi(r_n + s) - r_n - s \tag{10}$$

and the conclusion is immediate. (Note that $c_1 - 1 \neq 0$ since $|c_1| < 1$.) ∎

Theorem 2. E_k accelerates C_l along vertical paths. Further, if ϕ is an iteration function real on the real axis with $0 \leq \phi'(s) < 1$, s real, and r_0 is sufficiently small positive, then E_k is regular for **s** along all paths.

Proof

$$\frac{r_n^{(k)}}{r_n} \doteq \sum_{m=0}^{k} \frac{r_{n+m}}{r_n} \prod_{\substack{i=0\\ i\neq m}}^{k} \left(1 - \frac{a_{n+m+1}}{a_{n+i+1}}\right)^{-1} \rightarrow \sum_{m=0}^{k} \rho^m \prod_{\substack{i=0\\ i\neq m}}^{k} (1 - \rho^{m-i})^{-1} = 0. \tag{11}$$

The second part of the theorem, being a straightforward application of Theorem 5.2(2), is left to the reader. ∎

If **s** is divergent in the way $s_{n+1}/s_n \rightarrow \rho$, $|\rho| > 1$, then E_k "decelerates" **s**, i.e., $s_n^{(k)}/s_n \rightarrow 0$. This can be shown by using Theorem 1.4(1).

Moreover, E_k sums the series $\sum \lambda^n$ exactly, $\lambda \neq 0, 1$, i.e., $s_n^{(k)} = (1 - \lambda)^{-1}$, $n \geq 0$, $k \geq 1$.

5.3.2. Overholt's Procedure

This procedure (Overholt, 1965) deals with the iteration sequence directly. Let

$$s_{n+1} = \phi(s_n), \tag{1}$$

where ϕ is analytic in a suitably large neighborhood of s, $\phi(s) = s$, $|\phi'(s)| < 1$. Then

$$s_{n+1} - s = \phi'(s)(s_n - s) + \tfrac{1}{2}\phi''(s)(s_n - s)^2 + \cdots \tag{2}$$

or

$$r_{n+1} = b_1 r_n + b_2 r_n^2 + \cdots, \tag{3}$$

and inverting gives

$$r_n \doteq (1/b_1)r_{n+1} + b_2' r_{n+1}^2 + \cdots. \tag{4}$$

Theorem 1. Let

$$s_n^{(k+1)} = \frac{a_{n+k+1}^{k+1} s_{n+1}^{(k)} - a_{n+k+2}^{k+1} s_n^{(k)}}{a_{n+k+1}^{k+1} - a_{n+k+2}^{k+1}}, \qquad n, k \geq 0, \tag{5}$$

$$s_n^{(0)} = s_n, \qquad n \geq 0,$$

and let the denominator of (5) never be zero. Then

$$s_n^{(k)} = s + O(r_n)^{k+1}, \qquad n \to \infty. \tag{6}$$

Proof. We show instead that $s_{n+1}^{(k)} = s + O(r_{n+k})^{k+1}$, which, by (3), amounts to the same thing. Assume for some fixed $k \geq 0$ that

$$s_{n+1}^{(k)} = s + c_1(r_{n+k})^{k+1} + c_2(r_{n+k})^{k+2} + \cdots. \tag{7}$$

(3) gives

$$a_{n+k+1} = (b_1 - 1)r_{n+k} + b_2(r_{n+k})^2 + \cdots \tag{8}$$

so

$$\begin{aligned} a_{n+k+2} &= (b_1 - 1)r_{n+k+1} + b_2(r_{n+k+1})^2 + \cdots \\ &= (b_1 - 1)b_1 r_{n+k} + c r_{n+k}^2 + \cdots, \end{aligned} \tag{9}$$

by substitution of (3). Thus

$$(a_{n+k+2}/a_{n+k+1})^{k+1} = b_1^{k+1} + d r_{n+k} + \cdots. \tag{10}$$

Now letting $n \to n - 1$ in (7) and substituting (4) with $n \to n + k - 1$ gives

$$s_n^{(k)} = s + (c_1/b_1^{k+1})(r_{n+k})^{k+1} + \cdots, \tag{11}$$

and combining (7), (10), and (11) gives the inductive proof that establishes (6) (actually more, since we have shown that $s_n^{(k)}$ has a complete asymptotic expansion in powers of r_n). ■

Corollary. Overholt's procedure (5) is accelerative for iteration sequences of the prescribed type (provided $s_n^{(k)}$ is defined, $n, k \geq 0$).

It is easy to show that Overholt's procedure is regular for $\mathscr{C}_l$ but not accelerative, since members of $\mathscr{C}_l$ do not necessarily have the asymptotic representation $s_{n+1} \sim s + a_1 r_n + a_2 r_n^2 + \cdots$. In fact the algorithm on the basis of numerical evidence seems inferior to the GBW algorithm, as the following example points out.

As with the GBW algorithm, the Overholt method sums $\sum \lambda^n$ exactly, $\lambda \neq 0, 1$.

Table I

k	s_k	t_k	$s_0^{(k)}$ GBW	$s_0^{(k)}$ Overholt	$t_0^{(k)}$ GBW	$t_0^{(k)}$ Overholt
0	1	1	1	1	1	1
1	1.538461	1.7	1.370814	1.370814	1.333492	1.333492
2	1.295019	0.930700	1.368870	1.368884	1.375589	1.389681
3	1.401825	1.746142	1.368809	1.368808	1.369853	1.375163
4	1.354209	0.857797	—	—	1.368604	1.358817
5	1.375298	1.789719	—	—	1.368785	1.364680
6	1.365930	0.786117	—	—	1.368801	1.382128
7	1.370086	1.827823	—	—	1.368804	1.375144

Example. Leonardo's equation (Scheid, 1968, p. 314)

$$f(x) = x^3 + 2x^2 + 10x - 20 = 0 \tag{12}$$

has one real root, $s = 1.368808107$. The equation can be rewritten in two ways,

$$x = \phi(x), \qquad x = \psi(x) \tag{13}$$

to yield, respectively, the sequences

$$s_{n+1} = 20/(s_n^2 + 2s_n + 10), \qquad t_{n+1} = (20 - 2t_n^2 - t_n^3)/10, \tag{14}$$

starting with $s_0 = t_0 = 1$. s_n converges, since $\phi'(s) = 0.4438$; t_n diverges ($\psi'(s) = -1.1096$). To these sequences application of the GBW algorithm and the Overholt algorithm produces sequences $s_n^{(k)}$, $t_n^{(k)}$ (see Table I).

None of the previous theorems indicates what will happen with $t_n^{(k)}$ for any path, since t_n diverges. In fact t_n has essentially two convergent subsequences, $t_{2n} \to \alpha$ and $t_{2n+1} \to \beta$, $\alpha = 0.548946478$, $\beta = 1.92318948$. Both the Overholt and GBW methods produce $s_n^{(k)}$ that converge to s along all paths. What is remarkable is that the GBW method produces a $t_n^{(k)}$ that sums diagonally to s while all paths in the Overholt method for $t_n^{(k)}$ seem to produce divergence or at best grave numerical instability. To the accuracy permitted by the 19 places carried, $t_0^{(11)}$ was found to be 1.5, $t_0^{(12)}$ to be negative.

5.3.3. The Phenomenon of Diagonal Convergence

A strange thing occurs when either the GBW or Overholt process is applied to an iterative sequence which diverges very rapidly. To illustrate, let s_n be given by

$$s_{n+1} = s_n^2 - 1, \qquad s_0 = 2. \tag{1}$$

If **s** could be summed, one would expect the limit to be $(1 + \sqrt{5})/2$. In fact, **s** diverges with such rapidity that none of the methods we have studied will

Table II

$s_{n+1} = s_n^2 - 1,\ s_0 = 1$

$s_n^{(k)}$				
2				
	1.75			
3		1.736111111		
	2.5		1.735916990	
8		2.493956044		1.735916943
	7.214285714		2.493954545	
63		7.214094181		2.493954545
	62.031017370		7.214094181	
3968		62.031017308		7.214094181
	3967.000503841		62.031017308	⋮
15745023		3967.000503842	⋮	
	15745022.000000120	⋮		
	⋮			

assign it a limit. The result of applying the GBW algorithm produces the Table II. Clearly, there is regularity along no path. But, also,

$$\lim_{k\to\infty} s_n^{(k)} = \alpha_n; \tag{2}$$

i.e., there is convergence along all diagonal paths. Interesting are the facts that $\alpha_i \neq \alpha_j$, $i \neq j$, and the extraordinary rapidity with which the above limit is reached. There seems to be no easy way to determine closed form values of α_n. But the cause of the failure of the αs to be equal can be pinpointed: a certain function fails to satisfy a uniqueness property under infinite interpolation. Recall that if $\mathbf{z}$ is a sequence of distinct complex numbers and $z_n \to \infty$, there is always an entire function f for which $f(z_n) = w_n$ for any $\mathbf{w} \in \mathscr{C}_S$ (Davis, 1963, Theorem 5.2.1). f is not, in general, unique. Recall that in our case $s_n^{(k)}$ is the value at $z = 0$ of the Lagrangian interpolant of degree k that takes the values s_{n+j} at the points a_{n+j+1}, $0 \leq j \leq k$. The Newton interpolation formula gives

$$s_n^{(k)} = s_n - a_{n+1}g_1(n) + a_{n+1}a_{n+2}g_2(n) - \cdots \tag{3}$$

to $k + 1$ terms, where the g_j are the divided differences

$$g_{k-1}(n) = \begin{vmatrix} s_n & 1 & a_{n+1} & \cdots & a_{n+1}^{k-2} \\ s_{n+1} & 1 & a_{n+2} & \cdots & a_{n+2}^{k-2} \\ \vdots & \vdots & \vdots & & \vdots \\ s_{n+k-1} & 1 & a_{n+k} & \cdots & a_{n+k}^{k-2} \end{vmatrix} \Bigg/ V_k(a_{n+1}, a_{n+2}, \ldots, a_{n+k}). \tag{4}$$

f is the function inverse to $\phi - I$ since $(\phi - I)(s_n) = s_{n+1} - s_n = a_{n+1}$. Clearly here the f taking the values s_k at a_{k+1}, $k \geq j$, is not unique. In this example the series (3) converges with such rapidity that the first two terms give a good estimate for α_n for, say, $n > 2$:

$$\alpha_n \approx s_n - a_{n+1}^2/\Delta a_{n+1}. \tag{5}$$

When $n = 3$, for instance, the right- and left-hand sides agree to nine significant figures.

5.3.4. Construction of Scalar Iteration Functions

One of the most important problems in applied mathematics is that of determining vectors in $\mathscr{B}$ satisfying $x = \phi(x)$ where $\phi: \mathscr{B} \to \mathscr{B}$. (Obviously determining a zero of $f: \mathscr{B} \to \mathscr{B}$ can be formulated in terms of this problem.) One of the earliest methods of attacking the problem—one treated in every elementary book on numerical analysis, at least for scalar equations—is the method of *simple iteration*: Starting with some initial value s_0 one establishes the sequence $s_{n+1} = \phi(s_n)$. The difficulty with this method is well known—it hardly ever converges—and the search has long been on for more effective iteration methods.

The literature on iteration processes is enormous. The books by Ostrowski (1966) and Traub (1964) give a good survey of classical material. Good sources of more recent work, in particular, on operator equations are Kantorovich and Akilov (1964), Ortega and Rheinboldt (1970), Ostrowski (1973), and, of course, *Mathematical Reviews*.

The usual way of constructing an iteration sequence is to employ the derivatives of ϕ, the Newton–Raphson method being the classical prototype of this kind of method. Recently authors have studied methods for scalar equations that do not require derivatives, e.g., generalizations of the Steffensen iteration process [see Wimp (1970); Esser (1975); King (1979)]. In fact, a virtually inexhaustible source of iteration sequences that do not require derivatives can be obtained using any one of the sequence transformations in this book, for $s_n = \phi(\phi(\cdots \phi(s_0))\ldots)$ requires only repeated evaluations of ϕ. There is, however, a more imaginative way of proceeding, and that is to make use of the general nonlinear iteration function G of Eq. 5.1(3). Consider the formula

$$s_{n+1} = G(s_n, \phi(s_n), \phi_2(s_n), \ldots, \phi_k(s_n)), \qquad n \geq 0, \tag{1}$$

where

$$\phi_{k+1}(x) = \phi(\phi_k(x)), \qquad \phi_0(x) = x. \tag{2}$$

Assume that G is homogeneous and translative and related to g by Eq. 5.1(3).

Theorem 1. Let ϕ be analytic at α, $\phi(\alpha) = \alpha$, $\phi'(\alpha) \neq 0, 1$, and let g be continuous in a neighborhood of $\phi'(\alpha)e$, $g(\phi'(\alpha)e) = 1/[1 - \phi'(\alpha)]$. Then if s_0 is sufficiently close to α, **s** converges, $s = \alpha$, and either $r_{n+1}/r_n \to 0$ or $r_n \equiv. 0$.

Proof

$$s_{n+1} = s_n + [\phi(s_n) - s_n]g\left[\frac{\phi_2(s_n) - \phi(s_n)}{\phi(s_n) - s_n}, \ldots, \frac{\phi_k(s_n) - \phi_{k-1}(s_n)}{\phi_{k-1}(s_n) - \phi_{k-2}(s_n)}\right]. \tag{3}$$

Now

$$\phi(x) = \alpha + \phi'(\alpha)(x - \alpha) + \cdots \tag{4}$$

so

$$\phi_k(x) = \alpha + [\phi'(\alpha)]^k(x - \alpha) + \cdots \tag{5}$$

and

$$\frac{\phi_k(x) - \phi_{k-1}(x)}{\phi_{k-1}(x) - \phi_{k-2}(x)} = \phi'(\alpha) + O(x - \alpha), \qquad x \to \alpha. \tag{6}$$

Thus as $s_n \to \alpha$ the right-hand side of (3) behaves as

$$\alpha + r_n\{1 + [\phi'(\alpha) - 1 + O(r_n)][g(\phi'(\alpha)e) + o(1)]\} = \alpha + r_n o(1). \tag{7}$$

Thus, given $\varepsilon > 0$, if s_0 is sufficiently close to α, all subsequent s_n will be in $N_\varepsilon(\alpha)$, the denominator of (6) will be nonzero for $x = s_n$, and g will be defined. Thus

$$r_{n+1} = r_n o(1), \qquad n \to \infty, \tag{8}$$

and this proves the theorem. ∎

The theorem shows that although the sequence **t** defined by $t_{n+1} = \phi(t_n)$ converges (if it converges at all) linearly, the sequence **s** converges hyperlinearly.

If ϕ is analytic in a neighborhood of α and

$$\phi(x) = \alpha + \sum_{r=0}^{\infty} c_r(x - \alpha)^{m+r}, \qquad |x - \alpha| < \rho, \quad m \geq 1, \quad c_0 \neq 0, \tag{9}$$

then ϕ is said to be an *iteration function of order m*. Given an iteration function ϕ of order 1, can an iteration of arbitrary order be constructed? The answer is, of course, yes. We discuss this matter in the context of finding a root of the equation $x = \phi(x)$, where ϕ is an iteration function of first order at α.

Let f be any function with a simple zero at α (e.g., $f = x - \phi$ if $\phi'(\alpha) \neq 1$), and let g be analytic at α, $g(\alpha) \neq 0$. In a neighborhood of α, we have the Laurent series

$$g/f = b_{-1}/(x - \alpha) + b_0 + b_1(x - \alpha) + \cdots. \tag{10}$$

Then

$$h_r = \frac{1}{r!}\frac{d^r}{dx^r}\left(\frac{g}{f}\right) = \frac{(-1)^r b_{-1}}{(x-\alpha)^{r+1}} + b_r + O(x-\alpha), \tag{11}$$

$$h_{r-1}/h_r = -(x-\alpha)[1 + O(x-\alpha)^r], \tag{12}$$

and so

$$\Phi_{r+1} = x + h_{r-1}/h_r \tag{13}$$

is an iteration function of order $r + 1$ at least. $g = r = 1$ gives Newton's method. In general, Φ_{r+1} will involve derivatives of f of orders up through r.

Analogs of certain sequence transformations can be used to obtain iteration functions that do not require the differentiation of ϕ.

The first method is an analog of the GBW transformation.

Theorem 2. Let ϕ be an iteration function of first order at α with $\phi'(\alpha)$ not a root of 1. Then the function $\Phi_k(\phi)$ defined recursively by

$$\Phi_{k+1}[\phi(x)] = \frac{(\phi - x)\Phi_k(\phi(\phi(x))) - (\phi_{k+2} - \phi_{k+1})\Phi_k(\phi(x))}{(\phi - x) - (\phi_{k+2} - \phi_{k+1})}, \tag{14}$$
$$k \ge 0, \quad \Phi_0(\phi) = \phi,$$

where ϕ_k is defined in (2), is an iteration function of order $k + 1$ at least.

Proof. Let $\phi'(\alpha) = c$, $c^j \neq 1$. Assume

$$\Phi_k(\phi) = \alpha + A_k(x-\alpha)^{k+1} + B_k(x-\alpha)^{k+2} + \cdots, \qquad 0 \le k \le K. \tag{15}$$

Then

$$\Phi_k(\phi(\phi)) = \alpha + A_k c^{k+1}(x-\alpha)^{k+1} + \cdots, \qquad 0 \le k \le K, \tag{16}$$

so

$$\Phi_{K+1}(\phi) = \alpha + \frac{(\phi - x)[\Phi_K(\phi(\phi)) - \alpha] - (\phi_{K+2} - \phi_{K+1})[\Phi_K(\phi) - \alpha]}{(\phi - x) - (\phi_{K+2} - \phi_{K+1})}$$
$$= \alpha + \frac{N_K}{D_K}, \tag{17}$$

$$N_K = (c-1)(x-\alpha)A_K c^{K+1}(x-\alpha)^{K+1} - (c^{K+2} - c^{K+1})(x-\alpha)A_K(x-\alpha)^{K+1}$$
$$+ O(x-\alpha)^{K+3} = O[(x-\alpha)^{K+3}], \tag{18}$$

while

$$D_K = (c-1)(x-\alpha) - (c^{K+2} - c^{K+1})(x-\alpha) + O(x-\alpha)^2$$
$$= (c-1)(1-c^{K+1})(x-\alpha) + O[(x-\alpha)^2], \tag{19}$$

and the leading term in (19) is not zero. Thus

$$\Phi_{K+1}(\phi) = \alpha + O[(x - \alpha)^{K+2}], \tag{20}$$

as was to be shown. ■

$k = 0$ gives the Steffensen iteration function

$$\Phi_1 = (x\phi_2 - \phi^2)/(\phi_2 - 2\phi + x). \tag{21}$$

Consider the iteration sequence

$$s_{n+1} = \phi(s_n) = s_n^2 - 1, \qquad s_0 = 1.5. \tag{22}$$

Here $\alpha = (1 + \sqrt{5})/2 = 1.618033989$. None of the usual sequence transformations works very well in summing this very rapidly divergent sequence. In fact, nearly all of them produce divergent sequences, an exception being the GBW transformation. Yet the Steffensen function produces a rapidly convergent sequence,

$$s_{n+1} = \frac{s_n\phi(\phi(s_n)) - \phi^2(s_n)}{\phi(\phi(s_n)) - 2\phi(s_n) + s_n}, \tag{23}$$

$\mathbf{s} = \{1.5, 1.6429, 1.6189, 1.6180344, \ldots\}$. That the original GBW transformation produces only mild convergence while the iteration function constructed by analogy to the GBW transformation produces very rapid convergence is not so paradoxical as it may seem. This is because the iteration function Φ_k uses much more precise information about the sequence (22) than the GBW sequence transformation, namely, the exact form of the iteration function ϕ.

The Overholt procedure also produces an iteration function.

Theorem 3. Let ϕ be an iteration function of first order at α with $\phi'(\alpha)$ not a root of 1. Then the function $\Psi_k(\phi)$ defined recursively by

$$\Psi_{k+1}[\phi(x)] = \frac{(\phi_{k+1} - \phi_k)^{k+1}\Psi_k[\phi(\phi(x))] - (\phi_{k+2} - \phi_{k+1})^{k+1}\Psi_k[\phi(x)]}{(\phi_{k+1} - \phi_k)^{k+1} - (\phi_{k+2} - \phi_{k+1})^{k+1}}, \tag{24}$$

$k \geq 0$, $\Psi_0(\phi) = \phi$, is an iteration function of order $k + 1$ at least.

Proof. Left to the reader. ■

For $k = 0$, this method also yields the Steffensen iteration function.

The third iteration procedure arises from the formal elimination of the constants c_r, $1 \leq r \leq k$, from the system of equations

$$\Upsilon = \phi_j + \sum_{r=1}^{k} c_r(\phi_{r+j} - \phi_{r+j-1}), \qquad 0 \leq j \leq k. \tag{25}$$

Some readers may recognize this as the same formal procedure that leads to the Schmidt sequence transformation, to be discussed at length in Chapter 6. One interprets (25) as $k + 1$ equations in the k unknowns c_r. As such, for consistency, the determinant of the augmented matrix of the system must vanish. This produces a determinantal equation that may be solved for Υ, which can be relabeled Υ_k, and

$$\Upsilon_k = \frac{\begin{vmatrix} x & \phi - x & \cdots & \phi_k - \phi_{k-1} \\ \phi & \phi_2 - \phi & \cdots & \phi_{k+1} - \phi_k \\ \vdots & & & \vdots \\ \phi_k & \phi_{k+1} - \phi_k & \cdots & \phi_{2k} - \phi_{2k-1} \end{vmatrix}}{\begin{vmatrix} 1 & \phi - x & \cdots & \phi_k - \phi_{k-1} \\ 1 & \phi_2 - \phi & \cdots & \phi_{k+1} - \phi_k \\ \vdots & & & \vdots \\ 1 & \phi_{k+1} - \phi_k & \cdots & \phi_{2k} - \phi_{2k-1} \end{vmatrix}}. \tag{26}$$

To analyze Υ_k it is convenient to rewrite it

$$\Upsilon_k = \alpha + |\Delta^{(i-1)}(\phi_{j-1})|_{k+1}/|\Delta^{(i+1)}(\phi_{j-1})|_k, \tag{27}$$

i.e., $1 \le i, j \le k + 1$ in the numerator determinant and $1 \le i, j \le k$ in the denominator determinant, and

$$\begin{aligned} \Delta^{k+1}(\phi_j) &= \Delta^k(\phi_{j+1}) - \Delta^k(\phi_j), \qquad k, j \ge 0, \\ \Delta^0(\phi_0) &= \Delta^0(x) = x - \alpha. \end{aligned} \tag{28}$$

We now need the following lemma.

Lemma. Let

$$D_m = |b_i(x_j)|, \qquad 1 \le i, j \le m, \tag{29}$$

where

$$b_i(x_j) = \sum_{r=0}^{\infty} c_{i-1,r} x^r, \qquad 0 \le |x_j| < \rho_j, \quad \rho_j > 0. \tag{30}$$

Let

$$x_j = O(\eta), \qquad \eta \to 0. \tag{31}$$

Then

$$D_m = |c_{i-1,j-1}| V_m(x_1, x_2, \ldots, x_m) + O(\eta^{m(m-1)/2+1}). \tag{32}$$

Proof. In what follows it will be convenient to let D_m be a generic notation not necessarily involving the same $b_i(x_j)$ wherever the symbol appears.

Proof is by induction. Assume (32) true for $1 \le m \le N - 1$. By the explicit formula for V_m [Appendix 1(4)], this implies

$$D_m = O(\eta^{m(m-1)/2}), \qquad 1 \le m \le N - 1; \tag{33}$$

$$D_N = |c_i(x_j)| + R_N, \qquad 1 \le i, j \le N,$$

$$c_i(x_j) = \sum_{r=0}^{N-1} c_{i-1,r} x_j^r = b_i(x_j) - d_i(x_j), \tag{34}$$

$$d_i(x_j) = \sum_{r=N}^{\infty} c_{i-1,r} x_j^r.$$

Then

$$\begin{aligned} |c_i(x_j)| &= |c_{i-1,j-1}||x_j^{i-1}| \\ &= |c_{i-1,j-1}| V_N(x_1, \ldots, x_N), \qquad 1 \le i, j \le N. \end{aligned} \tag{35}$$

The remainder R_N may be written

$$R_N = \sum_{r=1}^{N} \sum_{(u_1, u_2, \ldots, u_r) \in {}_N S_r} (x_{u_1} x_{u_2} \cdots x_{u_r})^N T_N(u_1, u_2, \ldots, u_r), \tag{36}$$

where ${}_n S_k$ is the set of combinations of $(1, 2, \ldots, n)$ taken k at a time and T_N is a determinant of order N containing $d_i(x_{u_j})/x_{u_j}^N$ in the columns u_j, $j = 1, 2, \ldots, r$, and $c_i(x_k)$ in the kth column if $k \ne u_j$. By Laplace's expansion (Aitken, 1956, p. 78) $T_N(u_1, u_2, \ldots, u_r)$ may be expanded by minors chosen from the columns u_j and their cofactors whose elements are chosen from the remaining $N - r$ columns. These latter are determinants of the form D_{N-r}, and each may be estimated as $\eta \to 0$ by (33). Thus (36) may be written

$$\begin{aligned} R_N &= \sum_{r=1}^{N} O[\eta^{rN} \eta^{(N-r)(N-r-1)/2}] \\ &= \sum_{r=1}^{N} O[\eta^{N(N-1)/2 + r(r+1)/2}] = O(\eta^{N(N-1)/2 + 1}). \end{aligned} \tag{36'}$$

This establishes the lemma for $m = N$. Since the result is true for $m = 1$, the proof is complete. ■

We now return to the analysis of Υ_k. $\Delta^{(k)}(x)$ may be written

$$\Delta^{(k)}(x) = \sum_{r=1}^{\infty} b_{kr}(x - \alpha)^r. \tag{37}$$

Let

$$\phi(x) = \alpha + \sum_{r=1}^{\infty} c_r(x - \alpha)^r. \tag{38}$$

Then the b_{kr} can be computed recursively from the c_r. For instance,

$$\begin{aligned} b_{11} &= c_1 - 1, & b_{1j} &= c_j, \\ b_{21} &= (c_1 - 1)^2, & b_{22} &= c_2(c_1 - 1)(c_1 + 2), \\ b_{31} &= (c_1 - 1)^3, & b_{32} &= c_2(c_1 - 1)^2(c_1^2 + 3c_1 + 3), \end{aligned} \tag{39}$$

Theorem 4. Let $|b_{i+1,j}|_k, |b_{i,j+1}|_k, 1 \le i,j \le k$, be nonzero and let c_1 not be a root of 1.

Then Υ_k is an iteration function of order $k + 1$.

Proof. For the numerator determinant in (27) we use the lemma with the identifications

$$c_{i-1,j} = b_{i-1,j+1}, \qquad x_j = \bar{\phi}_{j-1} = \phi_{j-1} - \alpha, \qquad m = k + 1, \quad \eta = x - \alpha, \tag{40}$$

and for the denominator determinant

$$c_{i-1,j} = b_{i+1,j+1}, \qquad x_j = \bar{\phi}_{j-1}, \qquad m = k, \qquad \eta = x - \alpha. \tag{41}$$

The result is

$$\Upsilon_k = \alpha + \bar{\phi}_k \left[\frac{|b_{i-1,j}|_{k+1} V_{k+1}(\bar{\phi}_0, \ldots, \bar{\phi}_k) + O((x - \alpha)^{k(k+1)/2+1})}{|b_{i+1,j}|_k V_k(\bar{\phi}_0, \ldots, \bar{\phi}_{k-1}) + O((x - \alpha)^{(k-1)k/2+1})} \right]. \tag{42}$$

Since $|b_{i-1,j}|_{k+1} = |b_{i,j+1}|_k$, the use of (5) gives

$$\Upsilon_k = \alpha + \frac{|b_{i,j+1}|_k}{|b_{i+1,j}|_k} c_1^{k(k+2)/2} \prod_{j=1}^{k} (c_1^j - 1)(x - \alpha)^{k+1}[1 + O(x - \alpha)] \tag{43}$$

and this proves the theorem. ■

This process also yields the Steffensen iteration function when $k = 1$.

5.3.5. *Iteration Functions in Abstract Spaces*

Iteration functions suitable for the solution of operator equations can be derived in a straightforward manner. Let $\phi: \mathscr{B} \to \mathscr{B}$. Let ϕ^* be any element from the dual of $\mathscr{B}$ and consider

$$\Upsilon_k = \frac{\begin{vmatrix} I & \phi - I & \cdots & \phi_{k+1} - \phi_k \\ \langle \phi^*, \phi \rangle & \langle \phi^*, \phi_2 - \phi \rangle & \cdots & \langle \phi^*, \phi_{k+2} - \phi_{k+1} \rangle \\ \vdots & \vdots & & \vdots \\ \langle \phi^*, \phi_k \rangle & \langle \phi^*, \phi_{k+1} - \phi_k \rangle & \cdots & \langle \phi^*, \phi_{2k+1} - \phi_{2k} \rangle \end{vmatrix}}{\begin{vmatrix} 1 & \langle \phi^*, \phi - I \rangle & \cdots & \langle \phi^*, \phi_{k+1} - \phi_k \rangle \\ 1 & \langle \phi^*, \phi_2 - \phi \rangle & \cdots & \langle \phi^*, \phi_{k+2} - \phi_{k+1} \rangle \\ \vdots & \vdots & & \vdots \\ 1 & \langle \phi^*, \phi_{k+1} - \phi_k \rangle & \cdots & \langle \phi^*, \phi_{2k+1} - \phi_{2k} \rangle \end{vmatrix}}. \tag{1}$$

For $k = 1$ this gives an interesting operator generalization of the Steffensen iteration function:

$$\Upsilon_1 = \frac{\langle \phi^*, \phi(\phi) \rangle I - \langle \phi^*, \phi \rangle \phi}{\langle \phi^*, \phi(\phi) - 2\phi + I \rangle}. \tag{2}$$

Thus an iteration sequence **s** for the solution of $(I - \phi)(x) = 0$ is given by

$$s_{n+1} = \frac{s_n \langle \phi^*, \phi[\phi(s_n)] \rangle - \phi(s_n) \langle \phi^*, \phi(s_n) \rangle}{\langle \phi^*, \phi[\phi(s_n)] - 2\phi(s_n) + s_n \rangle}, \qquad n \geq 0, \quad s_0 \in \mathscr{B}. \tag{3}$$

The generalization of the GBW iteration function to operator sequences can be accomplished by using the recursion formula:

$$\Phi_{k+1} = \frac{\langle \phi^*, \phi - I \rangle \Phi_k(\phi(\phi)) - \langle \phi^*, \phi_{k+2} - \phi_{k+1} \rangle \Phi_k(\phi)}{\langle \phi^*, \phi - I - \phi_{k+2} + \phi_{k+1} \rangle}. \tag{4}$$

For $k = 0$, this reduces to

$$\Phi_1 = \frac{\langle \phi^*, \phi - I \rangle \phi(\phi) - \langle \phi^*, \phi(\phi) - \phi \rangle \phi}{\langle \phi^*, \phi(\phi) - 2\phi + I \rangle}, \tag{5}$$

which is a generalization of the Steffensen iteration function different from (3). The generalization of the Overholt procedure, Eq. 5.3.4(24), proceeds along the same lines. For $k = 0$, it reduces to (5).

For related constructions of sequence transformations in abstract spaces, see Chapter 6.

Chapter 6 The Schmidt Transformation; The ε-Algorithm

6.1. Background

In this chapter we shall discuss the Schmidt transformation. Schmidt (1941) used the method to solve by iteration systems of linear equations. Schmidt's paper was neglected for some time, and the rediscovery of the method is really due to D. Shanks, who resurrected the algorithm and discussed its remarkable properties in a lengthy paper (1955). The paper was widely read and made an enormous impact.

However, there was one drawback to the use of the Schmidt transformation, specifically, the need to compute determinants of large order; in fact, the cumbersome formulation of the algorithm made it difficult to analyze from a theoretical point of view, in particular, to determine its regularity or accelerativeness for special sequence spaces. Wynn changed all that by the publication of a paper (1956) in which he showed the Schmidt transformation could be formalized as a simple (although nonlinear) lozenge algorithm. This formalization has come to be called the ε-algorithm and has been the subject of an enormous amount of research. Over 50 papers on the subject have appeared by Wynn, and at least 30 by Brezinski. For a fairly complete bibliography, the reader is referred to Brezinski's book (1977).

It should be pointed out that Schmidt's idea did not appear ex nihilo. (What mathematical idea has?) The general idea goes back at least to Jacobi (1846). A special case of the Schmidt algorithm called the δ^2-process is usually attributed to Aitken (1926), who used the method, not surprisingly, to accelerate iterates in the Bernoulli method for solving algebraic equations. However, Todd (1962) observed that the method attributed to Aitken goes back at least to Kummer (1837).

What are the properties of the Schmidt transformation? Briefly, it is nonlinear and nonregular. This should not be thought a drawback: It *is* regular for certain important sequences, for instance, $\mathscr{R}_{\mathrm{TM}}$. Furthermore, when it works, i.e., accelerates a sequence, its performance is often nothing short of spectacular. I have indicated that the improvement in convergence obtained by linear (Toeplitz) methods is generally restricted to exponential improvement. The Schmidt transformation has no such limitations. In fact, it sums sequences so grossly divergent that they are hopelessly beyond the reach of linear methods, for instance, the partial sums of $0! - 1! + 2! - 3! + 4! - \cdots$.

As the reader knows, Toeplitz methods are easily conceptualized for the case of vector sequences. It might be thought that the nonlinearity of the Schmidt transformation would prevent its extension to abstract spaces except, perhaps, for those spaces where elements have inverses. This is not true: Brezinski has shown how the Schmidt transformation can be generalized to topological vector spaces in a very natural way by utilizing the dual space. This generalization will be presented in a later chapter when we discuss the BH protocol. Further, at least for $\mathscr{R}_S$, the Schmidt transformation has an elegant geometric interpretation.

6.2. Derivation

The heuristic derivation of the Schmidt transformation proceeds from the consideration that many sequences in applied mathematics converge as though they were composed of their limit and a linear combination of exponential "transients." Thus for such a sequence

$$s_n = s + \sum_{r=1}^{k} a_r \gamma_r^n, \qquad n \geq 0, \quad \gamma_i \neq \gamma_j, \quad i \neq j, \quad \gamma_i \neq 0; \tag{1}$$

in other words, $\mathbf{s}$ is a member of $\mathscr{C}_{E^k}$.

We write (1) as

$$r_m = \sum_{r=1}^{k} a_r \gamma_r^m, \qquad m \geq 0. \tag{2}$$

Since r_m is a linear combination of k exponentials, there is a linear difference operator of order k with constants coefficients that annihilates r_m [see Milne-Thomson (1960)]. Thus

$$\sum_{i=0}^{k} b_i r_{m+i} = 0, \qquad b_0 b_k \neq 0. \tag{3}$$

Now

$$\begin{aligned} r_{m+i} &= r_m + \Delta r_m + \Delta r_{m+1} + \cdots + \Delta r_{m+i-1} \\ &= r_m + \Delta s_m + \Delta s_{m+1} + \cdots + \Delta s_{m+i-1}, \qquad i \geq 1. \end{aligned} \tag{4}$$

Substituting this in (3), dividing by $\sum b_i$ (this requires that **s** not be a constant sequence), and redefining constants gives

$$s = s_m + \sum_{r=1}^{k} c_r \, \Delta s_{m+r-1}, \qquad n \leq m \leq n + k. \tag{5}$$

This can be interpreted as $k + 1$ equations in the k constants c_r. For a solution to exist the column vector $\{s - s_m\}$ must lie in the column space of the coefficient matrix of the system, and this is true iff the following determinantal equation is satisfied:

$$\begin{vmatrix} s - s_n & \Delta s_n & \cdots & \Delta s_{n+k-1} \\ s - s_{n+1} & \Delta s_{n+1} & \cdots & \Delta s_{n+k} \\ \vdots & \vdots & & \vdots \\ s - s_{n+k} & \Delta s_{n+k} & \cdots & \Delta s_{n+2k-1} \end{vmatrix} = 0. \tag{6}$$

Solving for s gives

$$s = W_k(s_n)/W_k(1), \tag{7}$$

where

$$\mathscr{W}_k(z_n, s_n) \equiv \mathscr{W}_k(z_n) = \begin{bmatrix} z_n & \Delta s_n & \cdots & \Delta s_{n+k-1} \\ z_{n+1} & \Delta s_{n+1} & \cdots & \Delta s_{n+k} \\ \vdots & \vdots & & \vdots \\ z_{n+k} & \Delta s_{n+k} & \cdots & \Delta s_{n+2k-1} \end{bmatrix} \tag{8}$$

$$W_k(z_n) = |\mathscr{W}_k(z_n)|.$$

Equation (7) will be satisfied if $\mathbf{s} \in \mathscr{C}_{E^k}$, but not necessarily otherwise. Nevertheless, the formula can be used to *define* a lozenge algorithm when $W_k(1) \neq 0$, i.e.,

$$e_k(s_n) = s_n^{(k)} = W_k(s_n)/W_k(1), \qquad n, k \geq 0, \tag{9}$$

whose regularity for certain sequence spaces can be investigated. Equation (9) we shall call the Schmidt transformation.

Equations (5) shows that for sequences of the same kind as (1), s_n differs from its limit s by certain perturbation terms, namely, linear combinations of $\Delta s_n, \ldots, \Delta s_{n+k-1}$. Levin (1960) took this as his point of departure and defined a very general class of transformations of sequences whose terms could be considered as their limits plus perturbations of a more general character (see Chapter 9).

The following properties of the Schmidt algorithm are fairly obvious:

(i) e_k is translative and homogeneous,
(ii) $s_n^{(k)} = H_n^{(k+1)}(\mathbf{s})/H_n^{(k)}(\Delta^2\mathbf{s})$ [see Eq. 1.5(9)], and
(iii) e_k is *exact* ($s_n^{(k)} = s$) for any member of $\mathscr{C}_{E^m}$ when $k \geq m$, for all n.

6.3. Exactness Results

It is of interest to determine that class of sequences for which e_k is exact, i.e., sequences such that, for fixed $k \geq 1$, $s_n^{(k)} \equiv s, n \geq 0$. The reason is that generally speaking a method is most likely to be regular for classes of sequences that behave "similarly" to those sequences for which the method is exact. This idea will be particularly useful when we discuss an extension of the Schmidt transformation to sequences in topological vector spaces.

We begin with some definitions.

Definition. Let $\mathbf{y} \in \mathscr{C}_S$. Then $\mathbf{y} \in \mathscr{K}_k$, $k \geq 1$, means that $\mathbf{y}$ satisfies a homogeneous linear difference equation with constant coefficients,

$$\mathscr{P}_k(y_n) = c_0 y_n + c_1 y_{n+1} + \cdots + c_k y_{n+k} = 0, \qquad c_0 c_k \neq 0, \tag{1}$$

and satisfies no equation of lower order and that $\mathbf{y}$ contains no terms purely algebraic in n.

We say $\mathbf{y}$ is a *maximal solution* of Eq. (1) if $\mathbf{y}$ satisfies no lower-order equation. A *maximal basis* for (1) is a basis each of whose members is a maximal solution.

Examples

(1) $\{n\lambda^n\} \in \mathscr{K}_2$, $\lambda \neq 0, 1$;
(2) $\{\lambda_1^n + \lambda_2^n\} \in \mathscr{K}_2, \lambda_1 \neq \lambda_2, \lambda_1, \lambda_2 \neq 0, 1$;
(3) $\{\lambda^n\} \notin \mathscr{K}_2$;
(4) $\{n + \lambda^n\} \notin \mathscr{K}_3$.

In Example 4 the function satisfies a difference equation of order 3 and none lower, but violates the condition on algebraic terms.

Remark 1. *Any* solution $\mathbf{y}$ of $\mathscr{P}_k = 0$ whose initial values form a vector from the natural basis of $\mathscr{R}^k$ is a maximal solution.

Remark 2. A $\mathbf{y}$ containing algebraic terms is equivalent to $\sum c_r = 0$ [see Milne-Thomson (1960)].

The members of $\mathscr{K}_k$ are easy to characterize [see Milne-Thomson (1960, Chapter 13)]. They are linear combinations of the kind

$$\lambda_1^n(d_1 n^{m_1} + d_1' n^{m_1 - 1} + \cdots) + \lambda_2^n(d_2 n^{m_2} + d_2' n^{m_2 - 1} + \cdots) \\ + \cdots + \lambda_r^n(d_r n^{m_r} + d_r' n^{m_r - 1} + \cdots), \qquad (2)$$

where the λ_j are distinct, $\lambda_j \neq 0, 1, d_j \neq 0$ and $m_1 + m_2 + \cdots + m_r + r = k$.

Let

$$\mathscr{M}_k(s_n) = \begin{bmatrix} \Delta s_n & \cdots & \Delta s_{n+k-1} \\ \vdots & & \vdots \\ \Delta s_{n+k-1} & \cdots & \Delta s_{n+2k-2} \end{bmatrix} \qquad (3)$$

denote the $(k + 1, 1)$ minor of $W_k(s_n)$, $M_k = |\mathscr{M}_k|$.

Lemma 1. Let $\mathbf{s} \in \mathscr{K}_k$. Then neither $W_k(1)$ nor $M_k(s_n)$ vanishes for any value of n.

Proof. Only the first statement will be shown; proof of the second is similar.

Let

$$c_0 s_n + c_1 s_{n+1} + \cdots + c_k s_{n+k} = 0, \qquad c_0 c_k \neq 0. \qquad (4)$$

One can rewrite M_k

$$M_k(s_n) = \begin{vmatrix} 1 & s_n & \cdots & s_{n+k} \\ \vdots & \vdots & & \vdots \\ 1 & s_{n+k} & \cdots & s_{n+2k-1} \end{vmatrix}. \qquad (5)$$

The functions $\{1, s_n, \ldots, s_{n+k-1}\}$ are linearly independent. Otherwise, for some constants d_j not all zero,

$$d_0 + d_1 s_n + \cdots + d_k s_{n+k-1} \equiv 0, \qquad (6)$$

which means either that s_n either contains algebraic terms ($d_0 \neq 0$) or else satisfies a difference equation of order $<k$ ($d_0 = 0$). But these functions are solutions of the $(k + 1)$th-order equation

$$c_0 \, \Delta y_n + c_1 \, \Delta y_{n+1} + \cdots + c_k \, \Delta y_{n+k} = 0. \qquad (7)$$

Heymann's theorem (Milne-Thomson, 1960, p. 357), gives

$$M_k(s_{n+1}) = (-1)^k (c_0/c_k) M_k(s_n). \qquad (8)$$

But this means that M_k can vanish for no value of n unless it vanishes identically for $n > n_0$, which cannot be. ■

Lemma 2. If $\mathbf{s}$ satisfies an equation of the form (1) of order $r \leq k$ and contains algebraic terms, then $W_k(1) \equiv 0$.

Proof. Since $\sum c_j = 0$, the equation can be rewritten

$$a_0 \, \Delta s_n + a_1 \, \Delta s_{n+1} + \cdots + a_{r-1} \, \Delta s_{n+r-1} = 0, \qquad a_{r-1} = c_r \neq 0. \tag{9}$$

Thus Rank $\mathscr{W}_k(1) < k + 1$, and so $W_k(1) \equiv 0$. ■

Remark 3. To illustrate the hypothesis in Lemma 1 that t_n must be free from algebraic terms, take $t_n = n + \lambda^n$, $\lambda \neq 0, 1$. Then $\mathbf{t}$ satisfies an equation of minimal order 3, yet $W_3(1) \equiv 0$.

The next result will be of use later on.

Lemma 3. Let $\mathscr{P}_k(y) = 0$ with $c_k = 1$. Then for any $\mathbf{t} \in \mathscr{C}_S$, $\mathscr{P}_k(t_n)$ is as given by Eq. (10) for any solution $\mathbf{s}$ of $\mathscr{P}_k = 0$ for which the denominator is defined, in particular, for $\mathbf{s}$ a maximal solution.

$$\mathscr{P}_k(t_n) \equiv (-1)^k \begin{vmatrix} t_n & s_n & \cdots & s_{n+k-1} \\ t_{n+1} & s_{n+1} & \cdots & s_{n+k} \\ \vdots & \vdots & & \vdots \\ t_{n+k} & s_{n+k} & \cdots & s_{n+2k-1} \end{vmatrix} \Bigg/ \begin{vmatrix} s_n & \cdots & s_{n+k-1} \\ \vdots & & \vdots \\ s_{n+k-1} & \cdots & s_{n+2k-2} \end{vmatrix} \tag{10}$$

Proof. Left to the reader.

Theorem. $s_n^{(k)}$ is defined for all $n \geq 0$ and $s_n^{(k)} \equiv s$ iff $\mathbf{r} \in \mathscr{K}_k$.

Proof. We can write

$$s_n^{(k)} - s = W_k(r_n, r_n)/W_k(1, r_n). \tag{11}$$

$\Rightarrow$: If $s_n^{(k)} \equiv s$, we must have for some a_j not all zero

$$a_0 r_n + a_1 \, \Delta r_n + \cdots + a_k \, \Delta r_{n+k-1} = 0, \qquad n \geq 0, \tag{12}$$

or, rewriting,

$$c_0 r_n + c_1 r_{n+1} + \cdots + c_k r_{n+k} = 0. \tag{13}$$

Differencing gives

$$c_0 \, \Delta r_n + c_1 \, \Delta r_{n+1} + \cdots + c_k \, \Delta r_{n+k} = 0, \tag{14}$$

and if $c_0 = 0$ or $c_k = 0$, rank $\mathscr{W}_k(1, r_n) < k + 1$, so $W_k \equiv 0$ and $s_n^{(k)}$ is not defined. This means the minimal order of the difference equation (13) is k. By Lemma 2, r_n can contain no algebraic terms. Thus $\mathbf{r}_n \in \mathscr{K}_k$.

$\Leftarrow$: $W_k(1, r_n) \neq 0$ and

$$c_0 r_n + c_1 r_{n+1} + \cdots + c_k r_{n+k} = 0, \qquad c_0 c_k \neq 0. \tag{15}$$

Rewriting gives

$$c_0' r_n + c_1' \, \Delta r_n + \cdots + c_k' \, \Delta r_{n+k-1} = 0, \qquad c_k' = c_k \neq 0, \tag{16}$$

so Rank $\mathscr{W}_k(r_n, r_n) < k + 1$ and $W_k(r_n, r_n) \equiv 0$. ■

Example ($k = 1$, the Aitken δ^2-process). e_1 is defined and exact when and only when $s_n = s + c\lambda^n$, $c \neq 0$, $\lambda \neq 0, 1$.

Example ($k = 2$). e_2 is defined and exact iff either $s_n = s + c_1\lambda_1^n + c_2\lambda_2^n$, $c_1c_2 \neq 0$, $\lambda_1, \lambda_2 \neq 0, 1$, $\lambda_1 \neq \lambda_2$, or $s_n = s + (c_1 n + c_2)\lambda^n$, $c_1 \neq 0$, $\lambda \neq 0, 1$.

6.4. The Effect of e_k on Certain Series

The work of this section will illustrate the remarks of the previous section. It turns out that e_k works well on (renders more rapidly convergent) sequences that behave as exponential sequences and does poorly on sequences that contain terms algebraic in n.

By elementary row and column manipulations,

$$W_k(s_n) = \begin{vmatrix} s_n & \Delta s_n & \cdots & \Delta^k s_n \\ \vdots & \vdots & & \vdots \\ \Delta^k s_n & \Delta^{k+1} s_n & \cdots & \Delta^{2k} s_n \end{vmatrix}, \tag{1}$$

and

$$W_k(1, s_n) = W_{k-1}(\Delta^2 s_n, \Delta^2 s_n), \qquad k \geq 1. \tag{2}$$

Since W_k can be written

$$W_k(s_n) = \begin{vmatrix} s_n & \cdots & s_{n+k} \\ \vdots & & \vdots \\ s_{n+k} & \cdots & s_{n+2k} \end{vmatrix}, \tag{3}$$

it follows that, for $s_n = \lambda^n p_n$,

$$W_k = \lambda^{(k+1)(n+k)} W_k(p_n, p_n). \tag{4}$$

Now let

$$p_n = n^\theta(c_0 + c_1/n + \cdots). \tag{5}$$

Since $\Delta^2 \lambda^n p_n = \lambda^n(\lambda - 1)^2 p_n^*$,

$$\begin{aligned} W_k(1, \lambda^n p_n) &= W_{k-1}(\lambda^n(\lambda - 1)^2 p_n^*, \lambda^n(\lambda - 1)^2 p_n^*), \\ &= (\lambda - 1)^{2k} \lambda^{k(n+k-1)} W_{k-1}(p_n^*, p_n^*), \\ p_n^* &= n^\theta(c_0 + c_1'/n + \cdots). \end{aligned} \tag{6}$$

By Lemma 1.7(2) and the representation (1),

$$W_k(p_n, p_n) \approx c_0^{k+1} \times \begin{vmatrix} n^\theta & \cdots & \theta(\theta-1)(\theta-k+1)n^{\theta-k} \\ \vdots & & \vdots \\ \theta(\theta-1)\cdots(\theta-k+1)n^{\theta-k} & \cdots & \theta(\theta-1)\cdots(\theta-2k+1)n^{\theta-2k} \end{vmatrix}$$

$$= c_0^{k+1} n^{(k+1)(\theta-k)} A_k \prod_{j=1}^{k} (-1)^j(-\theta)_j, \tag{7}$$

$$A_k = \begin{vmatrix} 1 & \theta & \cdots & \theta(\theta-1)\cdots(\theta-k+1) \\ \vdots & \vdots & & \vdots \\ 1 & \theta-k & \cdots & (\theta-k)\cdots(\theta-2k+1) \end{vmatrix},$$

and differencing rows gives $A_k = \prod_{j=1}^{k} (-1)^j j!$ and

$$W_k(p_n, p_n) \approx c_0^{k+1} n^{(k+1)(\theta-k)} \prod_{j=1}^{k} (-\theta)_j j!. \tag{8}$$

We combine these results as a theorem.

Theorem 1. Let s_n have the asymptotic expansion

$$s_n \sim s + \lambda^n n^\theta \sum_{r=0}^{\infty} \frac{c_r}{n^r}, \qquad c_0 \neq 0. \tag{9}$$

Then for $\lambda \neq 1$, $\theta \neq 0, 1, \ldots, k-1$,

$$s_n^{(k)} = s + \frac{c_0 \lambda^{n+2k} n^{\theta-2k} k!(-\theta)_k}{(\lambda-1)^{2k}}\left[1 + O\left(\frac{1}{n}\right)\right], \tag{10}$$

and thus, for fixed k, e_k accelerates convergence on vertical paths of all convergent sequences of the form (9).

For $\lambda = 1$, $\theta \neq 0, 1, \ldots, k-1$,

$$s_n^{(k)} = s + \frac{c_0 k! n^\theta}{(1-\theta)_k}\left[1 + O\left(\frac{1}{n}\right)\right], \tag{11}$$

and so e_k does not accelerate convergence.

For the effect of e_k on exponential sequences of the form $s + \sum c_k \lambda_r^n$, see Theorem 6.8(4).

We have already shown that e_1, which is Aitken's δ^2-process, is accelerative for $\mathscr{C}_l$ (see Section 5.1, Example 1); e_k is not accelerative for $\mathscr{C}_l$ when $k > 1$. The following argument, due to Smith and Ford (1979), shows what happens when $k = 2$. In the notation of Chapter 5,

$$e_2(s_n) = s_n + a_{n+1} g(\rho_{n+1}, \rho_{n+2}, \rho_{n+3}). \tag{12}$$

(The k in that chapter is here 4.) According to Theorem 5.1(4), $e_2(s_n)$ accelerates $\mathscr{A}_\rho$ iff g is continuous at (ρ, ρ, ρ) and has the value $1/(1-\rho)$ there.

But

$$g(x_1, x_2, x_3) = \frac{(1 + x_2)(x_2 - x_1) - x_2(x_3 - x_1)}{x_2(x_1 - 1)(x_3 - x_1) - (x_2 - x_1)(x_1x_2 - 1)}. \tag{13}$$

If $x_1 = \rho, x_2 = x, x_3 = y$, the denominator is zero on the curve

$$x(\rho - 1)(y - \rho) = (x - \rho)(x\rho - 1) \tag{14}$$

and the numerator on the curve

$$x(y - \rho) = (1 + x)(x - \rho). \tag{15}$$

The only x-values common to these curves must satisfy

$$\rho - 1 = (x\rho - 1)/(1 + x), \qquad \text{or} \qquad x = \rho. \tag{16}$$

Therefore, $y = \rho$ as well. We conclude that g is unbounded arbitrarily close to (ρ, ρ, ρ), so e_2 does not accelerate $\mathscr{A}_\rho$. e_k, in fact, is not regular for $\mathscr{A}_\rho$ when $k > 1$.

For totally monotone sequences, an interesting convexity property holds for $e_k(s_n)$. To establish it requires an inequality due to Bergstrom.

Lemma. Let A, B be positive definite matrices and let A_i, B_i denote the submatrices obtained by deleting the ith rows and ith columns. Then

$$|A|/|A_i| + |B|/|B_i| \leq |A + B|/|A_i + B_i|. \tag{17}$$

Proof. See Beckenbach and Bellman (1961, p. 67). ∎

Theorem 2. Let $\mathbf{s}, \mathbf{t} \in \mathscr{R}_{\mathrm{TM}}$. Then

$$e_k(s_n) + e_k(t_n) \leq e_k(s_n + t_n). \tag{18}$$

Proof. The case in which either $\mathbf{s}$ or $\mathbf{t}$ is the zero sequence is obvious. Assume neither is the zero sequence. Then neither of the distribution functions (see Section 1.5) for these functions can be equivalent to the zero function, and it is easy to show that W_k is positive definite in either case. The rest of the proof follows immediately from Eq. 6.4(1). ∎

A similar statement is possible when $\mathbf{s}, \mathbf{t} \in \mathscr{R}_{\mathrm{TO}}$.

6.5. Power Series and e_k; The Padé Table

When e_k is applied to the partial sums of a power series the result, obviously, is a rational approximation. This section investigates the nature of this approximation.

Definition. Let $s(z)$ be analytic at 0,

$$s(z) = \sum_{k=0}^{\infty} a_k z^k, \qquad a_0 \neq 0. \tag{1}$$

Let A be a polynomial of degree $\leq p$, B ($\not\equiv 0$) a polynomial of degree $\leq q$. If

$$s(z)B - A = O(z^{p+q+1}), \qquad z \to 0 \quad \text{in} \quad \mathscr{C}, \tag{2}$$

then the rational form A/B is called the p, q *Padé approximant* to $s(z)$ and written $[p/q]$. ■

When $[p/q]$ exists, it is unique when written in lowest terms. To show this, assume otherwise, i.e., that

$$s(z)B^* - A^* = O(z^{p+q+1}). \tag{3}$$

Then

$$AB^* - A^*B = O(z^{p+q+1}). \tag{4}$$

But the left-hand side is a polynomial of degree not exceeding $p + q$, yet is $O(z^{p+q+1})$. It must therefore vanish identically. Thus

$$A/B = A^*/B^*. \tag{5}$$

Now let

$$A = \sum_{j=0}^{p} \alpha_j z^j, \qquad B = \sum_{j=0}^{q} \beta_j x^j. \tag{6}$$

Equation (2) requires

$$(a_0 + a_1 z + \cdots)(\beta_0 + \beta_1 z + \cdots + \beta_q z^q) - (\alpha_0 + \alpha_1 z + \cdots + \alpha_p z^p) = O(z^{p+q+1}). \tag{7}$$

Then β_j must satisfy

$$\begin{aligned} a_{p+1}\beta_0 + a_p\beta_1 + \cdots + a_{p-1+1}\beta_q &= 0, \\ &\vdots \\ a_{p+q}\beta_0 + a_{p+q-1}\beta_1 + \cdots + a_p\beta_q &= 0, \qquad a_{-j} = 0. \end{aligned} \tag{8}$$

This is a homogeneous system of q equations in $q + 1$ unknowns and so always possesses nontrivial solutions. The α_j may then be computed from

$$\begin{aligned} \alpha_0 &= a_0\beta_0, \\ \alpha_1 &= a_1\beta_0 + a_0\beta_1, \\ &\vdots \\ \alpha_p &= a_p\beta_0 + a_{p-1}\beta_1 + \cdots + a_{p-q}\beta_q. \end{aligned} \tag{9}$$

A nontrivial solution with $\beta_0 \neq 0$ will exist iff $H_{p-q+1}^{(q)}(\mathbf{a}) \neq 0$. Thus if *all* the determinants $H_{p-q+1}^{(q)}(\mathbf{a}) \neq 0$, $p, q \geq 0$, there will exist for each p, q an A and B of degrees $\leq p$, $\leq q$, respectively, satisfying (2). This does not guarantee that A will be of *exact* degree p (although B will be of exact degree q since $\alpha_0 \neq 0$, $\beta_q \neq 0$). However, if the like determinants formed with the reciprocal series

$$\frac{1}{s(z)} = \sum_{k=0}^{\infty} b_k z^k \tag{10}$$

are nonvanishing, then A will also be of maximum degree. Such a series, or such a set of Padé approximants, Wall calls *normal*, although this terminology is not universal [cf. Henrici (1977, vol. 2, Chap. 12)].

With the approximants $[p/q]$ one can associate the following geometrical configuration known as the *Padé table for* $s(z)$:

$$\begin{matrix} [0/0] & [1/0] & [2/0] & \cdots \\ [0/1] & [1/1] & [2/1] & \cdots \\ [0/2] & [1/2] & [2/2] & \cdots \\ \vdots & \vdots & \vdots & \end{matrix}$$

The elements in the first row are simply the partial sums of the power series for s, $[p/0] = \sum_{j=0}^{p} a_j z^j$, those in the first column the partial sums of the series for $1/s$.

$[p/q]$ can always be obtained by brute force, i.e., by the above method of undetermined coefficients. However, this way of computing the Padé elements for a function tends to be very unstable numerically. Luke has found that in many cases double-precision arithmetic is not adequate for even moderate values of n (10 or so). Note, however, that the coefficient matrix of the system (8) is of Toeplitz type and thus can be inverted easily by using algorithms due to Trench (1964, 1965). (Toeplitz matrices are those having northwest-to-southeast diagonals constant.) Trench's algorithms require $O(n^2)$ operations, as compared with the usual methods, which characteristically require $O(n^3)$ operations. In fact, the Trench algorithms can be used to compute $e_k(s_n)$ systematically. [In Section 10.6 we show how a Trench algorithm can be used to compute a transformation that includes $e_k(s_n)$ as a special case.] To my knowledge, however, no work has yet been done on assessing the computational stability of Trench's methods. [For an excellent general discussion of methods for inverting Toeplitz and Hankel matrices, see Cornyn (1974).]

Closed-form expressions for the Padé elements are known for only a few special functions (other than rational functions) and then only for diagonal approximants $[p/p]$ or off-diagonal approximants, $[p/p-1]$ or $[p-1/p]$. The functions are special cases or confluent limits of the Gaussian hypergeometric function $F(1, b; c; z)$; see Luke's books (1969) for details. For certain other functions (see Section 2.5.3) the $[p-1, p]$ Padé may be computed systematically without solving equations, but since closed-form expressions for the Taylor coefficients of these functions are not known, the Padé can hardly be said to be closed form.

The following is due to Shanks (1955).

Theorem 1. Let $s_n = \sum_{j=0}^{n} a_j z^j$. Then $s_n^{(k)} = [n + k/k]$.

Note that it is possible for $W_k(1)$ to vanish identically. However, as Shank shows, this can happen only if at the same time $W_k(s_n) \equiv 0$. In such cases one defines $e_k(s_n) = e_{k-1}(s_n) = \cdots = e_{k-r}(s_n)$, where r is the smallest integer such that $W_{k-r}(s_n) \not\equiv 0$. With this understanding, the theorem still holds.

A number of theorems on Padé approximants can be translated directly into statements about the effect of e_k on partial sums of power series [see Gilewicz (1978, Section 6.3.1)]. (There are interesting results available on convergence in measure also. Here we shall be concerned only with uniform convergence.)

Theorem 2 (de Montessus de Balloire, 1902). Let $s(z)$ be meromorphic in N_R and have *exactly* k poles (each counted as often as its multiplicity) in N_R. Then

$$s_n^{(k)}(z) = s(z) + o(1), \qquad n \to \infty, \tag{11}$$

uniformly on every compact subset of N_R with the poles removed.

Proof. See de Montessus de Balloire (1902). ■

The most far-reaching result to date on the convergence of subsequences of $s_n^{(k)}(z)$ is due to Beardon (1968). The proof will not be given here.

Theorem 3. Let $s(z)$ be analytic at zero and meromorphic in a domain $\mathscr{D}$ containing zero. Then given any compact subset $\mathscr{K}$ of $\mathscr{D}$ with the poles removed there is some subsequence $s_{n_i}^{(k_i)}(z)$ that converges uniformly to $s(z)$ in $\mathscr{K}$.

These theorems cannot be strengthened in any obvious way. For instance, regarding Theorem 2, Perron (1929) has given an example of a function analytic in a disk N_ρ for which $s_n^{(1)}(z)$ (Aitken's δ^2-process) diverges over a set dense in N_ρ.

Several computational algorithms have been developed to relate the various components of the Padé table, e.g., schemes due to Wynn (the ε-algorithm, Section 6.2), Baker, Longman, and Pindor. For a good survey of these, see Gilewicz (1978, Section 7.3). Some of these schemes are computationally more efficient than the ε-algorithm [see Baker and Gammel (1970, Chap. 1)].

The previous results have touched on convergence only along vertical paths. Convergence along other paths is a much more difficult matter. The theorems to date—such as those of Chisholm–Gilewicz and Zinn–Justin—require troublesome hypotheses. One of these results will be given later.

As an example of what can happen, consider the diagonal path $[k/k]$, i.e., the path corresponding to $s_0^{(k)}(z)$. If s is given by a formal power series $\sum_{j=0}^{\infty} a_j z^j$, then it is known that the Padé approximant $[k/k] = s_0^{(k)}(z)$ is the $(2k+1)$th approximant to the (formal) continued fraction corresponding to s (when such exists); see Henrici (1977, vol. 2, Chap 12). It would be tempting to assert that for $s(z)$ analytic at 0, $s_0^{(k)}(z) \to s(z)$ in some neighborhood of 0. However, this is not, in general, the case, as shown by an example in Perron (1957, vol. 2, p. 158). Nor indeed, must $\sum a_j z^j$ converge for $s_0^{(k)}(z)$ to converge. Apparently, additional conditions must be placed on the coefficients a_j to ensure the convergence of the Padé table.

In a special but very important case, namely, when the a_j are the moments of some function ψ, many of these convergence questions can be answered satisfactorily. We shall tabulate below the most far-reaching results to date. (Since proofs in Padé theory tend to be very computational and to rely heavily on function theory, they have been omitted.)

In what follows, let $\psi \in \Psi^*$ and also Supp $\psi = [0, \infty)$. (When the latter is not true, the results can be refined in an obvious way.) Let

$$\begin{aligned} a_j &= \int_0^\infty t^j \, d\psi, \\ s(z) &= \int_0^\infty \frac{d\psi}{1 - zt}, \qquad z \notin (0, \infty), \\ s_n(z) &= \sum_{j=0}^{n} a_j z^j, \\ d\psi_j &= t^{j+1} \, d\psi, \qquad j \geq -1, \end{aligned} \tag{12}$$

and let $\{p_k^{(j)}(z)\}_{k=0}^{\infty}$ be the sequence of orthogonal polynomials associated with $d\psi_j$.

Theorem 4. $s_n^{(k)}$ is the ratio of two polynomials of exact degree $n+k$ and k, respectively. The denominator polynomial B is given by

$$B = z^k p_k^{(n)}(z^{-1}). \tag{13}$$

Thus all the poles of $s_n^{(k)}(z)$ lie in $[0, \infty)$.

Theorem 5

$$s_n^{(k)}(z) = \sum_{r=0}^{n} a_r z^r + z^{n+1} \sum_{r=1}^{k} \frac{\alpha_{kr}^{(n)}}{1 - z x_{kr}^{(n)}}, \tag{14}$$

where the αs and the xs are the weights and abscissas, respectively, of the Gaussian quadrature process based on $d\psi_n$.

Theorem 6. Let z be real negative.

(i) $(-1)^{n+1}[s_n^{(k+1)}(z) - s_n^{(k)}(z)] \geq 0$,
(ii) $(-1)^{n+1}[s_n^{(k+1)}(z) - s_{n+2}^{(k)}(z)] \geq 0$, and
(iii) $s(z) \leq s_0^{(k)}(z)$ when $s(z)$ converges.

Further, if $\psi(t) = c$, $t > \rho$, then, for $0 \leq z < 1/\rho$,

(iv) $s_{n+2}^{(k)}(z) \leq s_n^{(k+1)}(z) \leq s_{n+1}^{(k)}(z) \leq s_n^{(k)}(z) \leq s_n^{(k+1)}(z) \leq s(z)$.

Theorem 7. $s_n^{(k)}(z)$ converges along any path **P** to a function analytic in $\mathscr{C} - [0, \infty)$. If

$$\sum_{j=1}^{\infty} |a_j|^{-1/(2j+1)} = \infty, \tag{15}$$

all paths produce a common limit. Further, if (1) converges in some N_R, $R > 0$, $s_n^{(k)}(z)$ converges along any path **P** to $s(z)$ in $\mathscr{C} - [R, \infty)$.

In all cases above, convergence is uniform on compact subsets of the indicated region.

Theorem 8

$$r_n^{(k)}(z) = \frac{-z^{n+1}}{p_k^{(n)}(z^{-1})} \int_0^\infty \frac{p_k^{(n)}(t)\, d\psi_n}{1 - zt}, \qquad z \notin (0, \infty]. \tag{16}$$

Theorem 9. Let

$$a_j = O((2j+1)!\, R^{2j}) \tag{17}$$

for some $R > 0$. Then $s_n^{(k)}(z)$ converges to $s(z)$ as $k \to \infty$ uniformly on compact subsets of $\mathscr{C} - [0, \infty)$.

For Theorem 4, see Wall (1948, p. 388) and Allen *et al.* (1975); for Theorems 5, 8, and 9 see Allen *et al.* (1975); for Theorems 6 and 7 see Baker and Gammel (1970). [Actually, Theorem 9 follows directly from Theorem 5 by an application of Uspensky's result (1928) on quadrature formulas for infinite intervals.]

Example. If

$$d\psi = t^{\alpha} e^{-t}\, dt, \qquad \alpha > -1, \tag{18}$$

then

$$p_k^{(j)}(z) = L_k^{(j+\alpha+1)}(z). \tag{19}$$

We shall require the formula

$$\int_0^{\infty} e^{-t} t^{a} (1 - zt)^{b}\, dt = \frac{\Gamma(a+1)}{(-z)^{a+1}} \Psi\left(a+1, a+b+2; -\frac{1}{z}\right), \tag{20}$$

$$|\arg(-z)| < \pi, \quad \operatorname{Re} a > -1.$$

In the present case

$$s(z) = \frac{e^{-1/z}\Gamma(\alpha+1)}{(-z)^{\alpha+1}} \int_{-1/z}^{\infty} e^{-t} t^{-\alpha-1}\, dt. \tag{21}$$

[The integral is the incomplete Gamma function $\Gamma(-\alpha, -1/z)$.] Using the Rodrigues formula for the Laguerre polynomial in (16) and integrating k times by parts gives

$$r_n^{(k)}(z) = \frac{(-1)^n k!\,\Gamma(c)}{\Gamma(\alpha+1)(-z)^{\alpha+1}} \frac{\Psi(a, c; -1/z)}{\Phi(a, c; -1/z)} e^{-1/z}, \tag{22}$$

$$a = n + \alpha + k + 2, \quad c = n + \alpha + 2.$$

Now

$$\frac{\Psi(a, c; w)}{\Phi(a, c; w)} \approx \frac{2\pi a^{c-1} \exp(-4\sqrt{aw})}{\Gamma(a)\Gamma(c)}, \qquad a \to \infty, \quad |\arg w| < \pi \tag{23}$$

[see Slater (1960, p. 80)]. Thus

$$r_n^{(k)}(z) \approx [2\pi(-1)^n/(-z)^{\alpha+1}] \exp(-1/z) \exp(-4\sqrt{-k/z}), \qquad k \to \infty, \quad |\arg(-z)| < \pi. \tag{24}$$

Table I

k	$s_0^{(k)}$
2	0.615
4	0.598802
6	0.596816
8	0.596459999
10	0.596378884
12	0.596357234
14	0.596350734

This agrees, for $n = 0$, with a result given by Luke (1969, vol. 2, p. 200). [The formula (5.10) of Allen *et al.* (1975) seems to be wrong.]

For a numerical example, take $\alpha = 0$, $z = -1$, Then e_k in diagonal modes sums the highly divergent sequence

$$s_n = 0! - 1! + 2! - 3! + \cdots + (-1)^n n! = (\mathrm{FAC})_n \tag{25}$$

to the value

$$s = \int_0^\infty \frac{e^{-t}}{1+t}\,dt = 0.5963473611. \tag{26}$$

e_k is not regular for s_n in vertical modes. Let us tabulate some elements on the leading diagonal (see Table I). According to Eq. (24),

$$s_0^{(k)} - s \approx 2\pi \exp(1 - 4\sqrt{k}), \qquad k \to \infty. \tag{27}$$

Table II

$z = 0.85$		
s_n	$s_n^{(1)}$	$s_n^{(2)}$
1		
1.85	3.2	
2.37	2.79	2.67
2.60	2.71	2.695
2.67	2.70	2.6980
2.695	2.6985	2.69831
2.6979	2.698346	2.698329
2.6983	2.698331	2.69833035
2.698328	2.69833043	2.6983303913
2.698338	2.6983303937	2.6983303925

$z = 1.05$		
s_n	$s_n^{(1)}$	$s_n^{(2)}$
1		
2.05	−5.66	
3.36	−2.35	11.13
4.82	−0.54	−36.51
6.99	0.86	−5.88
10.38	2.23	−2.11
16.18	3.85	−0.18
27.09	6.05	
49.80		

e_k, particularly when computed by means of the ε-algorithm, is one of the few simple computational tools available for the numerical analytic continuation of an analytic function. What happens if one tries to continue a function beyond a natural boundary? Brezinski (1978) presents the example of

$$s(z) = \sum_{k=0}^{\infty} z^{k^2} \tag{28}$$

for which ∂N is a natural boundary. The results are given in Table II. The behavior of the algorithm seems to reflect the functional realities.

A series whose coefficients are given by (12) is called a *Stieltjes series.* If the Taylor series for $s(z)$ is not a Stieltjes series, it is still possible to say something about the convergence of $s_n^{(k)}$ on general paths, but restrictions, which may in practice be unverifiable, must be placed on the zeros of $W_k(s_n)$, $W_k(1)$. Let (1) hold near 0 with s_n its partial sums. The following result is due to Chisholm (1966).

Theorem 10. Let $s(z)$ be meromorphic in N_R. Let **P** be a path and let the zeros of $W_k(s_n)$, $W_k(1)$ have no accumulation point in N_σ for $\sigma < R$ and $(n, k) \in \mathbf{P}$. Then $s_n^{(k)}(z)$ converges to $s(z)$ uniformly in some nonempty region Δ, $0 \subset \Delta \subset N_\sigma$.

6.6. Geometrical Significance of the Schmidt Transformation

As Tucker (1973) has pointed out, e_k has an elegant geometrical interpretation. Let $\mathbf{s} \in \mathscr{R}_S$ and let

$$p_n = (s_n, s_{n+1}, \ldots, s_{n+k}) \tag{1}$$

be a sequence of points in $\mathscr{R}^{k+1}$. Denote by $\mathscr{L}$ the line through $(0, 0, \ldots, 0)$ and $(1, 1, \ldots, 1)$. Assume a unique plane can be passed through $p_n, p_{n+1}, \ldots, p_{n+k}$. Its equation is

$$\begin{vmatrix} x_1 & x_2 & \cdots & x_{k+1} & 1 \\ s_n & s_{n+1} & \cdots & s_{n+k} & 1 \\ \vdots & \vdots & & \vdots & \vdots \\ s_{n+k} & s_{n+k+1} & \cdots & s_{n+2k} & 1 \end{vmatrix} = 0. \tag{2}$$

If **s** converges, then

$$\lim_{n\to\infty} p_n = s(1, 1, \ldots, 1). \tag{3}$$

which lies on $\mathscr{L}$. If the plane intersects $\mathscr{L}$, say at $(t_n, t_n, \ldots, t_n)$, then it is reasonable to expect that t_n will be closer to s than any of the components of p_i, $n \leq i \leq n+k$.

Now,

$$\begin{vmatrix} t_n & t_n & \cdots & t_n & 1 \\ s_n & s_{n+1} & \cdots & s_{n+k} & 1 \\ \vdots & \vdots & & \vdots & \vdots \\ s_{n+k} & s_{n+k+1} & \cdots & s_{n+2k} & 1 \end{vmatrix}$$

$$= t_n \begin{vmatrix} 1 & 1 & \cdots & 1 & 0 \\ s_n & s_{n+1} & \cdots & s_{n+k} & 1 \\ \vdots & \vdots & & \vdots & \vdots \\ s_{n+k} & s_{n+k+1} & \cdots & s_{n+2k} & 1 \end{vmatrix}$$

$$+ \begin{vmatrix} 0 & 0 & \cdots & 0 & 1 \\ s_n & s_{n+1} & \cdots & s_{n+k} & 1 \\ \vdots & \vdots & & \vdots & \vdots \\ s_{n+k} & s_{n+k+1} & \cdots & s_{n+2k} & 1 \end{vmatrix}.$$

$$= 0. \tag{4}$$

Solving for t_n and performing obvious determinant manipulations shows

$$t_n = e_k(s_n). \tag{5}$$

Figure 1 illustrates $k = 1$ (Aitken's δ^2-process).

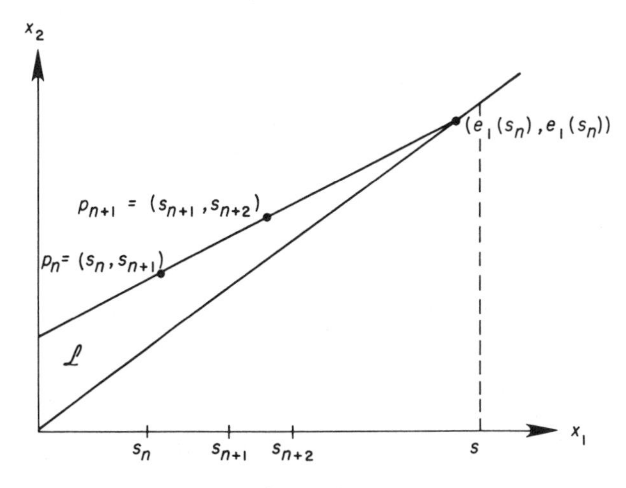

Fig. 1

6.7. The ε-Algorithm

The ε-algorithm, due to Wynn (1956b), is an economic computational procedure for calculating $e_k(s_n)$ without the necessity of evaluating determinants. It is a lozenge algorithm, actually a rhomboid of the kind 1.3(5). Not only does the ε-algorithm make the application of the Schmidt algorithm much more practical, but it helps to clarify some of the convergence properties of the latter, particularly in its application to monotone sequences.

Theorem 1. Let

$$\begin{aligned} \varepsilon_{k+1}^{(n)} &= \varepsilon_{k-1}^{(n+1)} + (\varepsilon_k^{(n+1)} - \varepsilon_k^{(n)})^{-1}, \qquad n, k \geq 0, \\ \varepsilon_{-1}^{(n)} &= 0, \qquad \varepsilon_0^{(n)} = s_n, \quad n \geq 0. \end{aligned} \tag{1}$$

(It is assumed that all quantities are defined.) Then

$$\varepsilon_{2m}^{(n)} = e_m(s_n), \qquad n, m \geq 0, \tag{2}$$

and

$$\varepsilon_{2m+1}^{(n)} = 1/e_m(\Delta s_n), \qquad n, m \geq 0. \tag{3}$$

Proof. The proof is quite computational and depends on an expansion of the ratio of determinants due to Schweins. The proof is sketched in Appendix 2. ■

The computational scheme for the ε-algorithm is as follows.

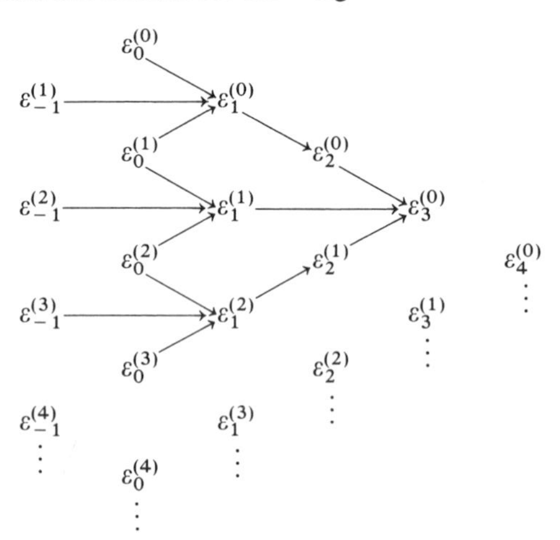

As a numerical example, take the iterative sequence considered by Wynn,

$$s_{n+1} = \tfrac{1}{4}(s_n^2 + 2), \qquad n \geq 0, \quad s_0 = 0. \tag{4}$$

Table III[a]

n	s_n	$\varepsilon_1^{(n)}$	$\varepsilon_2^{(n)}$	$\varepsilon_3^{(n)}$	$\varepsilon_4^{(n)}$
0	0.0				
		2			
1	0.5		0.5714285714		
		16		89.1111108	
2	0.5625		0.5851063830		0.5857434871
		60.2352941		1658.713	
3	0.5791015625		0.5857319781		0.5857857313
		211.05540		20262.3	
4	0.5838396549		0.5857818504		
		725.9366			
5	0.5852171856				

[a] The true root of $x^2 - 4x + 2$ is 0.5857864375

This sequence converges (very slowly) to the smaller zero of the Laguerre polynomial $L_2(x) = x^2 - 4x + 2$, i.e., $x = 0.5857864375$. Table III shows the effect of the ε-algorithm on s_n. It is apparent the convergence of $\varepsilon_{2k}^{(0)}$ is much superior to that of s_k. Note the odd entries, $\varepsilon_{2k+1}^{(n)}$, diverge as $n \to \infty$. This is generally the case. Tables IV and V show the effect of the algorithm on some other sequences.

$\varepsilon_k^{(n)}$ satisfies

$$\begin{aligned}
\varepsilon_{2m}^{(n)} &= H_n^{(m+1)}(\mathbf{s})/H_n^{(m)}(\Delta^2\mathbf{s}),\\
\varepsilon_{2m+1}^{(n)} &= H_n^{(m)}(\Delta^3\mathbf{s})/H_n^{(m+1)}(\Delta\mathbf{s}),\\
\varepsilon_{2m+2}^{(n)} - \varepsilon_{2m}^{(n)} &= -[H_n^{(m+1)}(\Delta\mathbf{s})]^2/H_n^{(m+1)}(\Delta^2\mathbf{s})H_n^{(m)}(\Delta^2\mathbf{s}).
\end{aligned} \tag{5}$$

The first two properties are obvious and the third is deducible using a determinantal expansion of the kind given in the appendix.

The two following theorems show the effect of $\varepsilon_k^{(n)}$ on sequences in $\mathscr{R}_{\mathrm{TM}}$ and $\mathscr{R}_{\mathrm{TO}}$.

Theorem 2. Let $\mathbf{s} \in \mathscr{R}_{\mathrm{TM}}$. Then

(i) $0 \le \varepsilon_{2m+2}^{(n)} \le \varepsilon_{2m}^{(n)}$, $n, m \ge 0$;
(ii) $\varepsilon_{2m+1}^{(n)} \le \varepsilon_{2m-1}^{(n)} \le 0$, $n, m \ge 0$; and
(iii) $\varepsilon_k^{(n+1)} \le \varepsilon_k^{(n)}$, $n, k \ge 0$.

Further the ε-algorithm is regular along any $[n, 2m]$ path for sequences such that $\mathbf{r} \in \mathscr{R}_{\mathrm{TM}}$ and $\lim_{n\to\infty} \varepsilon_{2m+1}^{(n)} = -\infty$.

Proof. The three inequalities are fairly straightforward, requiring the use of Theorem 1.6(6). They are left as an exercise. The rest of the proof is trickier. Assume without loss of generality that $s = 0$. Inequalities (iii) and (i) show

Table IV[a]

k	$(\text{LN } 2)_k$	$\varepsilon_k^{(0)}$	$(\text{PI}^2)_k$	$\varepsilon_k^{(0)}$	$(\text{EX } 3)_k$	$\varepsilon_k^{(0)}$
0	1	1	1	1	1	1
2	0.833333	0.7	1.361	1.45	7.72	−2
4	0.783333	0.693333	1.464	1.552	25	25
6	0.759524	0.693152	1.512	1.590	60.832	25
8	0.745635	0.693147332	1.540	1.609	128.196	25
10	0.736544	0.693147185	1.558	1.620	247.903	25
12	0.730134	0.693147180688	1.571	1.626	452.973	25
14	0.725372	0.693147180564	1.580	1.630	795.351	25

[a] $\ln 2 = 0.693147180560$, $\pi^2/6$ 1.644934.

$\varepsilon_{2m}^{(n)}$ is bounded and positive decreasing in n, and hence convergent. Putting $m = 0, 1, 2, \ldots$, in (i) shows vertical regularity.

Inequality (i) shows that $\varepsilon_{2m}^{(n)}$ is positive decreasing in m and bounded, and hence convergent. Let

$$t_n = \lim_{m \to \infty} \varepsilon_{2m}^{(n)}, \tag{6}$$

so that

$$0 \le t_n \le \varepsilon_{2m}^{(n)}. \tag{7}$$

Now,

$$0 \le \varepsilon_{2m+2}^{(n)} = \varepsilon_{2m}^{(n+1)} + 1/(\varepsilon_{2m+1}^{(n+1)} - \varepsilon_{2m+1}^{(n)}) \le \varepsilon_{2m}^{(n+1)} \tag{8}$$

since $\varepsilon_{2m+1}^{(n+1)} - \varepsilon_{2m+1}^{(n)} \le 0$ by (iii). Taking limits shows $t_n \le t_{n+1}$. But taking limits in (iii) with $k = 2m$ shows $t_{n+1} \le t_n$. Hence t_n is a constant, and letting $n \to \infty$ in (7) shows $t_n = 0$. This gives horizontal regularity.

It is an easy exercise to show that (iii) guarantees regularity for any path. ■

Table V[a]

k	$(\text{FAC})_k$	$\varepsilon_k^{(0)}$
2	2	0.666
4	20	0.615
6	620	0.602
8		0.598
10		0.597
12		0.596817

[a] $\varepsilon_{26}^{(0)} = 0.596353077$, $\int_0^\infty [e^{-t}/(t+1)]\, dt = 0.596347$

Theorem 3. Let $\mathbf{s} \in \mathscr{R}_{\text{TO}}$. Then

(i) $0 \le \varepsilon_{2m+2}^{(2n)} \le \varepsilon_{2m}^{(2n)}, \quad \varepsilon_{2m}^{(2n+1)} \le \varepsilon_{2m+2}^{(2n+1)} \le 0, \qquad n, m \ge 0;$

(ii) $(-1)^n \varepsilon_{2m+1}^{(n)} \le (-1)^n \varepsilon_{2m-1}^{(n)} \le 0, \qquad n, m \ge 0;$

(iii) $(-1)^n(\varepsilon_{2m}^{(n+1)} - \varepsilon_{2m}^{(n)}) \le (-1)^n(\varepsilon_{2m+2}^{(n+1)} - \varepsilon_{2m+2}^{(n)}) \le 0,$

$(-1)^n(\varepsilon_{2m+1}^{(n+1)} - \varepsilon_{2m+1}^{(n)}) \ge (-1)^n(\varepsilon_{2m-1}^{(n+1)} - \varepsilon_{2m-1}^{(n)}) \ge 0,$

$(-1)^n \varepsilon_{2m+2}^{(n)} \le (-1)^n \varepsilon_{2m}^{(n+2)}, \qquad n, m \ge 0.$

Further, the ε-algorithm is regular along any $[n, 2m]$ path for sequences such that $\mathbf{r} \in \mathscr{R}_{\text{TO}}$ and $\lim_{n\to\infty}(-1)^n \varepsilon_{2m+1}^{(n)} = \infty$.

Proof. Left to the reader. ∎

The very different behaviors of $\varepsilon_{2m}^{(n)}$ and $\varepsilon_{2m+1}^{(n)}$ can be appreciated from the following result, which describes the effect of ε_k on asymptotically exponential sequences.

Theorem 4. Let

$$s_n \sim s + \sum_{r=1}^{\infty} c_r \lambda_r^n, \qquad c_r \ne 0, \quad 1 > |\lambda_1| > |\lambda_2| > \cdots. \tag{9}$$

Then

$$\varepsilon_{2m}^{(n)} \sim s + c_{m+1}\lambda_{m+1}^n \prod_{j=1}^{m} \left[\frac{\lambda_{m+1} - \lambda_j}{1 - \lambda_j}\right]^2, \tag{10}$$

$$\varepsilon_{2m+1}^{(n)} \sim \frac{1}{c_{m+1}\lambda_{m+1}^n} \prod_{j=1}^{m} \left[\frac{1 - \lambda_j}{\lambda_{m+1} - \lambda_j}\right]^2, \tag{11}$$

Proof. Left to the reader, or see Golomb (1943). ∎

Thus e_k accelerates sequences of the form (9) exponentially,

$$(s_n^{(k)} - s_n)/(s_n - s) \approx C(\lambda_{k+1}/\lambda_1)^n. \tag{12}$$

6.8. The Stability of the ε-Algorithm

The execution of the algorithm involves division by small quantities, necessarily, in fact, if the algorithm is successful, since the reciprocal of $\varepsilon_{2m}^{(n+1)} - \varepsilon_{2m}^{(n)}$ is required to obtain $\varepsilon_{2m+1}^{(n)}$. This seems a sure recipe for numerical disaster: the diagonal convergence of the algorithm implying, so it seems, its numerical instability. However, $\varepsilon_{2m+1}^{(n)}$, even though large, is at the next stage the divisor, and so the import of cancellation of significant figures at any stage of the computation is unclear.

A stability analysis of the algorithm in its application to *general* sequences is a very difficult problem. It is not even possible at present to say much about how the computations should be arranged to minimize the accumulation of error. Some insight, however, can be gained by looking at the effect the algorithm has on specific sequences. Wynn (1966) has done this for the exponential sequence 6.8(9) and the Poincaré series

$$s_n \sim s + \sum_{r=1}^{\infty} c_r n^{-r}, \qquad c_1 \neq 0, \tag{1}$$

$$s_n \sim s + (-1)^n \sum_{r=1}^{\infty} c_r n^{-r}, \qquad c_1 \neq 0. \tag{2}$$

Let $\delta_k^{(n)}$ denote the relative error of $\varepsilon_k^{(n)}$ and consider the case of diagonal convergence. For Eq. 6.7(9) with $\lambda_r > 0$, one has

$$\delta_{2m+2}^{(n)} \sim -[\lambda_{m+1}/(1 - \lambda_{m+1})]^2 \delta_{2m}^{(n)}, \tag{3}$$

provided s is not too small. Thus if $\lambda_{m+1} > \frac{1}{2}$ the relative error is increased in moving from one even-numbered column to the next, considerably so if λ_{m+1} is close to one. If $\lambda_{m+1} \leq \frac{1}{2}$, however, the error is not magnified, and thus the process is numerically stable even though s_n is a monotone series. [Wynn shows that the ε-algorithm applied to (1) is disastrously unstable.]

For 6.7(9) with $\lambda_r < 0$, one has

$$\delta_{2m+2}^{(n)} \sim -[\lambda_{m+1}/(1 + \lambda_{m+1})]^2 \delta_{2m}^{(n)}, \tag{4}$$

and the absolute error is attenuated when one passes from one even-numbered column to the next. The ε-algorithm for this case is unconditionally stable, as it is for the sequence (2).

Obviously this brief discussion far from exhausts the subject. For more details and many examples, the reader should consult Wynn's papers (1959, 1961, 1963). What can be said in general is that the ε-algorithm tends to be numerically stable when used to compute limits of sequences whose terms oscillate about their limit, but numerically unstable when used on sequences that approach their limits monotonically.

6.9. Rational Analogs of the Formulas of Numerical Analysis

The ε-algorithm can be used in a straightforward manner to develop rational analogs of interpolation or quadrature formulas. For instance, in Newton's interpolation formula

$$N_h(f)(x) = \sum_{m=0}^{\infty} \frac{(-1)^m(-s)_m}{m!} h^m \Delta_h^j f_0, \qquad s = \frac{x - x_0}{h}, \quad f_j = f(x_0 + jh), \tag{1}$$

the partial sums may be used directly in the ε-algorithm. The resulting nonlinear interpolation formulas can be expected to be accurate for functions having poles in the vicinity of the interpolation interval.

For another example, take the first two terms of (1), let $x_0 \to x_n$, and integrate between x_n and $x_{n+1} = x_n + h$. We get the well-known formula that can be written

$$y_{n+1} - y_n = 0 + hf_n + \tfrac{1}{2}h\,\Delta f_n, \qquad f_n = y'_n. \tag{2}$$

Applying $e_1(s_0)$ to the above yields an approximation to $y_{n+1} - y_n$,

$$y_{n+1} \approx y_n + 2hy'^2_n/(3y'_n - y'_{n+1}). \tag{3}$$

Applying e_1 to a Taylor series gives

$$y_{n+1} \approx y_n + 2hy'^2_n/(2y'_n - hy''_n). \tag{4}$$

These formulas and more sophisticated ones have been studied by many writers (Lambert and Shaw, 1965, 1966; Shaw, 1967; Luke *et al.*, 1975).

As an example, take the nonlinear differential equation

$$y' = te^t y^2, \qquad y(0) = 1. \tag{5}$$

This equation has a movable singularity—a simple pole. The initial condition positions the pole at 1 and

$$y = e^{-t}/(1 - t). \tag{6}$$

Table VI demonstrates the result of applying (4) with $h = 0.1$. The error in $y(0.8)$ is about 2.7%.

A characteristic of these types of formulas is that they can be used to *predict* the location of the singularity. Thus the value h that causes the

Table VI

t	$y(t)$ from (3)	$y(t)$ true
0.1	1	1.00537
0.2	1.02518	1.02341
0.3	1.06322	1.05831
0.4	1.12506	1.11720
0.5	1.22472	1.21306
0.6	1.38980	1.37203
0.7	1.68526	1.65528
0.8	2.30981	2.24664
0.9	4.30434	4.06570

denominator of (3) to vanish yields, when added to the current value of t, an approximation to the pole:

$$\text{pole} \approx t_n + 2y'_n/y''_n. \tag{7}$$

The last five entries in the table result in the following sequence of approximants to the pole: 0.898, 0.950, 0.978, 0.991, 0.994.

Nonlinear formulas of great precision have been developed, including analogs of the Milne predictor–corrector formulas (see the references given in this section).

6.10. Generalizations of the ε-Algorithm

The ε-algorithm may be generalized to abstract spaces in several ways. One natural way to extend the method to sequences in topological vector spaces is by means of the Brezinski–Håvie protocol, discussed in Chapter 9.

Wynn (1962) gave an ingenious version of the method applicable to sequences in $\mathscr{C}^p$. For $x, y \in \mathscr{C}^p$, we can define the inner product in the usual way,

$$x = (x^{(1)}, x^{(2)}, \ldots, x^{(p)})^{\mathrm{T}}, \quad y = (y^{(1)}, y^{(2)}, \ldots, y^{(p)})^{\mathrm{T}}, \qquad x^{(i)}, y^{(i)} \in \mathscr{C},$$
$$(x, y) = \sum_{j=1}^{p} x^{(j)} \bar{y}^{(j)}. \tag{1}$$

Now let $\mathbf{s} \in \mathscr{C}_S^p$ and define

$$\varepsilon_{k+1}^{(n)} = \varepsilon_{k-1}^{(n+1)} + (\varepsilon_k^{(n+1)} - \varepsilon_k^{(n)})^*, \qquad n, k \geq 0;$$
$$\varepsilon_{-1}^{(n)} = 0, \quad \varepsilon_0^{(n)} = s_n, \qquad n \geq 0, \tag{2}$$

where

$$y^* = \bar{y}/(y, y). \tag{3}$$

y^* can be thought of as a pseudoinverse of y. When $p = 1$, of course, this reduces to the ε-algorithm in $\mathscr{C}_S$. Obviously the idea can be extended to any finite-dimensional complex inner product space or any real inner product space. In fact, the applications in Wynn's paper are to matrices.

An important class of sequences are those formed by iteration of the matrix equation

$$x = Ax + b, \tag{4}$$

where A is $p \times p$ with entries in $\mathscr{C}$, i.e., those sequences $\mathbf{s} \in \mathscr{C}_S^p$ defined by

$$s_{n+1} = As_n + b, \qquad n \geq 0. \tag{5}$$

Gekeler (1972) has shown that the effect of the vector ε-algorithm on the simple (Picard) iterative sequence $s_{n+1} = F(s_n)$ for determining fixed points of $F(x)$, F a real analytic function of the vector $x \in \mathscr{R}^p$, can be related to the effect of the algorithm on sequences of the form (5). To proceed we need a lemma, due to McLeod (1971), that is sort of a one-way version of Theorem 6.3 but a little more general than really required.

Lemma. Let $\mathbf{r} \in \mathscr{C}_S^p$ satisfy the irreducible equation

$$\begin{aligned} c_0 r_n + c_1 r_{n+1} + \cdots + c_k r_{n+k} &= 0, \qquad n \geq 0, \\ r_n = s_n - s, \qquad c_j \in \mathscr{R}, \quad c_0 &+ c_1 + \cdots + c_k \neq 0. \end{aligned} \tag{6}$$

Then if ε_i^j exists for all $i + j \leq 2k$,

$$\varepsilon_{2k}^{(0)} = s. \tag{7}$$

Remark. An equation of the kind considered is irreducible when it has no solutions in common with an equation of lower order (see Milne-Thomson, 1960, p. 366).

Proof. See Milne-Thomson (1960). ■

Note that, in contrast to the scalar case, here there is no simple formula for $\varepsilon_k^{(n)}$ in terms of determinants. In fact, the proof of this result is rather difficult. It has been conjectured but not proved that the given conditions are both necessary *and* sufficient for $\varepsilon_{2k}^{(0)} = s$. But this is not true even for scalar sequences, since the irreducibility of (6) is not equivalent to the statement, "$\mathbf{r}$ satisfies no such equation of lower order." Perhaps the correct iff statement will result when the irreducibility hypothesis is replaced by the statement in quotes.

Theorem. Let

$$s_{n+1} = As_n + b,$$

s_0 given, A $(p \times p)$ real, $b \in \mathscr{R}^p$, and $I - A$ invertible. Then if ε_i^j exists for all $i + j \leq 2k$,

$$\varepsilon_{2k}^{(n)} = s, \qquad n \geq 0, \tag{8}$$

where $s = (I - A)^{-1}b$ and k is the degree of the minimal polynomial of $s_0 - s$ [with respect to A; see Gantmacher (1959, vol. 1, p. 176)].

Proof. Let $p(\lambda) = \sum_{j=0}^{k} c_j \lambda^j$ be the minimal polynomial. Then

$$\sum_{j=0}^{k} c_j s_{n+j} = \left(\sum_{j=0}^{k} c_j\right) s + \left(\sum_{j=0}^{k} c_j A^{n+j}\right)(s_0 - s) = \left(\sum_{j=0}^{k} c_j\right) s \tag{9}$$

since $s_n = s + A^n(s_0 - s)$, or

$$\sum_{j=0}^{k} a_j r_{n+j} = 0. \tag{10}$$

Also, $\sum_{j=0}^{k} a_j \neq 0$, because otherwise we could write

$$p(\lambda) = (\lambda - 1)\sum_{j=0}^{k-1} c'_j \lambda^j, \tag{11}$$

and since $I - A$ is invertible, $p(\lambda)/(\lambda - 1)$ would be an annihilating polynomial for $s_0 - s$ of degree $k - 1$. ■

6.11. Fixed Points of Differentiable Functions

Let $\mathscr{B}_1, \mathscr{B}_2$ be two Banach spaces, $\mathscr{A}$ an open subset of $\mathscr{B}_1$, and f, g two continuous mappings of $\mathscr{A}$ into $\mathscr{B}_2$. f and g are said to be *tangent* at $x_0 \in \mathscr{A}$ if

$$\lim_{x \to x_0} \| f(x) - g(x)\| / \|x - x_0\| = 0. \tag{1}$$

Among all functions tangent at x_0 to f there is at most one mapping of the form $g\colon x \to f(x_0) + u(x - x_0)$ where u is linear [see Dieudonné (1969, Chap. 8)]. f is said to be *differentiable* at x_0 if such a mapping exists; u is then called the derivative of f at x_0 and written $f'(x_0)$. Higher-order derivatives are defined similarly, $f^{(m)}(x_0)$ being a m-linear continuous mapping of $\mathscr{B}_1 \times \mathscr{B}_1 \times \cdots \times \mathscr{B}_1$ (m times) into $\mathscr{B}_2$. If $f^{(m)}$ exists and is continuous at each point of $\mathscr{A}$, f is said to be an m-times continuously differentiable mapping of $\mathscr{A}$ into $\mathscr{B}_2$. If the segment joining x and $x + t$ is in $\mathscr{A}$, it can be shown that the generalized Taylor formula holds:

$$f(x + t) = f(x) + \sum_{j=1}^{m-1} \frac{1}{j!} f^{(j)}(x)t^{(j)} + \left(\int_0^1 \frac{(1 - \zeta)^{m-1}}{(m - 1)!} f^{(m)}(x + \zeta t)d\zeta\right)t^{(m)}, \tag{2}$$

where $t^{(k)}$ stands for $(t, t, \ldots, t)$ (k times). In particular, for every $\varepsilon > 0$, there is an $c > 0$ such that for $\|t\| < c$,

$$\left\| f(x + t) - \sum_{j=0}^{m} \frac{1}{j!} f^{(j)}(x)t^{(j)} \right\| \leq \varepsilon \|t\|^m. \tag{3}$$

For details and examples, see Dieudonné (1969).

Now consider the iterative scheme for determining a fixed point s of the function $f(x)$,

$$s_{n+1} = f(s_n). \tag{4}$$

If f is differentiable in some neighborhood of s and all members of $\mathbf{s}$ are sufficiently close to s, one may write

$$s - s_{n+1} = f'(s_n)(s - s_n) + O(\|s - s_n\|^2), \tag{5}$$

so *generally* the convergence of the process is poor and first order [see Definition 1.3(1iii)]. However, the vector ε-algorithm, as defined by Eq. 6.10(2) may be used to derive a quadratically (second-order) convergent process in the case that $\mathscr{B}_1 = \mathscr{B}_2 = \mathscr{R}^p$ and f is twice continuously differentiable on $\mathscr{A}$. In this case, of course, $f'(x)$ is a real $p \times p$ matrix. Let $s = f(s)$ and $Q_m(f'(s))$ be the set of vectors x for which m is the degree of the minimal polynomial of x (with respect to $f'(s)$). Define $\mathbf{s}$ as above and $\hat{\mathbf{s}}$ by

$$\hat{s}_n = s + [f'(s)]^n(s_0 - s). \tag{6}$$

Theorem. Let 1 not be an eigenvalue of $f'(s)$ and let $\varepsilon_i^{(j)}$, $\hat{\varepsilon}_i^{(j)}$, $i + j \le 2m$, exist for all s_0 sufficiently close to s with $s_0 - s \in Q_m(f'(s))$.

Put

$$\varepsilon_{2m}^{(0)} = G(s_0, s_1, \ldots, s_{2m}) = H(s_0). \tag{7}$$

Then the computational procedure

$$t_{i+1} = H(t_i), \qquad i \ge 0, \tag{8}$$

is, for t_0 sufficiently close to s and $t_0 - s \in Q_m(f'(s))$, quadratically convergent to s.

Proof. See Gekeler (1972).

Gekeler assumes that f is analytic (possesses a p-tuple series absolutely convergent in a neighborhood of s), but this seems not to be required. The author gives some interesting numerical examples.

The ε-algorithm has an obvious generalization to sequences of square matrices. For rectangular matrices, the generalized matrix inverse of Moore (1920) and Penrose (1955) is applicable, and many writers have investigated generalizations of the ε-algorithm applied to sequences of such matrices, particularly as they arise in the solution of linear systems of equations; see Pyle (1967), Wynn (1966, 1967), and Greville (1968).

For a numerical example of the theorem let $f\colon \mathscr{R}^4 \to \mathscr{R}^4$ be defined by

$$f(x) = s + A(x - s) + Q(x - s) \tag{9}$$

where

$$Q(w) = [-\tfrac{1}{2}(w_1^2 + w_1 w_4), -\tfrac{1}{2}w_2^2, -\tfrac{1}{2}w_3^2, -\tfrac{1}{2}(w_1 w_4 + w_4^2)]^{\mathrm{T}} \tag{10}$$

and

$$A = UDU^{-1}, \tag{11}$$

where D is the diagonal matrix (0.9, 0.8, 0.7, 0.6) and U is the orthogonal matrix

$$U = \frac{1}{2}\begin{bmatrix} 1 & 1 & 1 & 1 \\ 1 & 1 & -1 & -1 \\ 1 & -1 & 1 & -1 \\ 1 & -1 & -1 & 1 \end{bmatrix}. \tag{12}$$

It is easy to show that

$$f'(s) = A. \tag{13}$$

The mapping f is quadratic and

$$f(x) = s + f'(s)(x - s) + \frac{f''(s)}{2}(x - s)^{(2)}. \tag{14}$$

s turns out to be $(1, 1, 1, 1)^{\mathrm{T}}$. With the initial vector $t_0 = (2, 2, 2, 2)^{\mathrm{T}}$, the method described in the theorem produces iterates t_i for which

$$\|t_1 - t_0\| = 1.2 \times 10^{-2}, \qquad \|t_2 - t_1\| = 1 \times 10^{-5}. \tag{15}$$

Gekeler, who gives other examples, states that the method seems to produce the best results when the Jacobian matrix of the system $s = f(s)$ is symmetric.

Chapter 7 Aitken's δ^2-Process and Related Methods

7.1. Aitken's δ^2-Process

The most famous example of the Schmidt algorithm is a method usually attributed to Aitken (1926) but that is, in fact, much older. The method, which is discussed in most books on numerical analysis, results on taking $k = 1$ in e_k. We use the following notation: $A(\mathbf{s}) = \bar{\mathbf{s}}$, where

$$\bar{s}_n = \begin{cases} \dfrac{s_n s_{n+2} - s_{n+1}^2}{s_{n+2} - 2s_{n+1} + s_n} = s_n + a_{n+1}(1 - \rho_{n+1})^{-1}, & \rho_{n+1} \neq 1 \\ s_n, & \rho_{n+1} = 1. \end{cases} \tag{1}$$

This definition guarantees that $\bar{\mathbf{s}}$ always exists.

The following results are obvious:

(i) A is defined and regular for all convergent sequences having the property

$$|a_{n+1}/a_n - 1| \geq \delta, \qquad \delta > 0, \quad n \geq 0, \tag{2}$$

(thus A is regular when applied to the partial sums of convergent real alternating series);

(ii) if A is not regular for $\mathbf{s} \in \mathscr{C}_C$, then some subsequence of $\boldsymbol{\rho}$ has limit 1.

Thus A is regular for $\mathscr{C}_l$, which was established earlier. If $\mathbf{s}$ is a sequence of the form

$$a_n \sim n^\theta[c_0 + c_1/n + \cdots], \qquad c_0 \neq 0, \quad \operatorname{Re} \theta < 0, \tag{3}$$

then A is regular for $\mathbf{s}$ but does not accelerate $\mathbf{s}$; see Theorem 6.4(1). (This is a logarithmically convergent sequence.) A, however, is not regular.

Example 1 (Lubkin, 1952). Let $\mathbf{s}$ be defined by the partial sums of

$$s = 1 + \tfrac{1}{2} - \tfrac{1}{3} - \tfrac{1}{4} + \tfrac{1}{5} + \tfrac{1}{6} - \cdots, \tag{4}$$

so

$$s_0 = 1, \qquad s_1 = 1 + \tfrac{1}{2}, \qquad \ldots, \qquad s = \pi/4 + \tfrac{1}{2}\ln 2. \tag{5}$$

We can write

$$\bar{s}_n = s_n - (\Delta s_n)^2/\Delta^2 s_n \tag{6}$$

and

$$\bar{s}_{2m} = s_{2m} + \frac{(-1)^m(2m+3)}{(2m+2)(4m+5)}, \tag{7}$$

$$\bar{s}_{2m+1} = s_{2m+1} + (-1)^m \frac{(2m+4)}{(2m+3)}. \tag{8}$$

Thus A is not regular for $\mathbf{s}$. Note that $\bar{\mathbf{s}}$ contains essentially three distinct convergent subsequences. One of these, $\bar{s}_{2m}$, converges to s. This is no coincidence.

Theorem 1 (Tucker). Let $\mathbf{s} \in \mathscr{C}_C$. Then some subsequence of $\bar{\mathbf{s}}$ converges to s.

Proof. Suppose no subsequence of $\bar{\mathbf{s}}$ converges to s. This means $a_n \neq. 0$, $\rho_n \neq. 1$. Now

$$\bar{s}_n - s_{n+1} =. v_{n+1} =. -(a_{n+1}^{-1} - a_{n+2}^{-1})^{-1}. \tag{9}$$

Thus the assumption holds iff no subsequence of v_n converges to zero. This means

$$|a_n^{-1} - a_{n+1}^{-1}|^{-1} >. B \qquad \text{for some} \quad B > 0 \tag{10}$$

or

$$|a_n^{-1} - a_{n+1}^{-1}| <. 1/B. \tag{11}$$

But then, by Theorem 1.5(1), $\mathbf{s}$ diverges, a contradiction. ■

Corollary 1. If $\mathbf{s}$ and $\bar{\mathbf{s}} \in \mathscr{C}_S$, then $s = \bar{s}$.

Corollary 2. If $\mathbf{s}$ is such that $\bar{\mathbf{s}}$ is properly divergent (i.e., $|\bar{s}_n| \to \infty$), then $\mathbf{s}$ diverges.

Proof. If $\mathbf{s}$ were convergent some subsequence of $\bar{\mathbf{s}}$ would be convergent, a contradiction. ■

Tucker (1967, 1969), has obtained several sets of conditions that ensure that A accelerates convergence. These conditions, generally speaking, amount to restricting $\mathbf{s}$ to $\mathscr{C}_l$ or else are reformulations of the condition $(\Delta s_n)^2/(r_n \Delta^2 s_n) \approx 1$. Brezinski's result, Theorem 1.7(5), results in a criterion for certain real sequences.

Theorem 2. Let $\mathbf{a}$ be ultimately positive and monotone decreasing, and $\Delta(a_n/\Delta a_n) = o(1)$. Then $\mathbf{s}$ converges and A accelerates the convergence of $\mathbf{s}$.

Proof. Direct application of Theorem 1.7(5). ■

There are two useful results describing the effect of A on the partial sums of power series. Let $\bar{s}_n(z)$ be the effect of A on

$$s_n(z) = \sum_{k=0}^{n} a_k z^k \tag{12}$$

and

$$s(z) = \sum_{k=0}^{\infty} a_k z^k, \qquad |z| < \delta. \tag{13}$$

We know (see Section 6.5) that the analyticity of $s(z)$ in N_R does *not* guarantee the convergence of $\bar{s}_n(z)$. However, the following results, whose proofs are omitted, provide some information.

Theorem 3 (Tucker, 1969). Let $|\rho_n| \leq . \rho < 1$ and let $\bar{s}_n(1)$ converge more rapidly than $s_n(1)$. Then $\bar{s}_n(z)$ converges more rapidly than $s_n(z)$ [to $s(z)$] for each z such that $0 < |z| < 1/\rho$.

Theorem 4 (Beardon, 1968). Let $s(z)$ be analytic and bounded in N_R. Then there is some subsequence $\bar{s}_{n_i}(z)$ that converges to $s(z)$ uniformly on every compact subset of N_R.

Aitken's δ^2-process also accelerates convergence of hyperlinearly convergent series, i.e., series of the form $\rho_n = o(1)$, $\rho_n = r_{n+1}/r_n$. However, comparatively speaking, the method does not work as well on these sequences as on linearly convergent ones. Reich (1970) observes that it is more logical to compare $\bar{s}_n$ with s_{n+2} rather than s_n, since the computation of $\bar{s}_n$ involves s_{n+2}. For linearly convergent sequences, one still has $\bar{r}_n/r_{n+2} \to 0$. The examples $s_n = n^{-n}$ or 2^{-n^2} show this is *not* true for hyperlinear sequences.

Conditions for A to accelerate convergence of the infinite product $\prod (1 + a_n)$ (i.e., accelerate convergence of the sequence of partial products) have been investigated by Tucker. One result is that if $|\rho_n| \leq . \rho < 1$ and $a_n \neq -1, n \geq 0$, then A accelerates the convergence of $\prod (1 + a_n)$ iff $\Delta\rho_n \to 0$.

A sums divergent exponential sequences in certain cases.

Theorem 5. Let $\mathbf{s}$ be real, divergent, and bounded, and let

$$\rho_n = -1 + o(1),$$

ρ_n eventually monotone. Then A sums $\mathbf{s}$.

Proof. See Goldsmith (1965). ■

Example 2. A sums the series $\sum(-1)^n$ (to $\frac{1}{2}$).

There are ways of modifying the δ^2-process when the original is ineffective, for instance, in such problems as determining by the power method eigenvalues of a matrix which are close together. Iguchi (1975, 1976) discusses means of doing this and gives many examples.

7.2. The Lubkin W-Transform

This is a transformation introduced by Lubkin (1952) that is sort of an iteration of the Aitken δ^2-process. The formula is

$$s_n^* = s_n + a_{n+1}(1 - \rho_{n+1})/(1 - 2\rho_{n+1} + \rho_n\rho_{n+1}), \qquad \rho_n = a_{n+1}/a_n. \quad (1)$$

The work of Chapter 5 shows immediately that the W-transform is accelerative for $\mathscr{C}_l$. Further, for any ρ, W is regular in a sequence space slightly larger than $\mathscr{A}_\rho$. To make this precise, consider $D_n = 1 - 2\rho_{n+1} + \rho_n\rho_{n+1}$ and let $\rho_n = \rho - \delta_n$, $|\delta_n| \le \lambda$, $\lambda > 0$.

Then

$$\begin{aligned} D_n &= |(\rho - 1)^2 - \rho(\delta_n + \delta_{n+1}) + 2\delta_{n+1} + \delta_n\delta_{n+1}| \\ &> |\rho - 1|^2 - 2\lambda(|\rho| + 1) - \lambda^2. \end{aligned} \quad (2)$$

Then $D_n > 0$ for $\lambda \le \lambda^*$,

$$\lambda^* = [|\rho - 1|^2 + (|\rho| + 1)^2]^{1/2} - (|\rho| + 1). \quad (3)$$

This proves the following result.

Theorem 1. The W-transform is defined and regular for all convergent sequences $\mathbf{s}$ having the property

$$|a_{n+1}/a_n - \rho| \le \lambda^*, \qquad 0 < |\rho| < 1, \quad (4)$$

λ^* as above.

Theorem 2. s_n^* is defined and $s_n^* = s$, $n \geq 1$, iff s_n has the form

$$s_n = s + K \prod_{j=1}^{n} \left(\frac{ja + b + 1}{ja + b} \right), \qquad n \geq 1, \tag{5}$$

where $K \neq 0$, $a \neq 1$, $ja + b \neq 0, -1, 1, j \geq 1$.

Proof. Notice that s_n^* may be written

$$s_n^* = \Delta^2 \left(\frac{s_{n-1}}{\Delta s_{n-1}} \right) \Big/ \Delta^2 \left(\frac{1}{\Delta s_{n-1}} \right), \qquad n \geq 1. \tag{6}$$

The requirement $s_n^* \equiv s$ means s_n satisfies a first-order linear difference equation that may be solved by the usual techniques. Cordellier (1977) is responsible for the clever observation (6). Setting $a = 0$ shows W is exact for exponential sequences, convergent or not. ∎

There is a close connection between accelerativeness for the Aitken δ^2-process and W.

Theorem 3. Let A (resp. W) accelerate **s**. Then W (resp. A) accelerates **s** iff

$$\left(1 - \frac{a_{n+2}}{a_{n+1}} \right)^2 \approx \left(1 - 2 \frac{a_{n+2}}{a_{n+1}} + \frac{a_{n+2}}{a_n} \right). \tag{7}$$

Proof. Immediate by the use of Theorem 1.5(4). Note in accordance with my convention [see Definition 7.1(1)] A or W may be undefined for a (finite) number of values of n. ∎

For more results of the W-transform, see Lubkin (1952) and Tucker (1967, 1969). To close the discussion, observe that for a large class of logarithmically convergent sequences, the W-transform is accelerative, whereas the A-process is not.

Theorem 4. Let

$$a_n \sim n^{\theta}[c_0 + c_1/n + c_2/n^2 + \cdots], \qquad \operatorname{Re} \theta < -1. \tag{8}$$

Then W accelerates s_n. In fact

$$\frac{s_n^* - s}{s_n - s} = \frac{-2n^{\theta+1}}{(\theta + 1)} [1 + O(n^{-1})]. \tag{9}$$

Proof. Left to the reader. [Note that the denominator of (1) is $\theta(\theta + 1) \times n^{-2} + O(n^{-3})$.] ∎

7.3. Related Algorithms

A number of variants and generalizations of the δ^2-process have been given. In Aitken's process one assumes s_n converges as

$$s_{n+1} - s \approx \lambda(s_n - s). \tag{1}$$

Samuelson (1945) assumed

$$s_{n+1} - s \approx \lambda(s_n - s)^2. \tag{2}$$

Replacing n by $n + 1$ and eliminating λ from the two equations produces a quadratic equation for s qua $\bar{s}_n$. Ostrowski (1966) assumed more generally

$$s_{n+1} - s \approx \lambda(s_n - s)^m, \qquad m \geq 2, \tag{3}$$

and proposed the scheme

$$\bar{s}_n = s_{n+1} + (|a_{n+1}|^{m+1}/|a_n|^m)\operatorname{sgn}(s_{n+1} - s). \tag{4}$$

Of course, there can be a problem in determining the appropriate sign above.

Jones's method (1976) includes the δ^2-method and takes

$$\bar{s}_n = s_n - \Delta s_n/(d - 1), \tag{5}$$

where d is a root of

$$z^m + z^{m-1} + \cdots + z^2 + z - \Delta s_{n+1}/\Delta s_n = 0. \tag{6}$$

It is an easy matter to show the procedure is exact ($\bar{s}_n = s$) when s_n satisfies

$$s_{n+1} - s = \lambda(s_n - s)^m, \tag{7}$$

for some $\lambda \in \mathscr{C}$, $m \geq 1$. The selection of the correct root of (5) is not really a difficult matter—Jones has a discussion of this. The procedure is intended to be used on sequences which converge or diverge hyperlinearly, for instance, if one takes $m = 2$ in (5), (6) will sum the sequence

$$s_{n+1} = s_n^2 - 1, \qquad s_0 = 2, \tag{8}$$

to its "correct" value, $(1 + \sqrt{5})/2$. Note, however, the method is *not* accelerative for $\mathscr{C}_l$ since (6) with $\Delta s_{n+1}/\Delta s_n$ replaced by ρ has a root $= \rho$ iff $m = 1$.

All the above methods, however, have severe, perhaps fatal, computational deficiencies. If s_n converges *linearly*, one is better off using a column of the ε-algorithm to sum $\mathbf{s}$. If $\mathbf{s}$ converges *hyperlinearly*, why use an acceleration method at all? It is my experience that one picks up in $\bar{s}_n$ at most an extra significant figure or so over those present in s_{n+2}, which is used to compute $\bar{s}_n$. Finally, if $\mathbf{s}$ diverges hyperlinearly, severe loss of significance problems are encountered. If $m > 2$, it is unlikely these can be overcome even on the largest computers.

Iguchi (1975, 1976) discusses a generalization of the δ^2-process based on

$$\bar{s}_n = s_{n+2} + (s_{n+2} - s_n) \sum_{k=1}^{m} \left(\frac{a_{n+2}}{a_{n+1}} \right)^{2k}, \qquad m \geq 1 \tag{9}$$

($m \to \infty$ gives the δ^2-process).

Chapter 8 Lozenge Algorithms and the Theory of Continued Fractions

8.1. Background

In Chapter 6 it was shown how the Schmidt algorithm, when applied to the partial sums of a power series, produced the upper half of the Padé table. Since the diagonal Padé elements are the $(2n + 1, 2n + 1)$ approximants of the continued-fraction representation of the function defined by the power series, it seems clear some formal connection must exist between the Schmidt transformation, i.e., the ε-algorithm, and the theory of continued fractions. In fact, the ε-algorithm is just one of several computational formats relating various elements of the Padé table.

This chapter shows how two algorithms, the η-algorithm and the ε-algorithm, can be derived from the theory of continued fractions. The theory is both elegant and satisfying because it establishes a deep connection between an algorithm derived purely algebraically and certain important ideas in function theory.

The analysis in this chapter will depend heavily on material by Wall (1948) and Henrici (1977, Vol 2, Chapter 12).

8.2. The Quotient–Difference Algorithm; The η-Algorithm

This section considers a procedure due to Rutishauser, who developed it and explored its application in a series of books and papers [see e.g., Rutishauser (1954, 1957)]. We shall not deal extensively with the properties of the quotient–difference (q–d) algorithm here, but use it primarily as a tool for obtaining the other lozenge algorithms, the η- and ε-algorithms.

A formal (not necessarily convergent) power series

$$U = a_0 + a_1 z + a_2 z^2 + \cdots \tag{1}$$

and a formal continued fraction of the kind

$$K = \frac{a_0}{1-}\,\frac{q_1 z}{1-}\,\frac{e_1 z}{1-}\,\frac{q_2 z}{1-}\,\frac{e_2 z}{1-}\cdots \tag{2}$$

are said to *correspond* to each other if the nth approximant $P_n(z)/Q_n(z)$ of K, with

$$P_0 = 0, \qquad P_1 = a_0, \qquad P_2 = a_0, \qquad P_3 = a_0(1 - e_1 z), \qquad \ldots \tag{3}$$

and

$$Q_0 = 1, \qquad Q_1 = 1, \qquad Q_2 = 1 - q_1 z, \qquad Q_3 = 1 - e_1 z - q_1 z, \qquad \ldots, \tag{4}$$

if expanded in powers of z, satisfies

$$U - P_n/Q_n = O(z^n). \tag{5}$$

It is not clear that such a correspondence need exist. But the following theorem states when this happens.

Theorem 1. For U, there is at most one corresponding K. There is exactly one such K if and only if the Hankel determinants satisfy

$$H_n^{(k)}(\mathbf{a}) = \begin{vmatrix} a_n & a_{n+1} & \cdots & a_{n+k-1} \\ a_{n+1} & a_{n+2} & \cdots & a_{n+k} \\ \vdots & \vdots & & \vdots \\ a_{n+k-1} & a_{n+k} & \cdots & a_{n+2k-2} \end{vmatrix} \neq 0, \qquad n \geq 0, \quad k \geq 1. \tag{6}$$

Proof. See Henrici (Vol. 2, p. 518). ■

The q–d algorithm provides a systematic way of obtaining $\{q_n\}$ and $\{e_n\}$ from $\{a_n\}$. We assume the condition $H_n^{(k)} \neq 0$ of the previous theorem holds, but for the present the development is purely formal and no assumptions are made about convergence.

The *even* part of K is

$$K_{\mathrm{E}} = a_0\left[\frac{1}{1 - q_1 z -}\,\frac{q_1 e_1 z^2}{1 - z(q_2 + e_1) -}\,\frac{q_2 e_2 z^2}{1 - z(q_3 + e_2) -}\cdots\right]. \tag{7}$$

Its approximants are P_{2n}/Q_{2n}. The *odd* part of K is

$$K_{\mathrm{O}} = a_0\left[1 + \frac{q_1 z}{1 - z(q_1 + e_1) -}\,\frac{q_2 e_1 z^2}{1 - z(q_2 + e_2) -}\cdots\right]. \tag{8}$$

Its approximants are P_{2n+1}/Q_{2n+1}. Now consider a sequence of functions $\{U_i(z)\}$ that have continued fraction developments (2) with corresponding coefficients a_0^i, $\{q_n^{(i)}\}$, $\{e_n^{(i)}\}$, and

$$U_k = a_k + zU_{k+1}, \quad U_0 \equiv U, \qquad \text{so} \quad U_k = a_k + a_{k+1}z + a_{k+2}z^2 + \cdots. \tag{9}$$

Equating (7) for $k+1$ with (8) for k gives

$$a_0^{(k)}\left[1 + \frac{q_1^{(k)}z}{1 - z(q_1^{(k)} + e_1^{(k)}) -}\;\frac{q_2^{(k)}e_1^{(k)}z^2}{1 - z(q_2^{(k)} + e_2^{(k)}) -}\cdots\right]$$

$$= a_k + a_0^{(k+1)}z\left[\frac{1}{1 - zq_1^{(k+1)} -}\;\frac{q_1^{(k+1)}e_1^{(k+1)}z^2}{1 - z(q_2^{(k+1)} + e_1^{(k+1)}) -}\cdots\right]. \tag{10}$$

For the sake of the formal development, assume these fractions terminate. Then a uniqueness argument [see Wall (1948, Chapter IX)] can be invoked to show they are equal coefficient by coefficient. The result is

$$q_n^{(k)} + e_n^{(k)} = e_{n-1}^{(k+1)} + q_n^{(k+1)}, \qquad k \geq 0, \quad n \geq 1; \tag{11}$$

$$e_n^{(k)}q_{n+1}^{(k)} = q_n^{(k+1)}e_n^{(k+1)}, \qquad k \geq 0, \quad n \geq 1. \tag{12}$$

To obtain starting values, observe that $a_0^{(k)} = a_k$, so

$$q_1^{(k)} = a_{k+1}/a_k \qquad \text{and} \qquad e_0^{(k)} = 0, \quad k \geq 0. \tag{13}$$

Equation (9) shows

$$U = a_0 + a_1z + a_2z^2 + \cdots + a_Nz^N + z^{N+1}U_{N+1}, \tag{14}$$

for any N. But taking N sufficiently large shows

$$q_n^{(0)} = q_n \qquad e_n^{(0)} = e_n. \tag{15}$$

The q–d scheme may be arranged as follows:

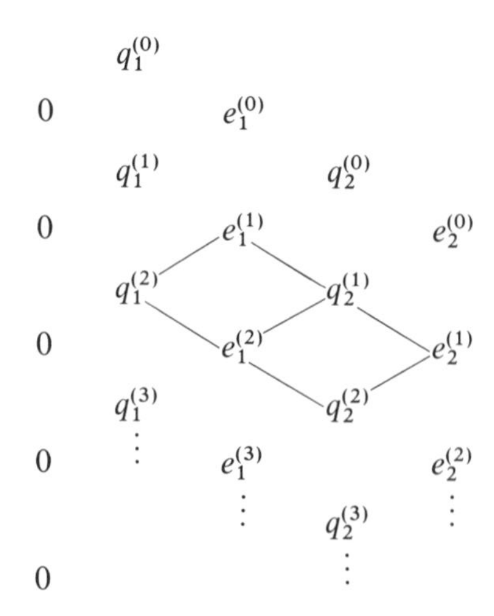

Table I

	$q_1^{(k)}$	$e_1^{(k)}$	$q_2^{(k)}$	$e_2^{(k)}$	$q_3^{(k)}$
	1				
0		$-\frac{1}{2}$			
	$\frac{1}{2}$		$\frac{1}{6}$		
0		$-\frac{1}{6}$		$-\frac{1}{6}$	
	$\frac{1}{3}$		$\frac{1}{6}$		$\frac{1}{10}$
0		$-\frac{1}{12}$		$-\frac{1}{10}$	
	$\frac{1}{4}$		$\frac{3}{20}$		
0		$-\frac{1}{20}$			
	$\frac{1}{5}$				
0					

The quantities in each formula constitute the four corners of a lozenge or rhombus, and one moves out in the table using first (11), then (12), and then repeating, as indicated in the above array.

As an example, take $U = e^z$. The results are given in Table I.

Thus

$$e^z = \frac{1}{1-}\,\frac{z}{1+}\,\frac{\frac{1}{2}z}{1-}\,\frac{\frac{1}{6}z}{1+}\,\frac{\frac{1}{6}z}{1-}\,\frac{\frac{1}{10}z}{1+}\cdots, \tag{16}$$

which is, apart from an equivalence transformation, the known continued fraction for e^z. In this case, the quantities $q_n^{(k)}$, $e_n^{(k)}$ may be written in closed form

$$\begin{aligned} q_n^{(k)} &= (n+k-1)/(k+2n-2)(k+2n-1), \qquad && k \geq 0, \quad n > 1; \\ e_n^{(k)} &= -n/(k+2n-1)(k+2n), && k \geq 0, \quad n \geq 1. \end{aligned} \tag{17}$$

One way of making the q–d formulas easier to use is to label each quantity by its direction from the *center* of the lozenge: E represents east, etc. Then (11) and (12) become

$$\mathrm{E} = \mathrm{W} + \mathrm{S} - \mathrm{N}, \qquad \mathrm{E} = \mathrm{WS/N}. \tag{18}$$

It can be shown that

$$q_n^{(k)} = H_{k+1}^{(n)} H_k^{(n-1)} / H_k^{(n)} H_{k+1}^{(n-1)}, \qquad e_n^{(k)} = H_k^{(n+1)} H_{k+1}^{(n-1)} / H_k^{(n)} H_{k+1}^{(n)}. \tag{19}$$

Thus the q–d recursion relations induce a recurrence relation sometimes attributed to Aitken (1931) but, in fact, known to Hadamard (1892). Of course the ε-algorithm also makes a statement, by means of Eq. 6.8(5), about Hankel determinants.

Theorem 2. Define $H_n^{(0)} = 1$. Then

$$[H_n^{(k)}(\Delta \mathbf{a})]^2 - H_n^{(k)}(\Delta^2 \mathbf{a}) H_n^{(k)}(\mathbf{a}) + H_n^{(k+1)}(\Delta \mathbf{a}) H_n^{(k-1)}(\Delta^2 \mathbf{a}) = 0, \quad n \geq 0, \quad k \geq 1; \tag{20}$$

$$H_n^{(k)}(\mathbf{a})^2 - H_{n-1}^{(k)}(\mathbf{a}) H_{n+1}^{(k)}(\mathbf{a}) + H_{n-1}^{(k+1)}(\mathbf{a}) H_{n+1}^{(k-1)}(\mathbf{a}) = 0, \quad n, k \geq 1. \tag{21}$$

Remark. (21) can be shown independently of the q–d algorithm by using Sylvester's expansion; see Section A.3 of the Appendix.

Bauer [Bauer (1959, 1965); Bauer *et al.* (1963)] seems to be the first to trace the connection between the q–d algorithm and the ε-algorithm. The basic idea is to convert the continued fraction K, which is equivalent to the formal power series U, into a *Euler* continued fraction

$$K' = \frac{a_0}{1-}\,\frac{\rho_1}{1+\rho_1-}\,\frac{\rho_2}{1+\rho_2-}\,\frac{\rho_3}{1+\rho_3-}\cdots. \tag{22}$$

Under appropriate conditions [see Wall (1948 p. 17, Theorem 2.1)] this continued fraction is equivalent to the infinite series

$$U' = a_0\left(1 + \sum_{r=1}^{\infty} \rho_1\rho_2\cdots\rho_r\right) \tag{23}$$

in the sense that the nth numerator of (22) is equal to the sum of the first n terms of (23) and the nth denominator is 1.

The η-algorithm establishes a correspondence between the terms of the above series and the coefficients a_j of U. The ε-algorithm results on interpreting the η-algorithm for sequences.

In what follows all convergence considerations are disregarded, since these are thoroughly discussed in Chapter 6.

The required conversion of the continued fraction K' to K depends on shameless algebraic trickery. Recall that the denominators of K satisfy

$$Q_{2m}(z) = -q_m z Q_{2m-2}(z) + Q_{2m-1}(z), \qquad m \geq 1. \tag{24}$$

$$Q_{2m+1}(z) = Q_{2m}(z) - e_m z Q_{2m-1}(z), \qquad m \geq 1. \tag{25}$$

Let λ be an arbitrary complex parameter. Write

$$q_m = \frac{Q_{2m-1}(\lambda) - Q_{2m}(\lambda)}{\lambda Q_{2m-2}(\lambda)}, \qquad m \geq 1, \tag{26}$$

$$e_m = \frac{Q_{2m}(\lambda) - Q_{2m+1}(\lambda)}{\lambda Q_{2m-1}(\lambda)}, \qquad m \geq 1. \tag{27}$$

Defining

$$g_m(\lambda) = Q_{m+1}(\lambda)/Q_m(\lambda), \quad m \geq 0, \qquad g_0(\lambda) \equiv 1, \tag{28}$$

(note $g_0(\lambda) = 1$), we also have

$$g_m(\lambda)g_{m+1}(\lambda) = Q_{m+2}(\lambda)/Q_m(\lambda), \qquad m \geq 0. \tag{29}$$

Using the formula (2.2) in Wall shows

$$\rho_m(z, \lambda) = \frac{zQ_{m-1}(z)[Q_m(\lambda) - Q_{m+1}(\lambda)]}{\lambda Q_{m+1}(z)Q_{m-1}(\lambda)}, \qquad m \geq 1. \tag{30}$$

Then

$$K(z) = K'(z) = \frac{a_0}{1-}\,\frac{\rho_1(z, \lambda)}{1 + \rho_1(z, \lambda)-}\,\frac{\rho_2(z, \lambda)}{1 + \rho_2(z, \lambda)-}\cdots. \tag{31}$$

Now let $\lambda = z$:

$$K(\lambda) = K'(\lambda) = \frac{a_0}{1-}\,\frac{\rho_1}{1 + \rho_1-}\,\frac{\rho_2}{1 + \rho_2-}\cdots, \tag{32}$$

where

$$\rho_m \equiv \rho_m(\lambda, \lambda) = [1 - g_m(\lambda)]/g_m(\lambda), \qquad m \geq 1. \tag{33}$$

Also, from (26) and (27),

$$q_m = [g_{2m-2}(\lambda)/\lambda][1 - g_{2m-1}(\lambda)], \qquad m \geq 1, \tag{34}$$

$$e_m = [g_{2m-1}(\lambda)/\lambda][1 - g_{2m}(\lambda)], \qquad m \geq 1. \tag{35}$$

Assume, as with the q–d algorithm, that both K and K' are used with two different functions, U_k and U_{k+1}, associated with quantities $g_m^{(k)}$, $q_m^{(k)}$, and $e_m^{(k)}$. Applying the q–d algorithm to K and using (34) and (35) with all quantities superscripted by k gives for $n = 1$ in (11)

$$(g_0^{(k+1)}/\lambda)(1 - g_1^{(k+1)}) = (g_0^{(k)}/\lambda)(1 - g_1^{(k)}) + (g_1^{(k)}/\lambda)(1 - g_2^{(k)}) \tag{36}$$

or

$$g_1^{(k+1)} = g_1^{(k)}g_2^{(k)}. \tag{37}$$

For $n = 1$, (12) gives

$$\frac{g_0^{(k+1)}}{\lambda}(1 - g_1^{(k+1)})\frac{g_1^{(k+1)}}{\lambda}(1 - g_2^{(k+1)}) = \frac{g_1^{(k)}}{\lambda}(1 - g_2^{k})\frac{g_2^{(k)}}{\lambda}(1 - g_3^{(k)}), \tag{38}$$

which when combined with (37) gives

$$(1 - g_1^{(k+1)})(1 - g_2^{(k+1)}) = (1 - g_2^{(k)})(1 - g_3^{(k)}). \tag{39}$$

Continuing this process gives a *lozenge algorithm* for the computation of $g_m^{(k)}$,

$$g_{2m-1}^{(k)}g_{2m}^{(k)} = g_{2m-2}^{(k+1)}g_{2m-1}^{(k+1)}, \tag{40}$$

$$(1 - g_{2m}^{(k)})(1 - g_{2m+1}^{(k)}) = (1 - g_{2m-1}^{(k+1)})(1 - g_{2m}^{(k+1)}), \qquad k \geq 0, \quad m \geq 1, \tag{41}$$

with starting values

$$g_0^{(k)} = 1, \qquad g_1^{(k)} = 1 - \lambda a_{k+1}/a_k. \tag{42}$$

To derive the η-algorithm, let

$$\eta_m^{(k)} = a_k \lambda^k \prod_{r=1}^{m} \rho_r^{(k)}, \qquad \eta_0^{(k)} = 1, \tag{43}$$

with

$$\rho_r^{(k)} = [1 - g_r^{(k)}(\lambda)]/g_r^{(k)}(\lambda). \tag{44}$$

Then, since $\rho_m^{(0)} = \rho_m$, we have from (32) and (23)

$$U(\lambda) = \eta_0^{(0)} + \eta_1^{(0)} + \eta_2^{(0)} + \cdots, \tag{45}$$

and so we have defined a series transformation of $U(\lambda)$. Of course, the $\eta_m^{(k)}$ satisfy lozenge relationships. For instance, let

$$S = (\eta_{2m-1}^{(k)} + \eta_{2m}^{(k)})/(\eta_{2m-2}^{(k+1)} + \eta_{2m-1}^{(k+1)}). \tag{46}$$

Substituting (43) in the above, factoring $\eta_{2m-1}^{(k)}$ from the numerator and $\eta_{2m-2}^{(k+1)}$ from the denominator, and pairing off factors by 2s using (44) and the $g_m^{(k)}$ recursion relationships yields

$$S = \frac{a_k(1 - g_1^{(k)})g_1^{(k+1)}g_0^{(k+1)}}{a_{k+1}\lambda g_1^{(k)}g_2^{(k)}} = \frac{a_k(1 - g_1^{(k)})}{\lambda a_{k+1}} = 1, \tag{47}$$

and this provides the first of two relationships. The iterates in the *η-algorithm* are defined by

$$\eta_{2m-1}^{(k)} + \eta_{2m}^{(k)} = \eta_{2m-2}^{(k+1)} + \eta_{2m-1}^{(k+1)}, \tag{48}$$

$$\frac{1}{\eta_{2m}^{(k)}} + \frac{1}{\eta_{2m+1}^{(k)}} = \frac{1}{\eta_{2m-1}^{(k+1)}} + \frac{1}{\eta_{2m}^{(k+1)}}, \qquad k, m \geq 0, \tag{49}$$

with starting values

$$1/\eta_{-1}^{(k)} = 0, \qquad \eta_0^{(k)} = a_k \lambda^k, \qquad k \geq 0. \tag{50}$$

The derivation of the second relationship above is straightforward.

By the η-algorithm we mean the summation of the sequence defined by

$$s_n = a_0 + a_1\lambda + \cdots + a_n\lambda^n \tag{51}$$

in terms of

$$\bar{s}_n = \eta_0^{(0)} + \eta_1^{(0)} + \cdots + \eta_n^{(0)}. \tag{52}$$

The computational scheme is as follows:

$$
\begin{array}{lllll}
a_0 = \eta_0^{(0)} & & & & \\
 & \eta_1^{(0)} & & & \\
a_1\lambda = \eta_0^{(1)} & & \eta_2^{(0)} & & \\
 & \eta_1^{(1)} & & \eta_3^{(0)} & \\
a_2\lambda^2 = \eta_0^{(2)} & & \eta_2^{(1)} & & \eta_4^{(0)} \\
 & \eta_1^{(2)} & & \eta_3^{(1)} & \vdots \\
a_3\lambda^3 = \eta_0^{(3)} & & \eta_2^{(2)} & \vdots & \\
 & \eta_1^{(3)} & \vdots & & \\
a_4\lambda^4 = \eta_0^{(4)} & \vdots & & & \\
\vdots & & & &
\end{array}
$$

Symbolically the η-algorithm may be written

$$
\begin{gathered}
\mathrm{N} + \mathrm{E} = \mathrm{W} + \mathrm{S} \Rightarrow \mathrm{E} = \mathrm{W} + \mathrm{S} - \mathrm{N}, \\
\frac{1}{\mathrm{N}} + \frac{1}{\mathrm{E}} = \frac{1}{\mathrm{W}} + \frac{1}{\mathrm{S}} \Rightarrow \mathrm{E} = \left(\frac{1}{\mathrm{W}} + \frac{1}{\mathrm{S}} - \frac{1}{\mathrm{N}}\right)^{-1},
\end{gathered} \tag{53}
$$

the formulas being applied alternately.

Often it is convenient to take $\lambda = 1$ in the algorithm. As an example, consider the divergent series

$$
0! - 1! + 2! - 3! + 4! - \cdots, \qquad a_k = (-1)^k k!. \tag{54}
$$

The η table is as follows:

$$
\begin{array}{cccccc}
1 & & & & & \\
 & -\frac{1}{2} & & & & \\
-1 & & \frac{1}{6} & & & \\
 & \frac{2}{3} & & -\frac{2}{21} & & \\
2 & & -\frac{1}{6} & & \frac{4}{91} & \\
 & -\frac{3}{2} & & \frac{6}{52} & & -\frac{6}{221} \\
-6 & & \frac{3}{10} & & -\frac{4}{91} & \\
 & \frac{24}{5} & & -\frac{8}{35} & & \\
24 & & -\frac{4}{5} & & & \\
 & -20 & & & & \\
-120 & & & & &
\end{array}
$$

The original series is therefore transformed into the series

$$1 - \tfrac{1}{2} + \tfrac{1}{6} - \tfrac{2}{21} + \tfrac{4}{91} - \tfrac{6}{221} + \cdots, \tag{55}$$

whose first six terms provide the sum 0.5882352 (cf. Example 6.5).

The ε-algorithm results from interpreting the η-algorithm as a sequence rather than a series transformation. Let

$$\varepsilon_0^{(0)} = 0, \qquad \varepsilon_{2m}^{(k)} = \sum_{r=0}^{k-1} \eta_0^{(r)} + \sum_{r=0}^{2m-1} \eta_r^{(k)}. \tag{56}$$

Then the starting value

$$\varepsilon_0^{(k)} = \sum_{r=0}^{k-1} \eta_0^{(r)} = \sum_{r=0}^{k-1} a_r \lambda^r \tag{57}$$

is the kth partial sum of the original series and

$$\varepsilon_{2m}^{(0)} = \sum_{r=0}^{2m-1} \eta_r^{(0)} \tag{58}$$

is the $2m$th partial sum of the transformed series.

We find that

$$\begin{aligned}
\varepsilon_{2m}^{(k+1)} - \varepsilon_{2m}^{(k)} &= \eta_0^{(k)} + \sum_{r=0}^{2m-1} \eta_r^{(k+1)} - \sum_{r=0}^{2m-1} \eta_r^{(k)} \\
&= \eta_{2m}^{(k)} + \sum_{r=0}^{2m-1} \eta_r^{(k+1)} - \sum_{r=1}^{2m} \eta_r^{(k)} \\
&= \eta_{2m}^{(k)} + \sum_{r=0}^{k-1} (\eta_{2r}^{(k+1)} + \eta_{2r+1}^{(k+1)} - \eta_{2r+1}^{(k)} - \eta_{2r+2}^{(k)})
\end{aligned} \tag{59}$$

or

$$\varepsilon_{2m}^{(k+1)} - \varepsilon_{2m}^{(k)} = \eta_{2m}^{(k)}, \tag{60}$$

and so the odd partial sums of the transformed series are given by

$$\varepsilon_{2m}^{(1)} = \sum_{r=0}^{2m} \eta_r^{(0)}. \tag{61}$$

Now let

$$\varepsilon_{2m+1}^{(k)} = \sum_{r=0}^{2k} \frac{1}{\eta_r^{(k)}}, \tag{62}$$

and for convenience, let $\varepsilon_{-1}^{(k)} = 0$.

One can show, as above, that

$$\begin{aligned}
\varepsilon_{2m+1}^{(k+1)} - \varepsilon_{2m+1}^{(k)} &= 1/\eta_{2m+1}^{(k)}, \\
\varepsilon_{2m+1}^{(k)} - \varepsilon_{2m-1}^{(k+1)} &= 1/\eta_{2m}^{(k)}, \\
\varepsilon_{2m+2}^{(k)} - \varepsilon_{2m}^{(k+1)} &= \eta_{2m+1}^{(k)}.
\end{aligned} \tag{63}$$

Applying to these formulas the iteration rules for the η-algorithm shows that the following hold for both even and odd subscripts:

$$(\varepsilon_n^{(k+1)} - \varepsilon_n^{(k)})(\varepsilon_{n+1}^{(k)} - \varepsilon_{n-1}^{(k+1)}) = 1, \qquad n, k \geq 0, \tag{64}$$

with

$$\varepsilon_{-1}^{(k)} = 0, \qquad \varepsilon_0^{(k)} = \sum_{r=0}^{k-1} a_r \lambda^r. \tag{65}$$

Chapter 9 | Other Lozenge Algorithms and Nonlinear Methods

9.1. A Multiparameter ε-Algorithm

The ε-algorithm may be considered one of a class of lozenge algorithms that depend on an arbitrary fixed sequence $\mathbf{y} \in \mathscr{C}_S$. The ε-algorithm results on choosing $\mathbf{y} = \{c\}$.

For any sequence $\mathbf{w} \in \mathscr{C}_S$, define the linear operator $R: \mathscr{C}_S \to \mathscr{C}_S$ by

$$\{R(\mathbf{w})\}_n = R(w_n) = \Delta(w_n/y_n). \tag{1}$$

R may be iterated by means of the rule

$$R^{k+1}(w_n) = R^k[\Delta(w_n/y_n)], \qquad R^0(w_n) = w_n, \tag{2}$$

so that $R^2(w_n) = \Delta\{\Delta(w_n/y_n)/y_n\}$, etc. Note that if $y_n = c$, then $R^k(w_n) = \Delta^k w_n/c^k$.

Now define

$$\varepsilon_{2m}^{(n)} = f_m(w_n) = \frac{\begin{vmatrix} w_n & \cdots & w_{n+m} \\ Rw_n & \cdots & Rw_{n+m} \\ \vdots & & \vdots \\ R^m w_n & \cdots & R^m w_{n+m} \end{vmatrix}}{\begin{vmatrix} y_n & \cdots & y_{n+m} \\ Rw_n & \cdots & Rw_{n+m} \\ \vdots & & \vdots \\ R^m w_n & \cdots & R^m w_{n+m} \end{vmatrix}}, \tag{3}$$

$$\varepsilon_{2m+1}^{(n)} = 1/f_m(Rw_n). \tag{4}$$

Theorem 1. $\varepsilon_k^{(n)}$ satisfies

$$\begin{aligned} \varepsilon_{k+1}^{(n)} &= \varepsilon_{k-1}^{(n+1)} + y_n(\varepsilon_k^{(n+1)} - \varepsilon_k^{(n)})^{-1}, \qquad n, k \geq 0, \\ \varepsilon_{-1}^{(n)} &= 0, \quad \varepsilon_0^{(n)} = w_n/y_n, \qquad n \geq 0. \end{aligned} \tag{5}$$

Proof. The proof is the same as that for the ε-algorithm and is left to the reader. ■

The most useful case occurs on choosing

$$w_n = s_n y_n \tag{6}$$

and defining

$$e_k(s_n) = f_k(w_n) = f_k(s_n y_n). \tag{7}$$

Then $\varepsilon_0^{(n)} = s_n, n \geq 0$. The transformation e_k is translative and homogeneous.

Exactness theorems for this transformation are, of course, more difficult than those for the ε-algorithm because of the nature of the operator R. However, some information is available; see Brezinski (1977, pp. 111ff.).

Example. $k = 1$ provides an interesting generalization of the Aitken δ^2-process:

$$\begin{aligned} e_1(s_n) = \bar{s}_n &= \frac{y_n s_n \Delta s_{n+1} - y_{n+1} s_{n+1} \Delta s_n}{y_n \Delta s_{n+1} - y_{n+1} \Delta s_n} \\ &= s_n + a_{n+1}\left(1 - \frac{a_{n+2} y_n}{a_{n+1} y_{n+1}}\right)^{-1}. \end{aligned} \tag{8}$$

The following convergence criterion is immediate.

Theorem 2. Let $\mathbf{s} \in \mathscr{C}_C$ and $a_{n+2} y_n / a_{n+1} y_{n+1}$ be bounded away from 1. Then $\bar{s}_n \to s$.

By means of Theorem 1.4(4) a connection may be established between the above transformation and the Aitken δ^2-process.

Theorem 3. Let the Aitken δ^2-process accelerate the convergence of $\mathbf{s}$. Then e_1 accelerates the convergence of $\mathbf{s}$ iff $(1 - y_n/y_{n+1})/(1 - a_{n+1}/a_{n+2})$ is a null sequence.

In particular e_1 will accelerate the convergence of some member of $\mathscr{C}_l$ iff $\mathbf{y}$ is ultimately "nearly constant" ($y_n/y_{n+1} \to 1$). It might be thought that $\mathbf{y}$ could be chosen to make this new method effective on logarithmically convergent series for which the Aitken δ^2-process (in fact, for which the ε-algorithm also) is ineffective. However, to make the process accelerative for $\mathscr{C}_{l'}$, it seems $\mathbf{y}$ must reflect some a priori knowledge about $\mathbf{s}$.

9.2. The ρ-Algorithm

Recall how the extrapolation deltoid of Section 3.3 was obtained. We assumed that a function f existed with the property that, at the distinct points x_j, $f(x_j) = s_j$. $s_n^{(k)}$ was then defined to be the Lagrangian polynomial through the points (x_j, s_j), $0 \leq j \leq k$, extrapolated to 0.

An approach based on the Thiele interpolation formula [see Milne-Thomson (1960, Chapter V)] yields a lozenge algorithm that is very similar to the algorithm of the previous section. Again let f be tabulated at x_j, $f(x_j) = s_j$. The reciprocal differences $\rho_k^{(n)}$ are defined recurrently by

$$\begin{aligned} \rho_{k+1}^{(n)} &= \rho_{k-1}^{(n+1)} + (x_{n+k+1} - x_n)(\rho_k^{(n+1)} - \rho_k^{(n)})^{-1}, \qquad n, k \geq 0, \\ \rho_{-1}^{(n)} &= 0, \quad \rho_0^{(n)} = s_n, \qquad n \geq 0. \end{aligned} \tag{1}$$

If f is a rational function of x of the form

$$f(x) = (sx^k + a_1 x^{k-1} + \cdots + a_k)/(x^k + b_1 x^{k-1} + \cdots + b_k), \tag{2}$$

then $\rho_{2k}^{(n)} = s$; i.e., the algorithm is exact; otherwise when $\mathbf{x}$ is a real sequence with $x_n \to \infty$ $\rho_{2k}^{(0)}$ will be the value of the interpolating rational function extrapolated to ∞. Note that $\rho_k^{(n)}$ is translative and homogeneous. [For a justification of these statements, see Milne-Thomson (1960).]

The obvious choice of $\mathbf{x}$ to make for logarithmically convergent sequences of the form $s_n \sim s + n^{\theta}(c_0 + c_1/n + \cdots)$ is $x_n = (n + 1)$.

Naturally, for any sequence $\mathbf{x}$ that is appropriate to use in the extrapolation deltoid of Section 3.3, $\{x_n^{-1}\}$ can be used in the ρ-algorithm. A very interesting nonlinear analog of the Romberg integration procedure results on choosing $x_n = 4^n$:

$$\begin{aligned} \rho_{k+1}^{(n)} &= \rho_{k-1}^{(n+1)} + 4^n(4^{k+1} - 1)(\rho_k^{(n+1)} - \rho_k^{(n)})^{-1}, \\ \rho_{-1}^{(n)} &= 0, \qquad n \geq 0, \\ \rho_0^{(n)} &= \frac{1}{2^n} \sum_{k=0}^{2^n}{}'' f\left(\frac{k}{2^n}\right), \qquad n \geq 0. \end{aligned} \tag{3}$$

$\rho_{2k}^{(0)}$ corresponds to extrapolating a rational approximation for the error in the trapezoidal formula, i.e., the right-hand side of

$$\rho_0^{(n)} - \int_0^1 f(x)dx \sim \frac{c_1}{4^n} + \frac{c_2}{4^{2n}} + \frac{c_3}{4^{3n}} + \cdots. \tag{4}$$

Table I shows the result of applying the ρ-algorithm to the example of Section 3.1.

Note that, as with the Romberg algorithm, diagonal convergence is not the most desirable. $\rho_2^{(6)}$ is as accurate as any other entry in the table and

Table I

ρ-Algorithm: $s = \int_0^1 (x + 0.05)^{-1}\,dx = \ln 21 = 3.044522437723$

n	$\rho_0^{(n)}$	$\rho_2^{(n)}$	$\rho_4^{(n)}$	$\rho_6^{(n)}$	$\rho_8^{(n)}$
0	10.476				
1	6.147	3.435 $(=\rho_2^{(0)})$			
2	4.219	3.133	3.053		
3	3.438	3.058	3.045	3.044541	
4	3.161	3.046	3.044546	3.044522697	3.044522425
5	3.760	3.044586	3.044522818	3.044522425	
6	3.053	3.044524	3.044522427		
7	3.047	3.044522424			
8	3.045				

clearly safer to use than entries on its right. This is probably due to the monotone character of s_n $(=\rho_0^{(n)})$.

9.3. The θ-Algorithm

The theorems and examples of the previous chapter show that the Schmidt transformation is highly effective on sequences in $\mathscr{C}_l$ but not at all on those in $\mathscr{C}_{l'}$. This is hardly surprising considering the way the transformation was derived. While it is true the Levin t and u transformations are more effective for $\mathscr{C}_{l'}$, these methods are time and space consuming as far as their computer implementation is concerned. It would be nice to have a method useful for $\mathscr{C}_{l'}$ that has some of the computational simplicity of the ε-algorithm.

The following derivation is very freewheeling, and the reader is asked to suspend any skepticism, accepting for the moment that the assumptions made are realistic and reasonable. The resulting method, which has acquired the label of the θ-algorithm, is in certain vertical modes of convergence an extraordinarily useful transformation. It is proclaimed by Smith and Ford (1979) in their exhaustive survey to be one of the three best (along with the Levin t and u transformations) across-the-board methods for accelerating the convergence of arbitrary sequences.

Let us modify the ε-algorithm slightly by writing

$$\theta_{k+1}^{(n)} = \theta_{k-1}^{(n+1)} + \beta_k(\theta_k^{(n+1)} - \theta_k^{(n)})^{-1}, \tag{1}$$

β_k to be determined. Generally speaking, if $\theta_{2k}^{(n)}$ is convergent as $n \to \infty$, $\theta_{2k+1}^{(n)}$ will be divergent. Since we are not really interested in the size of $\theta_{2k+1}^{(n)}$, assume that $\beta_{2k} \equiv 1$. Taking k odd gives

$$\theta_{2m+2}^{(n)} = \theta_{2m}^{(n+1)} + \beta_{2m+1}(\theta_{2m+1}^{(n+1)} - \theta_{2m+1}^{(n)})^{-1}. \tag{2}$$

Table II

θ-Algorithm, $\theta_k^{(0)}$

k	$s_n = (\mathrm{GAM})_n$	$s_n = (\mathrm{LN}\ 2)_n$	$s_n = (\mathrm{FAC})_n$ (divergent)	$s_n = (\mathrm{PI}^2)_n$	$s_n = (\mathrm{IT}\ 2)_n$	$s_n = (\mathrm{LUB})_n$	$s_n = (\mathrm{RT})_n$
2	0.577621	0.694	0.615	1.639	1.330	1.583	0.606
4	0.577455	0.693149	0.596984	1.644935	1.326	0.911	0.604901
6	0.577215433	0.6931471814	0.596357	1.644934082	1.305	−2.757	0.604898645
8	0.577215664	0.693147180560	0.596347258	1.644934067	1.307	4.898	0.604898643422
10	0.577215663314	0.693147180560	0.596347252	1.644934067	1.383	112.535	0.604898643422

Now if we wish $\theta_{2m+2}^{(n)}$ to converge more rapidly than $\theta_{2m}^{(n+1)}$, we should pick

$$\beta_{2m+1} = -\lim_{n\to\infty}(\theta_{2m+1}^{(n+1)} - \theta_{2m+1}^{(n)})\theta_{2m}^{(n+1)}. \tag{3}$$

Of course, we have no knowledge of what this limit is. We could allow β_{2m+1} to depend on n and pick $\beta_{2m+1} = -(\theta_{2m+1}^{(n+1)} - \theta_{2m+1}^{(n)})\theta_{2m}^{(n+1)}$, but this leads to nothing. Let us take differences with respect to n of (2) and then divide by $\Delta\theta_{2m}^{(n+1)}$. Then

$$\Delta\theta_{2m+2}^{(n)}/\Delta\theta_{2m}^{(n+1)} = 1 + (\beta_{2m+1}/\Delta\theta_{2m}^{(n+1)})\,\Delta(\Delta\theta_{2m+1}^{(n)})^{-1}, \tag{4}$$

and the accelerativeness requirement is now

$$\beta_{2m+1} = -\lim_{n\to\infty}\Delta\theta_{2m}^{(n+1)}/\Delta(\Delta\theta_{2m+1}^{(n)})^{-1}. \tag{5}$$

If we now allow β_{2m+1} to depend on n, i.e., to be equal to the quantity behind the limit, something nontrivial happens. The resulting algorithm can be expressed by the following equations:

$$\begin{aligned}
\theta_{-1}^{(n)} &= 0, \quad \theta_0^{(n)} = s_n, \qquad n \geq 0,\\
\theta_{2m+1}^{(n)} &= \theta_{2m-1}^{(n+1)} + (\theta_{2m}^{(n+1)} - \theta_{2m}^{(n)})^{-1}, \qquad m \geq 0,\\
\theta_{2m+2}^{(n)} &= \frac{\theta_{2m}^{(n+2)}\Delta\theta_{2m+1}^{(n+1)} - \theta_{2m}^{(n+1)}\Delta\theta_{2m+1}^{(n)}}{\Delta^2\theta_{2m+1}^{(n)}}, \qquad m \geq 0.
\end{aligned} \tag{6}$$

In particular,

$$\theta_2^{(n)} = s_{n+1} + \frac{a_{n+2}(1-\rho_{n+2})}{1 - 2\rho_{n+2} + \rho_{n+1}\rho_{n+2}}. \tag{7}$$

This is just s_{n+1}^* of the Lubkin W-transform [see Eq. 7.2(1)]. In this case the regularity and exactness results of Section 7.2 are applicable. Thus $\theta_2^{(n)}$ accelerates the convergence of sequences of the form $s_n \sim n^\theta(c_0 + c_1/n + \cdots)$, $\operatorname{Re}\theta < -1$.

Table II indicates that the θ-algorithm works well in diagonal modes of convergence on both logarithmic and linear sequences. Like other methods, however, it fails on hybrid sequences, such as $(\mathrm{LUB})_n$.

9.4. Implicit Summation: Logarithmically Convergent Sequences

This method, which is suitable for logarithmically convergent sequences, is radically different from any method previously discussed. The iterates $\bar{s}_n$ are not defined explicitly, but rather occur as a subsequence of the roots of certain polynomials.

Definition. Let **s** be a bounded complex sequence whose members are distinct. Let $\phi_k \neq 0$ be $o(1)$ and denote the zeros of the polynomial

$$\mu_n(t) = \sum_{k=0}^{n} \phi_k \prod_{\substack{j=0 \\ j \neq k}}^{n} \left(\frac{t - s_j}{s_k - s_j} \right) \tag{1}$$

by λ_{nk}, $1 \leq k \leq n$. If for some region $\mathscr{D} \supset [s_0, s_1, \ldots] \{\lambda_{nk}\}$ has exactly one limit point $\alpha \in \mathscr{D}$, then **s** is said to be *ϕ-summable to α.*

Thus $\bar{\mathbf{s}}$ is an equivalence class of sequences, namely, all sequences λ_{nm_n} converging to α. In practice, one selects a subsequence which converges rapidly.

Several facts about the method are immediately apparent. The method is homogeneous and translative in the sense that if **s** is ϕ-summable to α, $a\mathbf{s} + b$ is ϕ-summable to $a\alpha + b$, $a \neq 0$. Further, if a monotone sequence is to be ϕ-summable to α, all ϕ_k must be of one sign. To see this, suppose, e.g., $\phi_m < 0 < \phi_{m'}$. Then for n sufficiently large, $\mu_n(t)$ contains a root in the interval $(s_m, s_{m'})$, by Rolle's theorem. Thus some subsequence λ_{nm_n} will converge to some $\beta \in [s_m, s_{m'}]$, but clearly $\beta \neq \alpha$. This means that to apply the method to logarithmically convergent sequences requires choosing all the ϕ_k positive, say.

The technique is basically an extrapolation process, as Fig. 1 makes clear.

Tables III and IV show the effect of the method on the logarithmically convergent sequences $(\mathrm{PI}^2)_n$ and $(\mathrm{GAM})_n$. Both cases are based on the choice $\phi_k = 1/(4k + 9)$. Clearly, irrelevant roots are given to only three significant figures. The method seems to be a very powerful one for dealing with logarithmic convergence, one of the most powerful algorithms we have encountered.

The method presents no particular computational problems, at least for moderate values of n, since root-finding capabilities are standard software on today's computers. In the numerical examples developed so far, it has not been necessary to take n greater than 10 to achieve nine or ten significant figures.

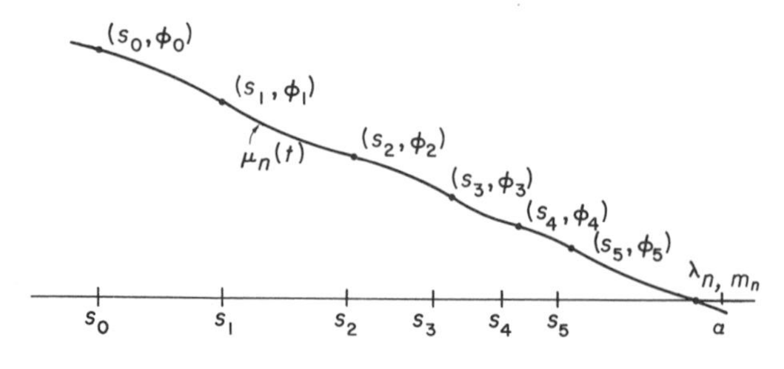

Fig. 1

Table III

Roots λ_{nk} of $\mu_n(t)$: $s_n = (\mathrm{PI}^2)_n$, $\pi^2/6 = 1.644934067$

n	1	2	3	4	5	6
	1.81	1.665887	1.647976	1.645328	1.644976	1.644938
		-1.30	$0.282 \pm 1.67i$	-0.512	$-4.22 \pm 0.975i$	$0.537 \pm 1.38i$
				$1.30 \pm 1.67i$	$1.81 \pm 1.47i$	$2.09 \pm 1.30i$
						-0.295

Table IV

Roots λ_{nk} of $\mu_n(t)$: $s_n = (\mathrm{GAM})_n$, $\gamma = 0.577215664$

n	1	2	3	4	5
	0.372	0.552003	0.572941	0.576504	0.577109
		1.89	$1.28 \pm 0.686i$	$0.825 \pm 0.714i$	$0.582 \pm 0.632i$
				1.55	$1.35 \pm 0.391i$

n	6	7	8	9	10
	0.577201	0.577213	0.577215482	0.577215647	0.577215663
	$0.444 \pm 0.548i$	$0.360 \pm 0.479i$	$0.304 \pm 0.422i$	$0.265 \pm 0.376i$	$0.238 \pm 0.338i$
	$1.10 \pm 0.552i$	$0.892 \pm 0.601i$	$0.734 \pm 0.600i$	$0.616 \pm 0.577i$	$0.529 \pm 0.547i$
	1.43	$1.33 \pm 0.266i$	$1.18 \pm 0.422i$	$1.03 \pm 0.506i$	$0.897 \pm 0.547i$
			1.36	$1.30 \pm 0.199i$	$1.19 \pm 0.336i$
					1.31

Theorem. Let the series

$$u = c_1 z + c_2 z^2 + \cdots, \qquad c_1 \neq 0, \tag{2}$$

converge for $|z| \leq 1$ and let the inverse function $z(u)$ be analytic and schlicht in N_R for some R.

Let

$$s_n = s + \sum_{r=1}^{\infty} \frac{c_r}{(n+1)^r}, \tag{3}$$

and suppose $[s_0 - s, s_1 - s, s_2 - s, \ldots, 0] \subset N_R$. Then if β is sufficiently close to 1, $\mathbf{s}$ is $(k+\beta)^{-1}$ summable to s.

Proof. Without loss of generality let $s = 0$. Defining

$$g(u) = \frac{z(u)}{1 + (\beta - 1)z(u)} \tag{4}$$

we have, since $z(u)$ is schlicht,

$$g(s_k) = \frac{1/(k+1)}{1 + (\beta - 1)/(k+1)} = \frac{1}{k+\beta} = \phi_k. \tag{5}$$

Thus β can be chosen sufficiently close to 1 so that g is analytic in any $\bar{N}_{R'}$ with $R' < R$.

Let $R' < R$ be such that $[s_0, s_1, \ldots] \subset N_{R'}$.

By Milne-Thomson (1960, p. 12),

$$g(t) - \mu_n(t) = \frac{(t - s_0)\cdots(t - s_n)}{2\pi i} \times \int_{\partial N_{R'}} \frac{g(u)\,du}{(u - s_0)\cdots(u - s_n)(u - t)}, \qquad t \in N_{R'}. \tag{6}$$

Since $|s_n| <. \varepsilon$,

$$|g(t) - \mu_n(t)| <. M(|t| + \varepsilon)^n/(R' - \varepsilon)^n, \qquad t \in N_{R'}, \tag{7}$$

for every $0 < \varepsilon < R'$. A direct argument based on this shows that $\mu_n(t) \to g(t)$ uniformly on compact subsets of N_R. Now $g(0) = 0$ and, by the schlichtness of $z(u)$, this is the only zero of g in N_R. By Hurwitz's theorem, $\{\lambda_{nk}\}$ has exactly one limit point in N_R, i.e., $\alpha = 0$. ∎

Example. Let

$$s_n = e^{-(n+1)^{-1}}. \tag{8}$$

Then $s = 1$,

$$z = -\ln(1 + u), \tag{9}$$

so $N_R = N$ and $s_j - s \in N$. In this case the permissible values of β can be determined. The denominator of (4) is zero iff $\ln(1 + u) = 1/(\beta - 1)$ or $u = e^{(\beta-1)^{-1}} - 1$. If we pick $1 < \beta < 1 + (\ln 2)^{-1}$, u will be >1 and hence the denominator of (4) will not vanish in N. s_n will then be $(k + \beta)^{-1}$-summable to 1.

Chapter 10 The Brezinski–Håvie Protocol

10.1. Introduction and Derivation; Sequences in a Banach Space

Let **s** be a complex convergent sequence approaching its limit s in the following manner:

$$s_n \sim s + \sum_{r=1}^{\infty} c_r f_r(n) \qquad \text{in} \quad J^+, \tag{1}$$

where $\{f_r(n)\}$ is an asymptotic scale. It is surprising that very often in practical problems the form of the functions f_r is known but the values of the coefficients A_r are not—or are at best, difficult to compute. As a simple example, suppose s is the value of an integral whose nth approximation by the trapezoidal rule is s_n. Then $f_r(n) = n^{-2r}$ but the A_r, which depend on the higher derivatives of the integrand, may be impossible to compute with any accuracy since, generally, only tabular values of the integrand are known. Many other examples are given in Chapter 3.

What the Brezinski–Håvie (BH) protocol does, loosely speaking, is to establish a deltoid algorithm that maps **s** into a sequence $\mathbf{s}^{(k)}$ where $s_n^{(k)}$ has a representation similar to (1), but where the first k terms are accounted for. Thus the convergence of **s** can be accelerated with no knowledge whatsoever of the coefficients A_r. The importance and usefulness of the method can hardly be exaggerated.

The idea of the general representation (1) was, apparently, first articulated by Levin (1973), although special cases, such as the Romberg scheme for integration, are classical and form the basis for many of the algorithms in previous chapters. The discovery of the deltoid computation of $s_n^{(k)}$ and its representation by a ratio of determinants, the real heart of the method, are due to Håvie (1979) and Brezinski, respectively. The latter was communicated to me privately in 1979.

We shall conduct the derivation of the algorithm for sequences in a Banach space. Although the formal aspects of derivation can be carried through for sequences in any topological vector space, interesting applications of the algorithm seem to be possible only when the underlying space is metrizable and the dual has a reasonable supply of functionals. For normed spaces the Hahn–Banach theorem guarantees the latter and, of course, the norm provides a convenient metric.

In what follows $\mathscr{B}$ will be a nontrivial real or complex Banach space with norm $\|\cdot\|$, and $\mathscr{B}^*$ will be its dual; **s** will be a sequence with members in $\mathscr{B}$, i.e., $\mathbf{s} \in \mathscr{B}_S$ and $\boldsymbol{\phi}$ a convergent (not to zero) sequence of functionals in $\mathscr{B}^*$, i.e., $\boldsymbol{\phi} \in \mathscr{B}^*_C$; $\phi_n \to \phi$. Assume ker $\phi_n \neq \mathscr{B}$. k will be an integer >0, for the moment fixed, and $\mathbf{f}_r = \{f_r(n)\} \in \mathscr{B}_S$, $\mathbf{f}_r \neq \mathbf{0}$, $1 \leq r \leq k$. Also, $f_r = o(1)$ as $n \to \infty$.

To say $\boldsymbol{\phi}$ is a *constant sequence* means that $\phi_n = \phi$, $n \geq 0$.

We seek to determine a family of transformations of **s**, $T_k(\mathbf{s}) = \mathbf{s}^{(k)}$, that are exact, i.e., $s_n^{(k)} \equiv 0$, when **s** is one of the sequences $\mathbf{f}_r$, $1 \leq r \leq k$. Obviously, there are many ways of doing this. One of the most natural ways, insofar as being in the spirit of the work of previous chapters, is to take

$$E_k(s_n) \asymp s_n^{(k)} = \begin{vmatrix} s_n & 0 \quad f_1(n) \cdots f_k(n) \\ \langle \phi_n, s_n \rangle & \\ \vdots & (C_n) \\ \langle \phi_{n+k}, s_{n+k} \rangle & \end{vmatrix} |C_n|^{-1}, \tag{2}$$

where

$$C_n = \begin{bmatrix} 1 & \langle \phi_n, f_1(n) \rangle & \cdots & \langle \phi_n, f_k(n) \rangle \\ \vdots & \vdots & & \vdots \\ 1 & \langle \phi_{n+k}, f_1(n+k) \rangle & \cdots & \langle \phi_{n+k}, f_k(n+k) \rangle \end{bmatrix}. \tag{3}$$

It is assumed here and in what follows that $|C_n| \neq 0$ for the values of n under consideration.

Note that E_k is not exact for constant sequences nor even for all members of $\text{Lin}\{\mathbf{f}_r\}$. Although it is possible to define a similar transformation that *is* exact for constant sequences, it does not have the very nice formal properties of the present one. At any rate, we may simply remark that E_k *is* exact for constants and members of $\text{Lin}\{\mathbf{f}_r\}$ whenever $\boldsymbol{\phi}$ is constant.

The following easily demonstrated theorem describes the effect of E_k on sequences of the form (1) when the right-hand side converges.

In what follows let

$$f_j^{(k)}(n) = \langle \phi_n, E_k(f_j(n)) \rangle, \qquad s_n^{(k)} = \langle \phi_n, s_n^{(k)} \rangle,$$
$$R_n^{(k)} = \langle \phi_n, r_n^{(k)} \rangle = \langle \phi_n, s_n^{(k)} - s \rangle. \tag{4}$$

Theorem 1. Let (1) hold where c_r is in the field of $\mathscr{B}$ and the series on the right converges absolutely, i.e., $\sum |A_r| \, \| f_r(n) \| < \infty$.

Then

$$s_n^{(k)} = E_k(s) + \sum_{r=k+1}^{\infty} c_r E_k[f_r(n)] \tag{5}$$

and

$$S_n^{(k)} = \langle \phi_n, E_k(s) \rangle + \sum_{r=k+1}^{\infty} c_r f_r^{(k)}(n). \tag{6}$$

Proof. Trivial. ■

Note $E_k(s) = s$ when $\boldsymbol{\phi}$ is constant.

To examine the convergence properties of E_k requires recursion formulas for $S_n^{(k)}$ and $R_n^{(k)}$.

Theorem 2

$$S_n^{(k+1)} = \frac{f_{k+1}^{(k)}(n) S_{n+1}^{(k)} - f_{k+1}^{(k)}(n+1) S_n^{(k)}}{f_{k+1}^{(k)}(n) - f_{k+1}^{(k)}(n+1)}, \quad n, k \geq 0, \qquad S_n^{(0)} = \langle \phi_n, s_n \rangle; \tag{7}$$

$$R_n^{(k+1)} = \frac{f_{k+1}^{(k)}(n) R_{n+1}^{(k)} - f_{k+1}^{(k)}(n+1) R_n^{(k)}}{f_{k+1}^{(k)}(n) - f_{k+1}^{(k)}(n+1)} + \frac{\Delta \langle \phi_n, s \rangle f_{k+1}^{(k)}(n)}{f_{k+1}^{(k)}(n) - f_{k+1}^{(k)}(n+1)}, \quad n, k \geq 0, \quad R_n^{(0)} = \langle \phi_n, r_n \rangle; \tag{8}$$

$$f_i^{(k+1)}(n) = \frac{f_{k+1}^{(k)}(n) f_i^{(k)}(n+1) - f_{k+1}^{(k)}(n+1) f_i^{(k)}(n)}{f_{k+1}^{(k)}(n) - f_{k+1}^{(k)}(n+1)}, \tag{9}$$

$i \geq 1, 0 \leq k \leq i-2,$

$f_i^{(0)}(n) = \langle \phi_n, f_i(n) \rangle, \qquad i \geq 1.$

Proof. Since

$$S_n^{(k)} = \begin{vmatrix} \langle \phi_n, s_n \rangle & \langle \phi_n, f_1(n) \rangle & \cdots & \langle \phi_n, f_k(n) \rangle \\ \vdots & \vdots & & \vdots \\ \langle \phi_{n+k}, s_{n+k} \rangle & \langle \phi_{n+k}, f_1(n+k) \rangle & \cdots & \langle \phi_{n+k}, f_k(n+k) \rangle \end{vmatrix}, \tag{10}$$

the first recursion formula is easily demonstrated by applying Sylvester's identity (Appendix, Section A.3). The proof for $R_n^{(k)}$ follows by the substitution $R_n^{(k)} = S_n^{(k)} - \langle \phi_n, s \rangle$, and the proof for $f_i^{(k)}(n)$ is straightforward. ■

A computational tableau for computing $f_k^{(k)}(n)$ will be discussed later along with the scalar case $\mathscr{B} = \mathscr{C}$.

Now consider a path **P**. Provided $r_n^{(k)} = s_n^{(k)} - s$ for n large on **P** does not approach too close, relatively speaking, to the kernel of ϕ, which we call $\mathscr{K}$, the convergence properties of $\langle \phi_n, r_n^{(k)} \rangle =: R_n^{(k)}$ can be related to those of $r_n^{(k)}$ in a simple way. Since the kernel of ϕ is a maximal linear subspace of $\mathscr{B}$, one might expect that the greater the dimensionality of $\mathscr{B}$, the less likely it is that $r_n^{(k)}$ can be relatively bounded away from $\mathscr{K}$, in other words, the less information $R_n^{(k)}$ can deliver about $r_n^{(k)}$. This expectation of failure is not entirely warranted, however. As an example will show, the functions $f_i(n)$ can often be chosen to keep $r_n^{(k)}$ far enough away from $\mathscr{K}$ for large classes of sequences.

To make these ideas more precise, we introduce the definitions

$$\mathscr{K} = \{g \in \mathscr{B} \mid \langle \phi, g \rangle = 0\}, \tag{11}$$

$$d(g, \mathscr{K}) = \inf_{h \in \mathscr{K}} \|g - h\|, \tag{12}$$

$$S = \langle \phi, s \rangle. \tag{13}$$

Theorem 3. For $n + k$ sufficiently large on **P**, let

$$d(r_n^{(k)}, \mathscr{K})/\|r_n^{(k)}\| \geq \delta > 0. \tag{14}$$

Then $s_n^{(k)}$ converges in norm to s on **P** iff $S_n^{(k)}$ converges to S on **P** in the two following cases:

(i) $\boldsymbol{\phi}$ constant, **P** any path;
(ii) $\boldsymbol{\phi}$ not necessarily constant, **P** a path (n, k) with $n \to \infty$.

Proof. For $n + k$ large,

$$\begin{aligned}\|\phi\| \, \|r_n^{(k)}\| &\geq |\langle \phi, r_n^{(k)} \rangle| = \|\phi\| \, d(r_n^{(k)}, \mathscr{K}) \\ &\geq \delta \|\phi\| \, \|r_n^{(k)}\|\end{aligned} \tag{15}$$

[see Jameson (1974, p. 188)].

Now observe that, since $\langle \phi_n, r_n^{(k)} \rangle = \langle \phi_n - \phi, r_n^{(k)} \rangle + \langle \phi, r_n^{(k)} \rangle$,

$$|\langle \phi_n, r_n^{(k)} \rangle| \geq \|r_n^{(k)}\| \{\|\phi\| \delta - \|\phi_n - \phi\|\}, \tag{16}$$

so it is clear we can find m_1 and m_2, $0 < m_1 < m_2$, such that

$$m_1 \|r_n^{(k)}\| \leq |\langle \phi_n, r_n^{(k)} \rangle| \leq m_2 \|r_n^{(k)}\|. \tag{17}$$

Further, $\langle \phi_n, r_n^{(k)} \rangle = S_n^{(k)} - S + \langle \phi - \phi_n, s \rangle$, and combining this with the above estimate shows

$$m_1 \|r_n^{(k)}\| - \|\phi_n - \phi\| \, \|s\| \leq |S_n^{(k)} - S| \leq m_2 \|r_n^{(k)}\| + \|\phi_n - \phi\| \, \|s\| \tag{18}$$

and the assertion of the second case follows on taking $n \to \infty$. ■

What is desired, of course, is information about the convergence of $s_n^{(k)}$ that does not require a knowledge of $S_n^{(k)}$. The following lemma is the link to a theorem that does this.

Lemma (Brezinski). Let $\mathbf{s} \in \mathscr{B}_C$, and for each k, let

$$\sigma_{nk} = f_{k+1}^{(k)}(n+1)/f_{k+1}^{(k)}(n) \tag{19}$$

be bounded away from $1, 0 \leq k \leq K$. Then $S_n^{(k)}$ converges to S on all vertical paths, $0 \leq k \leq K+1$.

Proof

$$R_n^{(k+1)} = R_{n+1}^{(k)} + \frac{\sigma_{nk}}{1-\sigma_{nk}}(R_{n+1}^{(k)} - R_n^{(k)}) + \frac{\Delta\langle\phi_n, s\rangle}{1-\sigma_{nk}}. \tag{20}$$

Since $\sigma_{nk}/(1-\sigma_{nk})$ is bounded, Eq. (20) furnishes an easy inductive proof on k that $R_n^{(k)} \to 0$. It is then trivial that $S_n^{(k)} \to S$. ∎

Paths where k is unbounded present a formidable problem. Aside from certain easy scalar cases, i.e., the summation by extrapolation formulas of Chapter 3, nothing has yet been accomplished on this problem.

We have now proved the following result.

Theorem 4. Let the hypothesis (14) of Theorem 3 hold and for each fixed $k, 0 \leq k \leq K$, let σ_{nk} be bounded away from 1. Then E_k is regular on all vertical paths, $0 \leq k \leq K+1$.

Example. Let $\mathscr{B} = \mathscr{R}^2$; $\phi_1 = \phi_2 = \phi$; $\langle\phi, (x, y)\rangle = y$ (so $\mathscr{K}$ is the x-axis). Let $k = 1$ and $f_1(n) = (\lambda_1^n, \lambda_2^n)$, $0 < |\lambda_1| < |\lambda_2| < 1$. Then $\sigma_{n0} = \lambda_2 \neq 1$. Let $s_n = (a_n, b_n)$, a_n, b_n null sequences (with no loss of generality). Then

$$r_n^{(1)} = r_n - [1/(\lambda_2 - 1)](\Delta b_n \sigma^n, \Delta b_n), \qquad \sigma = \lambda_1/\lambda_2, \tag{21}$$

$$\frac{d(r_n^{(1)}, \mathscr{K})}{\|r_n^{(1)}\|} = \left\{1 + \left[\frac{(\lambda_2 - 1)a_n - \Delta b_n \sigma^n}{\lambda_2 b_n - b_{n+1}}\right]^2\right\}^{-1/2}. \tag{22}$$

A sufficient condition for the regularity for E_1 in this case is that

$$(\lambda_2 - 1)a_n - \Delta b_n \sigma^n = o[\lambda_2 b_n - b_{n+1}]. \tag{23}$$

For instance, if $\Delta b_n = o(b_n)$, then it is sufficient that $a_n = o(b_n)$. E_k is then regular for this class of sequences.

10.2. The Case $\boldsymbol{\phi}$ Constant

For the case $\boldsymbol{\phi}$ = const the algorithm can be derived in the same way the Schmidt transformation was derived. One assumes that s_n behaves as

$$s_n = s + \sum_{r=1}^{k} c_r f_r(n). \tag{1}$$

Taking ϕ of Eq. (1), replacing n by m, $n \le m \le n + k$, and considering those equations and the above as $k + 2$ equations in the $k + 1$ unknowns $\langle \phi, s \rangle$, $c_1, c_2, \ldots, c_k$ produces the requirement

$$\begin{vmatrix} s_n - s & 0 & f_1(n) & \cdots & f_k(n) \\ \langle \phi, s_n \rangle & 1 & \langle \phi, f_1(n) \rangle & \cdots & \langle \phi, f_k(n) \rangle \\ \vdots & \vdots & \vdots & & \vdots \\ \langle \phi, s_{n+k} \rangle & 1 & \langle \phi, f_1(n+k) \rangle & \cdots & \langle \phi, f_k(n+k) \rangle \end{vmatrix} = 0, \tag{2}$$

but this is clearly equivalent to Eq. 10.1(2) when $s = s_n^{(k)}$, $\boldsymbol{\phi}$ = const.

Conversely, when s_n has the form (1), then E_k will be exact, $s_n^{(k)} \equiv s$, provided the algorithm is defined.

For the study of this algorithm, there are two modes of regularity or accelerativeness to consider. One pertains to weak convergence, i.e., convergence in the seminorm $|\langle \phi, \cdot \rangle|$. The other is the usual strong convergence, convergence in the norm. The regularity result below, though based on pretty specific properties of f_r, is often applicable.

Theorem 1. Let $\mathbf{s} \in \mathscr{B}_C$ and

$$\lim_{n \to \infty} \langle \phi, f_r(n+1) \rangle / \langle \phi, f_r(n) \rangle = b_r \ne 1, \qquad 1 \le r \le k, \tag{3}$$

where $b_i \ne b_j$, $i \ne j$.

Define

$$\eta_n = \langle \phi, r_{n+1} \rangle / \langle \phi, r_n \rangle \tag{4}$$

(which we assume exists for n sufficiently large) and denote by A_m the proposition

$$\text{“} \| f_m(n) \| / \langle \phi, f_m(n) \rangle = O(1). \text{”} \tag{5}$$

Then along any vertical path.

(i) if η_n is bounded, then E_k is regular for $\mathbf{s}$ in the seminorm $|\langle \phi, \cdot \rangle|$;
(ii) if $\eta_n \to b_j$ for some j, then E_k accelerates $\mathbf{s}$ in seminorm;
(iii) if η_n is bounded and A_m holds, $1 \le m \le k$, then E_k is regular for $\mathbf{s}$ in norm;

(iv) if $\eta_n \to b_j$ for some j, A_m holds, $1 \leq m \leq k$, $m \neq j$, and

$$\left\| r_n - \frac{\langle \phi, r_n \rangle f_j(n)}{\langle \phi, f_j(n) \rangle} \right\| \Big/ \|r_n\| = o(1), \tag{6}$$

then E_k accelerates **s** in norm.

Proof. All these statements are immediate consequences of the relationship

$$r_n^{(k)} \approx \frac{\langle \phi, r_n \rangle}{V_{k+1}(1, b_1, b_2, \ldots, b_k)} \times \begin{vmatrix} r_n/\langle \phi, r_n \rangle & 0 & f_1(n)/\langle \phi, f_1(n) \rangle & \cdots & f_k(n)/\langle \phi, f_k(n) \rangle \\ 1 & 1 & 1 & \cdots & 1 \\ \eta_n & 1 & b_1 & \cdots & b_k \\ \vdots & \vdots & \vdots & & \vdots \\ \eta_n \cdots \eta_{n+k-1} & 1 & b_1^k & \cdots & b_k^k \end{vmatrix}, \tag{7}$$

where V_k is a Vandermonde determinant (see Notation). Details are left to the reader with a hint: For (iv) subtract the first column of the determinant from the jth column. ∎

Example 1. Let

$$f_r(n) = x_n^r h, \qquad x_n \in \mathscr{R}^+, \quad h \in \mathscr{B}, \quad h \notin \mathscr{K}, \tag{8}$$

and let $x_{n+1}/x_n \to b$, $b \neq 1$, $b \neq 0$. Now, $\sigma_{nk} = x_{n+k+1}/x_n$. This provides a generalization to Banach spaces of the summation by extrapolation scheme of Section 3.3, but here there is no simple deltoid algorithm for the computation of $s_n^{(k)}$, only for $S_n^{(k)} = \langle \phi, s_n^{(k)} \rangle$.

Example 2. Let

$$f_r(n) = x_r^n h, \qquad x \in \mathscr{R}, \quad h \in \mathscr{B}, \quad h \notin \mathscr{K}, \tag{9}$$

and let the x_i be distinct numbers, none of which is 1. $\sigma_{nk} = x_{k+1}$ and this gives a generalization of the deltoid of Section 3.2. Although there is no deltoid in the general case, $s_n^{(k)}$ can be written out as a linear combination of the $f_i(n)$ with closedform coefficients that are Vandermonde determinants.

Both of these examples generalize the Romberg and Richardson procedures.

10.3. The Topological Schmidt Transformation

In the algorithm of Section 10.1 take $f_i(n) = \Delta s_{n+i-1}$. This gives a generalization of the Shanks–Schmidt transformation $e_k(s_n)$. The most useful case is when ϕ is constant. What results is

$$s_n^{(k)} = W_k(s_n, \langle \phi, \Delta s_n \rangle)/W_k(1, \langle \phi, \Delta s_n \rangle), \tag{1}$$

W_k as in Eq. 6.1(8).

Although the above is a nonlinear algorithm, its denominator is a scalar. Thus it requires nothing in the way of invertibility from the elements of $\mathscr{B}$. Obviously it retains the properties of homogeneity and translativity of the scalar algorithm.

The first question one asks is, How good is this abstract version of the Schmidt transformation? Insight into this very difficult question can be obtained by answering a simpler question: For what sequences is the algorithm exact? This question is easily resolved. Since, for the scalar algorithm, there is an intimate relation between exactness and regularity, one expects this relation to hold in other Banach spaces. We shall show that, roughly speaking, the topological E_k can be exact only for sequences that are linear combinations of fixed elements of $\mathscr{B}$, where all the dependence on n is restricted to the scalars in the linear combination. Unless $\mathscr{B}$ is finite dimensional, this is clearly a small class of sequences. Although more theoretical and numerical investigations are required, this probably means that E_k is regular for a disappointingly small subspace of $\mathscr{B}_C$.

In what follows Greek letters denote scalars. (Note that the arguments used carry through for any topological vector space.)

Theorem 1. Let k be fixed. Then $s_n^{(k)}$ is defined and $s_n^{(k)} \equiv s$, $n \geq 0$, iff $\langle \phi, r_n \rangle \in \mathscr{K}_k$ (see Section 6.3) and

$$s_n = s + \sum_{m=0}^{k-1} r_m \tau_n^{(m)}, \tag{2}$$

where $\tau_n^{(m)}$, $0 \leq m \leq k-1$, is a basis of solutions of the (scalar) equation $\mathscr{P}_k = 0$ satisfying

$$\tau_m^{(j)} = \delta_{mj}, \qquad 0 \leq m, j \leq k-1. \tag{3}$$

Proof

$\Rightarrow$: Let $\langle \phi, r_n \rangle = \xi_n$.

$$0 = s_n^{(k)} - s = r_n^{(k)} = W_k(r_n, \Delta\xi_n)/W_k(1, \Delta\xi_n). \tag{4}$$

Taking ϕ of both sides gives

$$0 \equiv \xi_n^{(k)} = W_k(\xi_n, \Delta\xi_n)/W_k(1, \Delta\xi_n). \tag{5}$$

By Theorem 6.3, the definition and exactness of $\xi_n^{(k)}$ imply $\xi_n \in \mathscr{K}_k$. Let $\{\tau_n^{(0)}, \tau_n^{(1)}, \ldots, \tau_n^{(k-1)}\}$ be a basis of $\mathscr{P}_k$ where $\mathscr{P}_k(\xi_n) = 0$, satisfying (2). Then

$$\xi_n = \sum_{m=0}^{k-1} \xi_m \tau_n^{(m)}. \tag{6}$$

But this means $\langle \phi, r_n - \sum r_m \tau_n^{(m)} \rangle \equiv 0$ or

$$r_n = \sum_{m=0}^{k-1} r_m \tau_n^{(m)} + v_n, \tag{7}$$

where $v_n \in \mathscr{K}$ (the kernel of ϕ). Putting $n = 0, 1, 2, \ldots, k-1$ shows $v_0, v_1, \ldots, v_{k-1} = 0$. Substituting (7) into (2) gives

$$0 = \sum_{m=0}^{k-1} r_m \frac{W_k(\tau_n^{(m)})}{W_k(1)} + \frac{W_k(v_n)}{W_k(1)}. \tag{8}$$

Now we have

$$\mathscr{P}_k(\Delta\xi_n) = \gamma_0 \Delta\xi_n + \gamma_1 \Delta\xi_{n+1} + \cdots + \gamma_k \Delta\xi_{n+k} = 0 \tag{9}$$

with, say, $\gamma_k = 1$. By Lemma 6.3(3) each term in the sum in (8) is proportional to $\mathscr{P}_k(\tau_n^{(m)})$ and hence equal to zero. Thus $W_k(v_n) = 0$, or v_n satisfies

$$\gamma_0 v_n + \gamma_1 v_{n+1} + \cdots + \gamma_k v_{n+k} = 0. \tag{10}$$

Since $v_0, \ldots, v_{k-1} = 0$, $v_n = 0$ for all n.

$\Leftarrow$: This part is trivial, since (2) shows $\mathscr{P}_k(r_n) \equiv 0$, and this is proportional to $W_k(r_n)$ $(\equiv 0)$. ■

The following, more of an observation than a theorem, is a useful negative criterion.

Theorem 2. Let $\langle \phi, r_n \rangle$ satisfy a homogeneous linear difference equation $\mathscr{P}_k = 0$ and let

$$\langle \phi, r_m \rangle = \eta_m, \qquad 0 \le m \le k-1, \tag{11}$$

where η_n is a nonmaximal solution of $\mathscr{P}_k = 0$. Then $s_n^{(k)}$ is not defined.

Example 1 $(k = 2)$. $\langle \phi, r_n \rangle \in \mathscr{K}_2$ if, for example,

$$\langle \phi, r_n \rangle = (c_1 n + c_2)\lambda^n, c_1 \neq 0, \lambda \neq 0, 1.$$

Then

$$\tau_n^{(0)} = \lambda^n(1 - n), \qquad \tau_n^{(1)} = n\lambda^{n-1}. \tag{12}$$

Thus E_2 will be exact for sequences of the form

$$s_n = s + r_0(1 - n)\lambda^n + r_1 n\lambda^{n-1}, \quad \lambda \neq 0, 1, \tag{13}$$

provided $\lambda r_0 - r_1 \notin \mathscr{K}$.

Example 2 ($k = 1$. Aitken's δ^2-process in a Banach space)

$$\bar{s}_n = s_n - (\langle \phi, \Delta s_n \rangle / \langle \phi, \Delta^2 s_n \rangle)\, \Delta s_n. \tag{14}$$

Theorem 3. Let $s \in \mathscr{B}_C$. Then $\bar{s}_n$ in (14) is defined and $\bar{s}_n \equiv s$ iff $r_0 \notin \mathscr{K}$ and $s_n = s + \lambda^n r_0$ for some $\lambda \in \mathscr{C}$, $\lambda \neq 0, 1$.

If $\mathbf{s} \in \mathscr{B}_C$, our previous work, Theorem 10.1(4), states that this algorithm converges provided

$$d(\bar{r}_n, \mathscr{K})/\|\bar{r}_n\| \geq \delta > 0 \tag{15}$$

and

$$\langle \phi, a_{n+2} \rangle / \langle \phi, a_{n+1} \rangle \notin .[\alpha, \beta], \qquad 0 < \alpha < 1 < \beta. \tag{16}$$

The acceleration properties of the algorithm are easily established.

Theorem 4. Let $\mathbf{s} \in \mathscr{B}_C$ and

$$\langle \phi, a_{n+2} \rangle / \langle \phi, a_{n+1} \rangle = \rho + o(1), \qquad 0 < |\rho| < 1. \tag{17}$$

Then $\bar{\mathbf{s}}$ converges to s more rapidly than $\mathbf{s}$ in the seminorm $|\langle \phi, \cdot \rangle|$.

If, further,

$$\|r_{n+1} - \rho r_n\| / \|r_n\| = o(1), \tag{18}$$

then $\bar{\mathbf{s}}$ converges more rapidly than $\mathbf{s}$ in norm.

Proof. By Theorem 1.4(1),

$$\langle \phi, r_{n+1} \rangle / \langle \phi, r_n \rangle = \rho + o(1), \tag{19}$$

so Theorem 10.2(1) may be applied. ■

Brezinski (1975) has studied this algorithm. It may be used to generalize the Padé table in the following way. Let $\mathbf{a} \in \mathscr{B}_S$ and $z \in \mathscr{C}$. Then we may write the formal power series

$$f(z) = \sum_{j=0}^{\infty} a_j z^j \tag{20}$$

with partial sums

$$s_n = \sum_{j=0}^{n} a_j z^j. \tag{21}$$

$s_n^{(k)}$ will then define a formal rational approximation to f whose numerator is a polynomial of degree $n + k$ in z with coefficients in $\mathscr{B}$ and whose denominator is a scalar polynomial of degree n. For details, see Brezinski (1975). Germain-Bonne (1978) has also studied this algorithm.

The topological Schmidt transformation provides a construction for iteration functions for the solution of operator equations. Let $f: \mathscr{B} \to \mathscr{B}$ and define

$$f_{k+1}(x) = f_k(f(x)), \qquad f_0(x) = x. \tag{22}$$

Take, in (1),

$$\begin{aligned} \Delta s_{n+k} &\to f_{k+1}(s_n) - f_k(s_n), && k \geq 0, \\ s_{n+k} &\to f_k(s_n), && k \geq 0. \end{aligned} \tag{23}$$

Thus the $k = 1$ case of Eq. (1) produces

$$s_{n+1} = s_n - \frac{\langle \phi, f(s_n) - s_n \rangle [f(s_n) - s_n]}{\langle \phi, f_2(s_n) - 2f(s_n) + s_n \rangle} \tag{24}$$

for the solution of $x = f(x)$. In contrast with often-used methods such as the generalized Newton iteration scheme, these formulas do not require the evaluation of the (Fréchet) derivatives of f.

There are a multitude of other ways the BH protocol can be used to construct iteration functions. One could take $\boldsymbol{\phi}$ const,

$$\begin{aligned} f_j(n + k) &= [f_k(s_n) - f_{k-1}(s_n)] \langle \phi, f_k(s_n) - f_{k-1}(s_n) \rangle^{j-1}, j \geq 1, k \geq 0, \\ s_{n+k} &\to f_k(s_n), \qquad k \geq 0, \end{aligned} \tag{25}$$

and replace $s_n^{(k)}$ by s_{n+1} on the left-hand side of Eq. 10.1(2).

10.4. The Scalar Case

When $\mathscr{B}$ is its scalar field and $\phi_n \equiv I$ (the identity), $S_n^{(k)} = s_n^{(k)}$ and there is a deltoid computational scheme for the computation of $S_n^{(k)}$. Then

$$s_n^{(k)} = \frac{\begin{vmatrix} s_n & f_1(n) & \cdots & f_k(n) \\ \vdots & \vdots & & \vdots \\ s_{n+k} & f_1(n+k) & \cdots & f_k(n+k) \end{vmatrix}}{\begin{vmatrix} 1 & f_1(n) & \cdots & f_k(n) \\ \vdots & \vdots & & \vdots \\ 1 & f_1(n+k) & \cdots & f_k(n+k) \end{vmatrix}} = E_k(s_n) \tag{1}$$

and the algorithm becomes

$$s_n^{(k+1)} = \frac{f_{k+1}^{(k)}(n)s_{n+1}^{(k)} - f_{k+1}^{(k)}(n+1)s_n^{(k)}}{f_{k+1}^{(k)}(n) - f_{k+1}^{(k)}(n+1)}, \quad n, k \geq 0; \quad s_n^{(0)} = s_n, \quad n \geq 0; \tag{2}$$

$$f_i^{(k+1)}(n) = \frac{f_{k+1}^{(k)}(n)f_i^{(k)}(n+1) - f_{k+1}^{(k)}(n+1)f_i^{(k)}(n)}{f_{k+1}^{(k)}(n) - f_{k+1}^{(k)}(n+1)},$$

$$i \geq 1, \quad 0 \leq k \leq i-2; \qquad f_i^{(0)}(n) = f_i(n), \qquad i \geq 1, \tag{3}$$

and $f_i^{(k)}(n) = E_k(f_i(n))$, i.e., the ratio (1) with s_n replaced by $f_i(n)$.

The computational tableau of the algorithm is as follows. The $s_n^{(k)}$ array is filled out in diagonal lines, the kth line being $\{s_{k-1}^{(0)}, s_{k-2}^{(1)}; s_{k-3}^{(2)}, \ldots, s_0^{(k-1)}\}$:

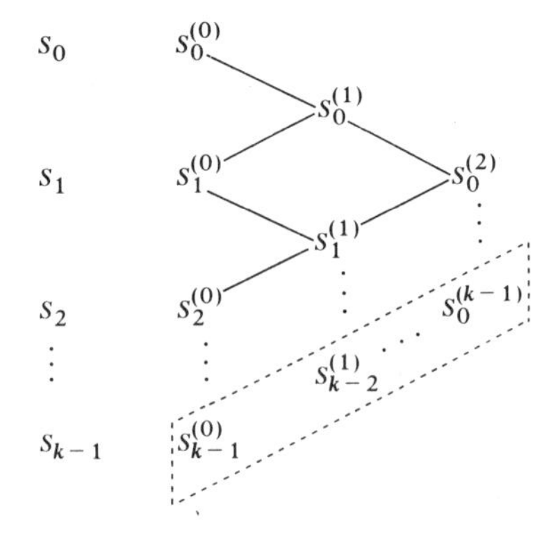

To compute this kth diagonal line, $k-1$ subsidiary arrays are needed. Each array has the following form:

*i*th Array

$$\begin{array}{lcl}
f_i^{(0)}(0) = f_i(0) & & \\
f_i^{(0)}(1) = f_i(1) & \ddots & \\
 & \ddots & f_i^{(i-1)}(0) \\
 & & f_i^{(i-1)}(1) \\
\vdots & & \vdots \\
 & & f_i^{(i-1)}(k-i-1) \\
f_i(k-2) & \cdot^{\cdot^{\cdot}} & f_i^{(i-1)}(k-i) \\
f_i(k-1) & \cdot^{\cdot^{\cdot}} &
\end{array}$$

For instance, to compute the diagonal $\{s_3^{(0)}, s_2^{(1)}, s_1^{(2)}, s_0^{(3)}\}$ the following three arrays are needed:

Array 1	Array 2		Array 3		
$f_1(0)$	$f_2(0)$		$f_3(0)$		
		$f_2^{(1)}(0)$		$f_3^{(1)}(0)$	
$f_1(1)$	$f_2(1)$		$f_3(1)$		$f_3^{(2)}(0)$
		$f_2^{(1)}(1)$		$f_3^{(1)}(1)$	
$f_1(2)$	$f_2(2)$		$f_3(2)$		$f_3^{(2)}(1)$
		$f_2^{(1)}(2)$		$f_3^{(1)}(2)$	
$f_1(3)$	$f_2(3)$		$f_3(3)$		

Clearly, the amount of computer storage necessary at the completion of the computation of the kth diagonal is $k(k + 1)(2k + 1)/6 \approx k^3/3$.

The computations are probably best done in the following order:

(a) Initialize $s_0^{(0)} = s_0, f_1^{(0)}(0) = f_1(0), f_1^{(0)}(1) = f_1(1)$.

(b) Assume the $(k - 1)$th $s_n^{(k)}$ diagonal has been filled out, $k \geq 2$.

(c) For $k > 2$ compute new ascending diagonals of each of the $k - 2$ arrays; i.e., DO, for $1 \leq i \leq k - 2$,

$$\text{generate} \quad f_i^{(0)}(k - 1) = f_i(k - 1);$$
$$\text{compute} \quad f_i^{(j)}(k - 1 - j) \quad \text{from (3)}, \qquad 1 \leq j \leq i - 1.$$

(d) For $k > 2$ fill out array $k - 1$, i.e., generate $f_{k-1}^{(0)}$ and DO, for $1 \leq i \leq k - 1$:

$$\text{generate} \quad f_{k-1}^{(0)}(i) = f_{k-1}(i);$$
$$\text{compute} \quad f_{k-1}^{(j)}(i - j) \quad \text{from (3)}, \qquad j = 1, 2, \ldots, i \quad (i - 1 \text{ if } i = k - 1).$$

(e) For $k \geq 2$ fill out the kth diagonal of $s_n^{(k)}$; i.e., generate s_{k-1} and compute $s_{k-1-i}^{(i)}$, $1 \leq i \leq k - 1$, from (2).

(f) Go back to (a).

Note that moving down one diagonal in the $s_n^{(k)}$ table necessitates adding one more subsidiary array.

The Brezinski–Håvie protocol for scalar sequences is undoubtedly the most elegant and flexible computational procedure yet discovered for the transformation of sequences.

The flexibility of the algorithm lies in the possible choices of $\mathbf{f}_r$. Generally speaking, the choices are made with a foreknowledge of the kinds of sequences one wishes to accelerate.

Before discussing particular cases of the scalar algorithm, let us gather together the convergence and acceleration results.

Theorem 1. Let, for each fixed k, σ_{nk} [Eq. 10.1(19)] be bounded away from 1. Then E_k is regular on all vertical paths.

Theorem 2. Let

$$s_n = s + \sum_{r=1}^{\infty} a_r f_r(n), \tag{4}$$

where for each n the series on the right converges absolutely. Then

$$s_n^{(k)} = s + \sum_{r=k+1}^{\infty} a_r f_r^{(k)}(n). \tag{5}$$

Proof. Follows from Theorem 10.1(1). This shows E_k is exact for constant sequences and $\mathrm{Lin}\{\mathbf{f}_1, \mathbf{f}_2, \ldots, \mathbf{f}_k\}$ when $\mathbf{f}_j$ is independent of $\mathbf{s}$. ■

For the next three theorems, the common hypothesis is

$$\lim_{n\to\infty} f_r(n+1)/f_r(n) = b_r \neq 1, \qquad 1 \leq r \leq k, \quad b_i \neq b_j, \quad i \neq j. \tag{6}$$

Recall $h_n = r_{n+1}/r_n = (s_{n+1} - s)/(s_n - s)$.

Theorem 3. Let $\mathbf{s} \in \mathscr{C}_C$.

(i) If $h_n = O(1)$, then E_k is regular along vertical paths.
(ii) If $h_n \to b_j$ for some j, then E_k is accelerative along vertical paths.

Our last two results require the following.

Lemma

$$f_j^{(k)}(n) \approx f_j(n)\frac{(b_j - b_1)\cdots(b_j - b_k)}{(1 - b_1)\cdots(1 - b_k)}, \qquad j > k \geq 1. \tag{7}$$

Proof. Left to the reader. ■

Theorem 4. Let

$$\lim_{n\to\infty} r_{n+1}^{(k)}/r_n^{(k)} = b_{k+1}, \qquad k \geq 0. \tag{8}$$

Then

$$\lim_{n\to\infty} r_n^{(k+1)}/r_{n+1}^{(k)} = 0, \qquad k \geq 0. \tag{9}$$

Proof. Left to the reader. ■

Theorem 5. For some λ let $|a_j| < \lambda^j$ and let (6) and

$$f_{r+1}(n) = o(f_r(n)) \tag{10}$$

hold uniformly in r. Let the representation (4) hold.

Then

$$\lim_{n\to\infty} r_n^{(k+1)}/r_{n+1}^{(k)} = 0, \qquad k \geq 0. \tag{11}$$

Proof. Requirement (10) guarantees the absolute and uniform convergence of (4) for $n > n_0$. It is easily seen that $r_n^{(k)} \approx a_{k+1} f_n^{(k+1)}$, and the previous theorem may be invoked. ■

Example 1

$$f_1(n) = \lambda^n, \quad s_n = \lambda^n/n, \qquad \lambda \in [-1, 1). \tag{12}$$

Then

$$s_n^{(1)} = \lambda^{n+1}/(\lambda - 1)n(n+1), \qquad s_n^{(1)}/s_{n+1} = 1/(\lambda - 1)n = o(1). \tag{13}$$

The conditions of Theorems 1, 3, and 4 are satisfied but not those of Theorems 2 and 5 since there is no $a_1 \neq 0$ such that $\lambda^n/n = a_1\lambda^n + \cdots$.

Example 2. If $f_k(n) = \Delta s_{n+k-1}$, the result is the Schmidt transformation but, interestingly, the algorithm for computing the transformation is *not* the ε-algorithm. Work remains to be done in assessing the relative computational advantages of the two algorithms.

10.5. The Levin Transformations

In 1973, Levin gave a general transformation of series that is enormously useful in numerical analysis and that has been the subject of a wide literature. Essentially, the transformation is a special case of the previous transformation, although Levin did not develop an algorithm for computing $s_n^{(k)}$ efficiently. There are a number of useful cases of his algorithm, and I will examine each of these in turn. After making the assumption 10.4(4) Levin effected the specialization

$$f_j(n) = x_n^{j-1}\zeta_n, \qquad x_m \neq x_n, \qquad m \neq n. \tag{1}$$

Expanding by minors, one finds

$$\begin{aligned} s_n^{(k)} &= \sum_{m=0}^{k} \frac{s_{n+m}}{\zeta_{n+m}} \pi_n^{(k,m)} \Bigg/ \sum_{m=0}^{k} \frac{1}{\zeta_{n+m}} \pi_n^{(k,m)}, \\ \pi_n^{(k,m)} &= \sum_{\substack{r=0 \\ r\neq m}}^{k} (x_{n+r} - x_{n+m})^{-1}, \qquad s_n^{(0)} = s_n, \end{aligned} \tag{2}$$

where it is assumed, of course, that all quantities are defined. Furthermore,

$$f_j^{(k)}(n) = \sum_{m=0}^{k} x_{n+m}^{j-1} \pi_n^{(k,m)} \bigg/ \sum_{m=0}^{k} \frac{1}{\zeta_{n+m}} \pi_n^{(k,m)}. \tag{3}$$

There are a number of ways to choose the x_n and ζ_n that make sense. One is as follows. If the sequence **s** converges rapidly enough and, say, the terms $|a_j|$ are ultimately monotone in such a way that we may write

$$s_n - s \approx -a_{n+1}, \tag{4}$$

then a good choice would seem to be $\zeta_n = a_{n+1}$. (h_n can be multiplied, of course, by any constant without affecting the algorithm.) For this choice of h_n, we wish to analyze the acceleration properties of the algorithm using the theorems of Chapter 5.

Let, as usual,

$$\rho_n = a_{n+1}/a_n. \tag{5}$$

Note that $s_n^{(k)}$ depends on $s_n, s_{n+1}, \ldots, s_{n+k+1}$ and is translative and homogeneous. Thus we need to consider the functions

$$(s_n^{(k)} - s_n)/a_{n+1} = g_n(\rho_{n+1}, \ldots, \rho_{n+k}). \tag{6}$$

An application of the Smith–Ford (1979) theorem gives

Theorem. In (2) let $\zeta_n = a_{n+1}$, $a_n \doteq\!\!\!\!/\; 0$. Let $\mathbf{s} \in \mathscr{A}_\rho$, $0 < |\rho| < 1$, and for some u_n let

$$\lim_{n\to\infty} \pi_n^{(k,m)}/u_n = \pi^{(k,m)}, \tag{7}$$

where

$$\sum_{m=0}^{k} \rho^{-m} \pi^{(k,m)} \neq 0. \tag{8}$$

Then $s_n^{(k)}$ accelerates the convergence of **s** along any vertical path.

Proof.

$$\begin{aligned} \frac{s_n^{(k)} - s_n}{a_{n+1}} &\doteq \sum_{m=1}^{k} \frac{(s_{n+m} - s_n)}{a_{m+n+1}} \pi_n^{(k,m)} \bigg/ \sum_{m=0}^{k} \frac{1}{a_{m+n+1}} \pi_n^{(k,m)} \\ &\doteq \sum_{m=1}^{k} \frac{(1 + \rho_{n+1} + \rho_{n+1}\rho_{n+2} + \cdots + \rho_{n+1} \cdots \rho_{n+m-1})\pi_n^{(k,m)}}{\rho_{n+1}\rho_{n+2} \cdots \rho_{n+m}} \\ &\quad \times \left(\sum_{m=0}^{k} \frac{1}{\rho_{n+1}\rho_{n+2} \cdots \rho_{n+m}} \pi_n^{(k,m)} \right)^{-1}. \end{aligned} \tag{9}$$

Thus

$$g_n(x_1, x_2, \ldots, x_k) = \sum_{m=1}^{k} \frac{(1 + x_1 + x_1x_2 + \cdots + x_1 \cdots x_{m-1})\pi_n^{(k,m)}}{x_1x_2 \cdots x_m} \times \left(\sum_{m=0}^{k} \frac{1}{x_1x_2 \cdots x_m} \pi_n^{(k,m)} \right)^{-1} \tag{10}$$

and

$$g_n(\rho e) \doteq \sum_{m=1}^{k} \frac{-1 + \rho^{-m}}{1-\rho} \pi_n^{(k,m)} \Bigg/ \sum_{m=0}^{k} \rho^{-m}\pi_n^{(k,m)}$$
$$\doteq \frac{1}{1-\rho} - \sum_{m=0}^{k} \pi_n^{(k,m)} \Bigg/ (1-\rho) \sum_{m=0}^{k} \rho^{-m}\pi_n^{(k,m)}. \tag{11}$$

By elementary properties of interpolation sums, $\sum \pi_n^{(k,m)} \equiv 0$. Thus

$$g_n(\rho e) \doteq 1/(1-\rho). \tag{12}$$

By uniform convergence,

$$g(\rho e) = 1/(1-\rho), \tag{13}$$

and this concludes the proof. ■

In this proof it was assumed $\mathbf{x}$ is independent of $\mathbf{s}$. However, $\mathbf{x}$ *may* depend on $\mathbf{s}$ if the dependency is such that the homogeneity and translativity of the transformation is maintained, e.g., $x_n = \Delta s_n$.

10.5.1. The t-Transform

In his analysis, Levin took $\zeta_n = a_n$, but, as Smith and Ford point out, $\zeta_n = a_{n+1}$ makes better sense and simplifies the convergence analysis. One natural choice of the x_n would result by assuming that $\mathbf{s}$ converges as

$$s = s_n + a_{n+1}v_n, \tag{1}$$

where v_n is a Poincaré asymptotic series, in other words, by taking

$$x_n = 1/(n+1). \tag{2}$$

In what follows, it is assumed that $a_n \not\doteq 0$. This choice and dividing numerator and denominator by common factors amounts to the choices

$$\zeta_n = a_{n+1} \qquad \text{and} \qquad \pi_n^{(k,m)} = (n+m+1)^{k-1}(-1)^m \binom{k}{m}$$

in Eq. 10.5(2). Then

$$t_k(s_n) \equiv s_n^{(k)} = \frac{\sum_{m=0}^{k} (s_{n+m}/a_{n+m+1})(n+m+1)^{k-1}(-1)^m\binom{k}{m}}{\sum_{m=0}^{k} (1/a_{n+m+1})(n+m+1)^{k-1}(-1)^m\binom{k}{m}}. \tag{3}$$

This is called the *Levin t-transform.*

Theorem 1. t_k, $k \geq 1$, is accelerative for $\mathscr{C}_l$ along any vertical path.

Proof. In Theorem 10.5 take $u_n = n^{k-1}$ and then $\pi^{(k,m)} = (-1)^m\binom{k}{m}$. We have

$$\sum_{m=0}^{k} \frac{1}{\rho^m} \pi^{(k,m)} = \left(1 - \frac{1}{\rho}\right)^k \neq 0. \quad \blacksquare \tag{4}$$

t_k turns out to be regular for any path for another large and important class of sequences.

Theorem 2. If $\mathbf{s} \in \mathscr{C}_C$ and $\mathbf{a}$ is alternating, $s_n^{(k)}$ is defined and converges to s along any path.

Proof. We can write

$$r_n^{(k)} = \sum_{m=0}^{k} r_{n+m}\mu_{mk},$$

$$\mu_{mk} \equiv \mu_{mk}(n) = \frac{(n+m+1)^{k-1}}{|a_{n+m+1}|}\binom{k}{m} \Bigg/ \sum_{r=0}^{k} \frac{(n+r+1)^{k-1}}{|a_{n+r+1}|}\binom{k}{r}. \tag{5}$$

First, assume $n \to \infty$ on $\mathbf{P}$. In this case, use

$$|r_n^{(k)}| \leq \sup_{m \geq n} |r_m| \sum_{m=0}^{k} \mu_{mk} = \sup_{m \geq n} |r_m| \to 0. \tag{6}$$

If n is bounded on $\mathbf{P}$, use Theorem 5.2(1). Conditions (i) and (ii) of that theorem are satisfied. We can majorize μ_{km} by throwing away all the terms in its denominator except the last, so

$$\begin{aligned} \mu_{km} &\leq \frac{|a_{n+k+1}|}{|a_{n+m+1}|}\left(\frac{n+m+1}{n+k+1}\right)^{k-1} \frac{k!}{m!(k-m)!} \\ &\approx \frac{|a_{n+k+1}|}{|a_{n+m+1}|} \frac{(n+m+1)^{k-1}}{m!\,e^{n+1}} k^{m-k+1}, \qquad k \to \infty, \quad n \text{ bounded}, \end{aligned} \tag{7}$$

so $\mu_{km} = o(1)$ in k along $\mathbf{P}$ and this establishes convergence along $\mathbf{P}$. $\blacksquare$

t does not work well on monotone sequences.

Theorem 3. Let $\operatorname{Re}\theta < -1$, $c_0 \neq 0$, and

$$a_n \sim n^\theta(c_0 + c_1/n + c_2/n^2 + \cdots). \tag{8}$$

Then

$$r_n^{(k)} \approx -k!n^{\theta+1}c_0/(\theta+1)(-\theta)_k, \qquad k \geq 0. \tag{9}$$

Proof. Exercise. ■

10.5.2. The u-Transform

The t-transform was designed to be used on rapidly convergent alternating series. The u-transform is designed for monotonic series and has the following heuristic basis. Consider

$$s_n = \sum_{k=0}^{n} \frac{1}{(k+1)^\alpha}, \qquad \alpha > 1. \tag{1}$$

Then, according to the work in Chapter 1 [Theorem 1.7(3)],

$$s_n - s \approx -(n+1)^{1-\alpha}/(1-\alpha), \tag{2}$$

or $s_n - s \approx C(n+1)a_{n+1}$. Since the above sequence is such a typical one, it makes sense to take

$$\zeta_n = (n+1)a_{n+1} \tag{3}$$

and, as before, $x_n = (n+1)^{-1}$. (This is not the precise choice Levin made—$\zeta_n = na_n$, $x_n = n^{-1}$—but seems preferable since the transform is now defined for all n.) The result is called the *Levin u-transform*:

$$u_k(s_n) = s_n^{(k)} = \frac{\sum_{m=0}^{k} (s_{n+m}/a_{n+m+1})(-1)^m(n+m+1)^{k-2}\binom{k}{m}}{\sum_{m=0}^{k} (1/a_{n+m+1})(-1)^m(n+m+1)^{k-2}\binom{k}{m}}. \tag{4}$$

Theorem 10.5 gives immediately the next result.

Theorem 1. The u_k transform, $k \geq 1$, is accelerative for $\mathscr{C}_l$ along any vertical path.

It is ironic that, despite its derivation, it has not been established that u_k is regular for monotone series. (In fact, I suspect this is not true.) Nevertheless, a result in the previous section continues to hold.

Theorem 2. If $\mathbf{s} \in \mathscr{C}_C$ and $\mathbf{a}$ is alternating, $s_n^{(k)}$ is defined and converges to s along any path.

For certain kinds of monotone series, however, the columns of the u-transform give excellent results. These are series whose general term has a Poincaré type of asymptotic expansion.

Theorem 3. Let $\operatorname{Re}\theta < -1$, $c_0 \neq 0$, and

$$a_n \sim n^{\theta}(c_0 + c_1/n + c_2/n^2 + \cdots). \tag{5}$$

Then for some $m \geq k$,

$$r_n^{(k)} \approx Cn^{\theta+1-m}. \tag{6}$$

Proof. Left to the reader. ■

If $\mathbf{s}$ behaves as

$$s_n \sim s + \lambda^n n^{\theta} \sum_{r=0}^{\infty} \frac{c_r}{n^r}, \qquad 0 < |\lambda| < 1, \tag{7}$$

then it is easily shown that

$$r_n^{(k)} \approx C\lambda^n n^{\theta-2k}, \qquad n \to \infty, \tag{8}$$

for both t_k and u_k. Thus the Levin transformations enhance exponential convergence algebraically.

Levin defined another transform, called the v-transform, by taking $e_1(s_n)$, the Aitken δ^2-iterate, as an estimate for s; i.e., in Eq. 10.4(1), $f_j(n) = [a_{n+1}/(\rho_n - 1)]x_n^{j-1}$, $x_n = (n+1)^{-1}$. Thus the u-transform is defined by

$$s_n^{(k)} = \frac{\sum_{m=0}^{k} (s_{n+m}(\rho_{n+m} - 1)/a_{n+m+1})(n+m+1)^{k-1}(-1)^m\binom{k}{m}}{\sum_{m=0}^{k} ((\rho_{n+m} - 1)/a_{n+m+1})(n+m+1)^{k-1}(-1)^m\binom{k}{m}}, \tag{9}$$

$$\rho_n = a_{n+1}/a_n.$$

Obviously this idea can be elaborated *ad absurdum*, since *any* sequence transformation can be used for ζ_n. Smith and Ford, however, think that v has special advantages, and consider it one of the best practical transformations.

10.5.3. Exactness Theorems for t and u

t and u turn out to be exact for a surprisingly large and varied class of sequences. To explore this matter, we first demonstrate an exactness result for a general case of the scalar algorithm 10.4(1).

Theorem 1. Let $k \geq 1$ and $f_j(n) = a_{n+1}g_j(n)$ where the $\mathbf{g}_j$ are linearly independent and independent of $\mathbf{s}$. Let r_n, $a_n \neq 0$ and let the denominator of Eq. 10.4(1) vanish for no value of n. Then $s_n^{(k)} \equiv s_n$, $n \geq 0$, iff

$$s_n = s + (s_0 - s) \prod_{r=1}^{n-1} (1 + \tau_r^{-1}), \qquad n \geq 0, \tag{1}$$

where $\boldsymbol{\tau}$ is a nontrivial member of $\text{Lin}[\mathbf{g}_1, \ldots, \mathbf{g}_k]$.

Furthermore if $g_1(n) \equiv 1$ and $g_j(n)$ is an asymptotic scale, the transform is exact for $\mathbf{s} \in \mathscr{C}_C$ only if

$$s_n = s + \lambda^n g_j(0) g_j(1) \cdots g_j(n-1) e^{n\varepsilon_n} \tag{2}$$

for some j, $1 \leq j \leq k$, and some $\boldsymbol{\varepsilon} \in \mathscr{C}_N$.

Proof

$\Leftarrow$:

$$s_n^{(k)} - s = \begin{vmatrix} r_n & f_1(n) & \cdots & f_k(n) \\ \vdots & \vdots & & \vdots \\ r_{n+k} & f_1(n+k) & \cdots & f_k(n+k) \end{vmatrix} \times \begin{vmatrix} 1 & f_1(n) & \cdots & f_k(n) \\ \vdots & \vdots & & \vdots \\ 1 & f_1(n+k) & \cdots & f_k(n+k) \end{vmatrix}^{-1} \tag{3}$$

Let

$$v_n = \prod_{r=0}^{n-1} (1 + \tau_r^{-1}). \tag{4}$$

Differencing (1) shows

$$a_{n+1} = (s_0 - s) v_n / \tau_n, \tag{5}$$

and if a_n is defined and nonzero, then $\tau_n \neq 0$, $s_0 \neq s$. Thus $r_n / a_{n+1} = \tau_n$ or $r_n = a_{n+1} \tau_n$ and $s_n^{(k)} - s = 0$, $n \geq 0$.

$\Rightarrow$: We must have

$$c_1 r_n + a_{n+1}[c_2 g_1(n) + \cdots + c_{k+1} g_k(n)] \equiv 0, \tag{6}$$

and by linear independence of the $\mathbf{g}_j$, $c_1 \neq 0$. This can be written

$$r_n - a_{n+1} \tau_n \equiv 0 \tag{7}$$

for some $\tau_n \in \text{Lin}[\mathbf{g}_1, \ldots, \mathbf{g}_k]$, or

$$r_{n+1} / r_n = (1 + \tau_n^{-1}), \tag{8}$$

and taking products gives (1).

To prove the second part of the theorem, write

$$1 + 1/\tau_n = 1 + [c_1 + c_2 g_2(n) + \cdots + c_k g_k(n)]^{-1}$$
$$= (1 + 1/c_1)(1 + \delta_n), \qquad \boldsymbol{\delta} \in \mathscr{C}_N, \quad c_1 \neq -1. \tag{9}$$

Note that $c_1 \neq 0$, since otherwise $\mathbf{s}$ is not convergent. Substituting (9) in (8), taking products, and using Theorem 1.4(2) gives (2) (with $j = 1$). The case $c_1 = -1$ is handled similarly. ■

Theorem 2 (Exactness for Euler Series). Let

$$s_n = s + \sum_{k=0}^{n} k^\alpha x^k, \qquad \alpha \in J^0, \quad x \neq 0. \tag{10}$$

Then for t_k, $s_n^{(k)} \equiv s$ for $x \neq 1$, $k \geq \alpha + 1$. For u_k, $s_n^{(k)} \equiv s$ for $k \geq \alpha + 2$.

Proof. We shall prove the first statement. According to (7), t is exact iff

$$\sum_{k=0}^{n} k^\alpha x^k = (n+1)^\alpha x^{n+1}\left[c_0 + \frac{c_1}{(n+1)} + \cdots + \frac{c_{k-1}}{(n+1)^{k-1}}\right] \tag{11}$$

for some constants $c_0, c_1, \ldots, c_{k-1}$. By the work in Section 1.7 this is certainly possible provided $x \neq 1$ and $k - 1 \geq \alpha$. (Notice that in this case the asymptotic expansion for r_n terminates.) ■

Theorem 3. If t_k is exact for some sequence $\mathbf{s}$, then u_m is exact for $\mathbf{s}$ when $m > k$.

If u_k is exact, u_m is exact, $m \geq k$. If t_k is exact, t_m is exact, $m \geq k$.

Proof. Trivial. ■

Theorem 4. Let $s_n = s + (s_0 - s)p_n$. u_1 is exact for $p_n = (a+1)_n/n!$; u_2 is exact for $p_n = (a+1)_n/(b+1)_n$ or c^n; u_3 is exact for the previous sequences and $p_n = c^{-n}(c+1)_n$, or $(b/a)^n(a+1)_n/(b+1)_n$.

t_1 is exact for c^n; t_2 is exact for $p_n = c^{-n}(c+1)_n$, c^n, or $(b/a)^n(a+1)_n/(b+1)_n$.

Proof. Left to the reader. (Assume that the parameters are such that all quantities are defined and denominators of t or u are never zero.) ■

10.5.4. Numerics

Diagonal modes of convergence seem to always be preferable with the Levin transforms. Table I shows the effect of t and u on typical sequences. t sums alternating series well but not monotone series; u sums both. Further, *both* sum the divergent $1 - 1! + 2! - 3! + \cdots$. In almost every case, the

Table I

$s_0^{(k)}$: Levin Transforms

	$s_n = (\text{GAM})_n$		$s_n = (\text{LN } 2)_n$	
k	t	u	t	u
2	0.645	0.577621	0.692	0.694
4	0.608	0.577268	0.693144	0.693161
6	0.594	0.577216	0.693147186	0.693147203
8	0.588	0.577215661	0.693147180584	0.693147180437
10	0.585	0.577215664926	0.6931471805598	0.693147180559951
	$\gamma = 0.577215664901533$		$\ln 2 = 0.693147180559945$	

$s_n = (\text{FAC})_n$ (divergent)		$s_n = (\text{PI}^2)_n$		$s_n = (\text{IT } 1)_n$	
t	u	t	u	t	u
0.571	0.615	1.521	1.639	1.369456	1.369734
0.595	0.598	1.586	1.644676	1.368860	1.368882
0.596399	0.596368	1.611	1.644931	1.368812	1.368813
0.596346	0.596341	1.624	1.644934081	1.368808438	1.368808559
0.596347283	0.596347823	1.630	1.644934067	1.368808134	1.368808143
$\int_0^\infty e^{-t}\,dt/(1+t) = 0.596347361$		$\pi^2/6 = 1.644934066846$		The root of $x^3 + 2x^2 + 10x - 20 = 0$ is 1.368808107	

performance of u is spectacular, and it is clear why Smith and Ford call it one of the three best all-round practical summation methods (along with the θ-algorithm and Levin v-transform). However, no method can do everything; apparently the t and u methods perform less satisfactorily on iteration sequences, convergent or divergent. True they both sum $(\text{IT } 1)_n$ well diagonally, but s_n itself is rapidly convergent also. For $(\text{IT } 2)_n$ both methods fail. For t and u, respectively,

$$s_0^{(15)} = 1.2596, \qquad s_0^{(15)} = 1.2606$$

and the data seem to indicate $s_0^{(k)}$ converges in either case, although it is not clear to what. (s_n has two limit points, 0.549 and 1.293.) The GBW algorithm is the only one I know of that will sum this kind of sequence.

The example of 7.1(4), $(\text{LUB})_n$, can be used to show neither t nor u is regular for any path $\mathbf{P} = (n, k)$, $k > 0$. For t the sequence $s_0^{(k)}$ is (1, 1.3 (exact), 3 (exact), 0.955, 0.876, 1.45, 1.316, ...) and for u it is (1, 1.214, 1.583, 0.870, 0.904, 1.917, ...).

10.6. Special Computational Procedures: The Trench Algorithm

Two of the drawbacks to the BH protocol are computing time and storage space. However, if a certain relationship prevails among the $f_j(n)$, a very efficient algorithm due to Trench (1965) for the inversion of finite Hankel matrices can be used to compute $s_n^{(k)}$. It is surprising that when $f_j(n) = \Delta s_{n+j-1}$, yielding the iterates of the Schmidt transformation, the result is even more efficient than the ε-algorithm for the computation of $s_n^{(k)}$, requiring only one-third as many operations.

Recall that $s_n^{(k)}$ results from solving the system

$$s_m = s + \sum_{r=1}^{k} c_r f_r(m), \qquad n \le m \le n + k. \tag{1}$$

Differencing gives

$$\Delta s_m = \sum_{r=1}^{k} c_r \, \Delta f_r(m), \qquad n \le m \le n + k - 1. \tag{2}$$

Now assume f_r has the property

$$f_r(n) = g(n + r - 1). \tag{3}$$

The system may be written

$$\Delta s_{m+n} = \sum_{r=0}^{k-1} c_{r+1} D_{r+m}, \quad 0 \le m \le k - 1, \qquad D_j \equiv \Delta g(n + j). \tag{4}$$

Define

$$H_k = [D_{i+j}], \qquad 0 \le i, j \le k. \tag{5}$$

The algorithm for the inversion of the H_k proceeds as follows. Let

$$H_k^{-1} = [b_{ij}^{(k)}], \qquad 0 \le i, j \le k, \tag{6}$$

and assume H_m^{-1} is known, $0 \le m \le k$ for some *fixed* n. (The algorithm generates a *diagonal* of $s_n^{(k)}$.)

Initialize as follows:

$$\begin{gathered} \gamma_{-1} = 0, \qquad \lambda_{-1} = 1; \\ u_{i,-2} = 0, \quad u_{0,-1} = 1, \quad u_{i,-1} = 0, \qquad i \ne 0; \\ u_{-1,i} = u_{i+1,i-1} = 0, \qquad u_{i+1,i} = 1. \end{gathered} \tag{7}$$

Then compute

$$\begin{gathered} \lambda_k = \sum_{j=0}^{k} D_{j+k} u_{j,k-1} \qquad \gamma_k = \sum_{j=0}^{k} D_{j+k+1} u_{j,k-1}, \\ u_{r,k} = (\lambda_{k-1}^{-1}\gamma_{k-1} - \lambda_k^{-1}\gamma_k) u_{r,k-1} + u_{r-1,k-1} - \lambda_{k-1}^{-1}\lambda_k u_{r,k-2}, \\ 0 \le r \le k. \end{gathered} \tag{8}$$

Next compute

$$\lambda_{k+1} = \sum_{j=0}^{k+1} D_{j+k+1} u_{jk}, \tag{9}$$

and finally

$$b_{ij}^{(k+1)} = \begin{cases} b_{ij}^{(k)} + \lambda_{k+1} u_{ik} u_{jk}, & 0 \le i \le j \le k+1 \\ b_{ji}^{(k+1)}, & 0 \le j < i \le k+1. \end{cases} \tag{10}$$

$s_n^{(k)}$ may be computed from (1):

$$s_n^{(k)} = s_n - [f_1(n), f_2(n), \ldots, f_k(n)] H_{k-1}^{-1} \begin{bmatrix} \Delta s_n \\ \Delta s_{n+1} \\ \vdots \\ \Delta s_{n+k-1} \end{bmatrix}. \tag{11}$$

Brezinski (1976) has shown that for the special case of the ε-algorithm, the computation in (11) can be bypassed and the algorithm becomes more compact.

An important fact is that

$$c_j = j\text{th component of } \{H_{k-1}^{-1}(\Delta s_n, \ldots, \Delta s_{n+k-1})^{\mathrm{T}}\}, \tag{12}$$

and thus if it is known that s_n has a complete asymptotic expansion of the kind

$$s_n \sim s + \sum_{r=1}^{\infty} c_r f_r(n),$$

then Eq. (12) gives a lozenge algorithm for the computation of each c_j, $j \ge 1$; i.e., just label the left-hand side of (12) $c_{j,n}^{(k)}$.

Chapter 11 The Brezinski–Håvie Protocol and Numerical Quadrature

11.1. Introduction; The G-Transform

The sequence transformations discussed in the previous chapters can be put to obvious use to compute approximate values of $s = \int_a^b g(x)\,dx$. If any sequence **s** of approximants to s has been obtained (for instance, by the application of some standard numerical quadrature rule to progressively finer subdivisions of $[a, b]$), then any acceleration method can be applied to **s**. We shall not belabor this obvious approach. Brezinski (1978) discusses this approach exhaustively and gives many numerical examples.

One problem in such an ad hoc approach is that there is usually no clear way of finding those functions for which the method yields exact answers and thus of characterizing the class of functions for which the method should be expected to provide good answers.

However, there is another more intuitive way of proceeding. The underlying philosophy of this method, which is applied to infinite integrals ($b = \infty$), was first set forth in a paper by Gray and Atchison (1967) and developed in subsequent papers by Gray, Atchison, and Clark. Their algorithm came to be known as the G-transformation, and is a special case—vertical convergence in the second column of the $s_n^{(k)}$ array—of the algorithm to be developed in this section. The latter method is one to which the BH protocol can easily be applied.

The derivation will be informal; convergence theorems will come later. Suppose we have a way of computing approximate values of

$$G(t) = \int_a^t g(x)\,dx, \qquad t \geq a, \tag{1}$$

for a sequence of values of t. Let ρ be a fixed number >0 and write

$$G_j = G(t + j\rho), \quad j \geq 0, \qquad G_0 = G(t),$$
$$g_j = g(t + j\rho), \tag{2}$$
$$I = \int_a^\infty g(x)\,dx = G_\infty.$$

We have

$$G(t) = I - \int_0^\infty g(x + t)\,dx, \qquad t \geq a. \tag{3}$$

Now assume that a quadrature formula with equally spaced nodes is available for the evaluation of the above integral; in fact, for the purposes of the derivation, we assume the integral can be represented *exactly* for all t by such a formula, so that

$$I - G_0 = \sum_{r=0}^{k-1} c_r g_r. \tag{4}$$

Replacing t by $t + m\rho$, $0 \leq m \leq k$, yields a system of $k + 1$ equations in the k unknowns $c_0, c_1, \ldots, c_{k-1}$. For consistency, the augmented determinant of the system must vanish and that determinant can be solved for I. For general integrands, of course, the result will no longer be exact, but we can use it to obtain an approximation $I_t^{(k)}$ to I that looks like

$$I_t^{(k)} = \begin{vmatrix} G_0 & g_0 & \cdots & g_{k-1} \\ \vdots & \vdots & & \vdots \\ G_k & g_k & \cdots & g_{2k-1} \end{vmatrix} \Bigg/ \begin{vmatrix} 1 & g_0 & \cdots & g_{k-1} \\ \vdots & \vdots & & \vdots \\ 1 & g_k & \cdots & g_{2k-1} \end{vmatrix}. \tag{5}$$

Thus

$$I_t^{(0)} = G(t), \qquad I_t^{(1)} = \frac{G(t)g(t + \rho) - G(t + \rho)g(t)}{g(t + \rho) - g(t)}, \ldots. \tag{6}$$

There are several ways the above formula can be used. One could assume, for instance, that G is tabulated at equally spaced points, $t_0 + jh$, $t_0 \geq a$, $j \geq 0$. The BH protocol can then be applied by taking

$$f_r(n) = g(t_0 + (n + r - 1)\rho), \qquad s_n = G(t_0 + n\rho), \tag{7}$$

and so $s_n^{(k)}$ yields an extrapolation to I in terms of the known values $G(t_0 + n\rho)$, $n \geq 0$. Of course, t_0 may be chosen larger than a, the advantage being that g may have a singularity near a and that the necessity of tabulating g near a can be avoided. Alternatively, one may take $t_0 = a$ and define, arbitrarily $f_1(0) = 0$ when g is singular at a.

Table I

k	$s_{10-k}^{(k)}$	k	$s_{10-k}^{(k)}$
0	−0.45939	6	−0.49968
1	−0.49006	7	−0.49979
2	−0.49634	8	−0.49986
3	−0.49829	9	−0.49990
4	−0.49909	10	−0.50003
5	−0.49948		

Example 1

$$\begin{aligned} g(x) &= -e^{\sqrt{x}}/2\sqrt{x}(1 + e^{\sqrt{x}})^2, \qquad x > 0, \\ G(t) &= (1 + e^{\sqrt{t}})^{-1} - \tfrac{1}{2}, \\ I &= \int_0^\infty g(x)\,dx = -\tfrac{1}{2}, \quad t_0 = a = 0, \quad \rho = 1. \end{aligned} \tag{8}$$

Here G is known explicitly, but, surprisingly, not more than ten or so tabular values are required to determine I to almost five places despite the fact that g is singular at zero. Thus we may assume, for the example $f_1(0) = 0$, that 11 values of G are known and tabulate the 11th ascending diagonal of $s_n^{(k)}$ (see Table I).

Example 2

$$\begin{aligned} g(x) &= -e^{-x}(x + 1)/x^2, \qquad x > 0, \\ G(t) &= e^{-t}/t - 1/e, \\ I &= \int_1^\infty g(x)\,dx = e^{-1} = 0.367879441, \qquad t_0 = a = \rho = 1. \end{aligned} \tag{9}$$

The sixth ascending diagonal is tabulated in Table II. In this example double precision (16 significant figures) was used, and $s_0^{(15)}$ is accurate to 16 significant figures. This indicates the method has great numerical stability, at least when applied to monotonic integrands.

Table II

k	$s_{5-k}^{(k)}$	k	$s_{5-k}^{(k)}$
0	−0.367466316	3	−0.367879339
1	−0.367863093	4	−0.367879477
2	−0.367878981	5	−0.367879363

Table III

k	$s_{12-k}^{(k)}$	k	$s_{12-k}^{(k)}$
0	1.04471	8	0.99996
2	0.99818	10	0.99967
4	1.00015	12	0.99996
6	0.99968		

Example 3. This has an oscillatory integrand, corresponding to

$$I = \int_0^{\infty} \frac{\sin x - x \cos x}{x^2}\, dx = 1. \tag{10}$$

Then

$$G(t) = 1 - (\sin t)/t. \tag{11}$$

Some elements on the 13th ascending diagonal are tabulated in Table III. The error in $s_0^{(15)}$ is 1×10^{-6}. Obviously, the algorithm was not designed for integrands that decay algebraically or logarithmically. For $\int_2^{\infty} x^{-1}(\ln x)^{-2}\, dx$, as another example, $s_0^{(15)} = 1.262$, while the true value of the integral is $1/\ln 2 = 1.443$.

We now look at the exactness problem for this algorithm.

Theorem 1. For some complex constant $d_0, d_1, \ldots, d_k$, let $\lambda \in \mathscr{C}_C$ be a sequence of roots with negative real parts of the exponential polynomial

$$H(\lambda) = d_0 + \lambda \sum_{r=0}^{k-1} d_{r+1} e^{\lambda r \rho}. \tag{12}$$

Then if

$$g(t) = \sum p_m(t) e^{\lambda_m t}, \tag{13}$$

where $p_m(t)$ is a polynomial of degree less than the multiplicity of λ_m, infinite sums being allowed subject to convergence conditions, the transformation (5) is exact for each t; i.e., $I_t^{(k)} \equiv I$, $t > a$, provided the denominator of (5) does not vanish.

Proof. Define

$$\mathscr{L}(f) = d_0 f(t) + \sum_{r=0}^{k-1} d_{r+1} f'(t + r\rho). \tag{14}$$

If g satisfies the equation $\mathscr{L}(g) = 0$, then, by integration between t and ∞,

$$d_0[G(t) - I] + \sum_{r=0}^{k-1} d_{r+1} g(t + r\rho) = 0, \qquad t > a, \tag{15}$$

so the numerator of the determinantal expression of $I_t^{(k)} - I$ will vanish. Let λ_0 be a root of $H(\lambda)$ of multiplicity m. We need only show $\mathscr{L}(t^j e^{\lambda_0 t}) = 0$ for $0 \le j \le m - 1$.

We can write

$$e^{\lambda t} H(\lambda) = c_0 e^{\lambda t} + \sum_{r=0}^{k-1} c_{r+1} \frac{d}{dt} e^{\lambda(t+r\rho)}, \tag{16}$$

so

$$0 = \frac{d^j}{d\lambda^j} e^{\lambda t} H(\lambda)|_{\lambda=\lambda_0} = c_0 t^j e^{\lambda_0 t} + \sum_{r=0}^{k-1} c_{r+1} \frac{d}{dt} (t + r\rho)^j e^{\lambda_0(t+r\rho)}$$
$$= \mathscr{L}(t^j e^{\lambda_0 t}), \qquad 0 \le j \le m - 1, \tag{17}$$

which was to be shown. ■

Corollary 1 ($k = 1$). Let $f \in L(0, \infty)$. Then $I_t^{(1)}$ is defined and exact iff $g(t) = Me^{-\alpha t}$, $M \ne 0$, Re $\alpha > 0$.

Corollary 2. For some complex constants $d_1, \ldots, d_k$, $d_1 + \cdots + d_k \ne 0$, let $\boldsymbol{\lambda}$ be a sequence of roots with negative real parts of

$$\sum_{r=0}^{k-1} d_{r+1} e^{\lambda r\rho} = 0. \tag{18}$$

Then $I_t^{(k)}$, $k \ge 1$, is defined and exact for

$$g(t) = \sum p_m(t) e^{\lambda_m t}, \tag{19}$$

where p_m is as in Theorem 1.

Proof. Completion of the proof, which requires Heymann's theorem to guarantee the nonvanishing of the denominator of $I_t^{(k)}$, is left to the reader; see Section 6.3. ■

The following result on accelerativeness is easy to demonstrate.

Theorem 2. Let $D(t)$ denote the denominator of $I_t^{(k)}$ and $M_r(t)$ the rth cofactor of the first column of D. Let M_r/D be bounded. Let I exist, g be bounded, and

$$(G_r - I)/(G_0 - I) = g_r/g_0 + o(1), \qquad 1 \le r \le k. \tag{20}$$

Then

$$\lim_{t\to\infty} \{(I_t^{(k)} - I)/[G(t) - I]\} = 0. \tag{21}$$

Proof. Left to the reader. ■

Let us take as an example the important case $k = 1$,

$$I_t^{(1)} = \frac{G(t + \rho) - G(t)g(t + \rho)/g(t)}{1 - g(t + \rho)/g(t)}. \tag{22}$$

If $g(t + \rho)/g(t) = \lambda + o(1)$, $0 < \lambda < 1$, the hypotheses of Theorem 2 are satisfied—in fact, in this case the conditions are necessary *and* sufficient for the accelerativeness of $I_t^{(1)}$; see Gray and Atchison (1967).

The algorithm is most suitable for integrands that behave exponentially. Obviously if $f = O(t^{-\alpha})$, the conditions of theorem are not satisfied; in fact, for $k = 1$, one has

$$I_t^{(1)} - I = G(t) - I + O(t^{1-\alpha}). \tag{23}$$

An algorithm suitable for cases in which f behaves algebraically can be obtained by making an exponential substitution in (2)–(6). This amounts to taking in the BH protocol

$$\begin{gathered} f_r(n) = t_0 \rho^{n+r-1} g(t_0 \rho^{n+r-1}), \\ s_n = G(t_0 \rho^n), \qquad t_0 \geq a \geq 1, \quad \rho > 1. \end{gathered} \tag{24}$$

However, these equations offer no clear computational advantage over (7), since tabular values of G for very large t are required.

An exactness theorem analogous to Theorem 1 is easily established for the new algorithm. Details are left to the reader. Theorem 2 remains unchanged. For the important case $k = 1$, these results show the algorithm is exact for functions $f(t) = Mt^{-\alpha}$, $M \neq 0$, $\operatorname{Re} \alpha > 1$, and accelerative if $f(t) = O(t^{-\alpha})$, $\operatorname{Re} \alpha > 1$. The papers by Gray, Atchison, and Clark detail many other properties of the $k = 1$ algorithm.

11.2. The Computation of Fourier Coefficients

Suppose it is required to compute the Fourier coefficients

$$I(m) = \int_0^1 f(x) \cos(2\pi m x)\, dx, \tag{1}$$

and that a sequence **s** of values of the trapezoidal sums

$$T_n = \frac{1}{n} \sum_{k=0}^{n}{}'' f\left(\frac{k}{n}\right) \tag{2}$$

is known. Further, assume that Romberg integration (Section 3.1) has been applied to s_n to produce a value of $I(0)$ accurate to as many figures as are required of $I(m)$.

The BH protocol, combined with a method due to Lyness (1970, 1971) can be used to attack this problem. To be accurate, we should speak of a "class" of methods, since Lyness's theory has a great deal of flexibility, which allows one to take advantage of additional data, i.e., a knowledge of the derivatives of f. Here only the simplest form of his algorithm will be used. (It seems a pity that Lyness's work, uncomplicated and beautifully ingenious, has received almost no attention from the authors of books on numerical analysis.)

Suppose f has the Fourier series development

$$f(y) = I(0) + 2\sum_{k=1}^{\infty} \int_0^1 f(x) \cos[2\pi k(x - y)]\, dx. \tag{3}$$

Let y assume the values j/m and sum from $j = 0$ to $m - 1$. The result may be expressed

$$2\sum_{k=1}^{\infty} I(km) = r_m, \qquad r_m = T_m - I(0). \tag{4}$$

[For details, see Luke (1969, Vol. II, p. 215).]

Now, the Möbius inversion formula (Hardy and Wright, 1959, p. 237) states that, subject to certain convergence conditions, the sum

$$G_m = \sum_{k=1}^{\infty} F_{k\cdot m}, \qquad m \geq 1, \tag{5}$$

may be inverted to yield

$$F_m = \sum_{k=1}^{\infty} \mu_k G_{k\cdot m}, \qquad m \geq 1, \tag{6}$$

where μ_k is the Möbius function,

$$\mu_k = \begin{cases} 1 & k = 1 \\ 0 & \text{if} \quad k \text{ has a square factor} \\ (-1)^r & \text{if} \quad k \text{ is the product of } r \text{ prime numbers.} \end{cases} \tag{7}$$

(The first ten values of μ_k are $+1, -1, -1, 0, -1, +1, -1, 0, 0, +1$). Lyness applied this formula to the sum (4) to obtain

$$I(m) = \frac{1}{2}\sum_{k=1}^{\infty} \mu_k r_{k\cdot m}. \tag{8}$$

This is the series from which we wish to compute $I(m)$. We show how the BH protocol can be applied to the partial sums of this series. Let

$$I_n(m) = \frac{1}{2}\sum_{k=1}^{n+1} \mu_k r_{k\cdot m}, \qquad n \geq 0, \tag{9}$$

and define

$$R_n \equiv I_n(m) - I(m) = \frac{1}{2} \sum_{k=n+2}^{\infty} \mu_k r_{k \cdot m}. \tag{10}$$

From the fact that

$$r_n \sim c_1/n^2 + c_2/n^4 + \cdots, \tag{11}$$

$$R_n \sim d_1 f_1(n) + d_2 f_2(n) + \cdots, \qquad f_j(n) = \sum_{k=n+2}^{\infty} \frac{\mu_k}{k^{2j}}. \tag{12}$$

However,

$$\frac{1}{\zeta(2j)} = \sum_{k=1}^{\infty} \frac{\mu_k}{k^{2j}}, \tag{13}$$

so

$$f_j(n) = \frac{1}{\zeta(2j)} - \sum_{k=1}^{n+1} \frac{\mu_k}{k^{2j}}, \qquad n \geq 0, \quad j \geq 1, \tag{14}$$

and to complete the BH protocol one takes

$$\begin{aligned} s_n &= I_n(m) = \frac{1}{2} \sum_{k=1}^{n+1} \mu_k r_{k \cdot m}, \\ r_n &= T_n - I(0), \\ T_n &= \frac{1}{n} \sum_{k=0}^{n}{}'' f\left(\frac{k}{n}\right), \qquad I(0) = \int_0^1 f(x)\,dx. \end{aligned} \tag{15}$$

[The numbers $\zeta(2j)$ are extensively tabulated; see e.g., Abramowitz and Stegun (1964).]

One would expect, based on the representation 10.4(1), that $r_k^{(n)} = O(n^{-2k-2})$, $n \to \infty$. (This has not been proved, of course.) The original series, Eq. (8), converges only as n^{-2}.

If f has derivatives, i.e., if the values of $c_1, c_2, \ldots, c_{2r+1}$, are known, these may be used in an obvious way to make the process even more efficient, with T_n minus the first several terms in the series (11) taken for T_n.

11.3. The tanh Rule

The basis of the tanh rule is the approximation of a doubly infinite integral by means of a trapezoidal approximating sum. Thus the quadrature process is similar to the methods based on cardinal interpolation. However, there is an important difference, one that changes completely the nature of the error term: The infinite sum is truncated at $\pm N(h)$. The problem is, how should N be chosen to obtain optimal results?

Following Schwartz (1969), we make a change of variable in the finite integral $\int_{-1}^{1} g(x)\,dx$. Let ψ be a reasonably smooth function that is monotone and maps $(-1, 1)$ into $(-\infty, \infty)$.

$$\int_{-1}^{1} g(x)\,dx = \int_{-\infty}^{\infty} g(\psi(t))\psi'(t)\,dt \approx h \sum_{r=-n}^{n} \psi'(rh)g(\psi(rh)). \tag{1}$$

How should ψ and $h \equiv h(n)$ be chosen? Schwartz suggested $\psi(t) = \tanh(\frac{1}{2}t)$ (hence the name "tanh rule") and $h = \pi\sqrt{2/n}$.

For integrands g in Hardy class H^2, Haber (1977) has computed the asymptotic form of the error norm and has shown that for the above choice of ψ, the choice of h is optimal. [The functions in the Hardy class H^2 are functions analytic in N for which $\int_0^{2\pi} |f(re^{i\theta})|^2\,d\theta$ is bounded as $r \to 1$.]

Let $g \in H^2$ and define

$$s(g) = \int_{-1}^{1} g(x)\,dx, \tag{2}$$

$$s_n(g) = h \sum_{r=-n}^{n} \frac{g(\tanh(nh/2))}{2\cosh^2(nh/2)}, \qquad h = \pi\sqrt{2/n}. \tag{3}$$

It can be shown that $s - s_n$ is a bounded linear functional on H^2. Haber found that

$$\|s_n - s\| \sim Ke^{-(\pi/\sqrt{2})\sqrt{n}}, \qquad K = (4\pi^2 + 8\ln 2)^{1/2}. \tag{4}$$

Note that this seems to be considerably inferior to the bound obtained for the trapezoidal rule in Section 3.4. However, there the sum is not truncated and the class of functions is smaller.

Haber's computations seem to indicate that a good choice for the BH protocol is

$$f_j(n) = e^{-(\pi/\sqrt{2})\sqrt{n}}/(n+1)^{(j-1)/2}. \tag{5}$$

The function $g(x) = (1 - x^2)^\alpha$ is in H^2 provided $\operatorname{Re}\alpha > -\frac{1}{2}$. Then $I = \Gamma(\alpha+1)\sqrt{\pi}/\Gamma(\alpha+\frac{3}{2})$ and

$$s_n = \begin{cases} h\sum_{k=0}^{n}{}' \left[\cosh\left(\dfrac{kh}{2}\right)\right]^{-2\alpha-2}, & n \ge 1 \\ \pi\sqrt{2/n}\left[\dfrac{1}{2} + \sum_{k=1}^{n}\left(\cosh\dfrac{k\pi}{\sqrt{2n}}\right)^{-2\alpha-2}\right], & n \ge 1, \end{cases} \tag{6}$$

and $s_0 = 0$.

Table IV

BH Protocol Applied to $I_\alpha = 2 \int_0^1 (1 - x^2)^\alpha \, dx$

	$\alpha = -\frac{1}{3}, I_{-1/3} = 2.587109559$		$\alpha = -\frac{1}{4}, I_{-1/4} = 2.396280467$	
k	s_k (tanh rule)	$s_0^{(k)}$	s_k (tanh rule)	$s_0^{(k)}$
2	2.611931003	2.266890051	2.440806880	2.048670072
4	2.586166070	2.563060233	2.399070105	2.371528094
6	2.586239244	2.586139159	2.396475368	2.395295728
8	2.586715520	2.587082817	2.396260717	2.396255106
8	2.586937436	2.587108878	2.396257569	2.396279761
12	2.587032111	2.587109544	2.396267876	2.396280440

Table IV displays s_k versus $s_0^{(k)}$, i.e., vertical versus diagonal, convergence for the choice (5) and the cases $\alpha = -\frac{1}{3}$ and $\alpha = -\frac{1}{4}$. Clearly, the BH protocol is a powerful tool to use in conjunction with the tanh rule.

Chapter 12 | Probabilistic Methods

12.1. Introduction

Historically, the construction of summability methods has been based on the philosophy and techniques of classical analysis. Actually, the problem of accelerating the convergence of a sequence is more at home in a probabilistic setting. A formulation in terms of prediction theory or recursion filtering, for instance, immediately suggests the minimization of the expectation $\{E(|\bar{r}_n|)\}$ of the transformed error sequence if the original sequence is interpreted as a sequence of random variables.†

By assuming certain distribution functions for the $\{s_n\}$ and performing this minimization, one is led naturally to a class of methods for transforming sequences.

Of course, the methods will depend on the parameters of the chosen distributions. If these parameters are unknown, any well-known estimation technique can be applied. Each estimation technique provides a different summation method.

Although the construction of summation methods has not traditionally been based on probabilistic techniques, the methods themselves have been put to extensive probabilistic use. For example, Chow and Teicher (1971) represent the strong law as a trivial special case of the following Toeplitz summability.

Let $\{X_n\}_{n=1}^{\infty}$ be independent identically distributed random variables with finite first moment. Suppose

$$a_n \geq 0, \qquad n \geq 0, \tag{1}$$

† Good sources for the theory of probability and stochastic processes needed in this chapter are Papoulis (1965) and Miller (1974).

and

$$s_n = \sum_{j=0}^{n} a_j, \qquad n \geq 0, \tag{2}$$

diverges. Define the transformed sequence $\{T_n\}$ by

$$T_n = s_n^{-1} \sum_{j=0}^{n} a_j X_j, \qquad n \geq 0. \tag{3}$$

If $T_n - C_n \to 0$ almost surely for some centering constants $\{C_n\}$, then $\{X_n\}$ is called a_n-summable with probability 1.

Note that the strong law is obtained by using

$$C_n = EX, \qquad n \geq 0, \tag{4}$$

the common mean of the underlying distribution, and

$$a_n = 1, \qquad n \geq 0. \tag{5}$$

The summation methods to be derived here are nonlinear and nonregular. They are simple to use. They are useful for summing classical series and also for summing "statistical" series whose terms are realizations of random sequences. Numerical examples of both kinds of applications are included here. The advantages the methods hold for statistical applications are clear: For series defined by complicated experiments in which obtaining data is difficult and expensive, the use of the proper summation method based on an appropriate probabilistic assumption can result in practical advantages.

Finally, we shall show that for one large and important class of sequences, the methods are regular, namely, the sequence space of partial sums of alternating series whose terms in absolute value are monotone decreasing. No other nonregular method has been shown to be regular for this sequence space.

12.2. Derivation of the Methods

To motivate our derivation, suppose that the series $\sum_{k=0}^{\infty} a_k$ is a realization of the following "experiment":

Let $\{x_k\}_{k=1}^{\infty}$ be a sequence of independent random variables with

$$E(x_k) = p, \qquad k \geq 1, \tag{1}$$

and

$$E(x_k^2) = q, \qquad k \geq 1, \tag{2}$$

where $|p| < 1$ and $q < \infty$.

Defining

$$a_k = a_0 \prod_{j=1}^{k} x_j, \qquad k \geq 1, \tag{3}$$

one finds that

$$E(a_k) = p^k a_0, \qquad k \geq 1. \tag{4}$$

Since

$$s = \sum_{k=0}^{\infty} a_k, \tag{5}$$

it follows that

$$E(s_n) = a_0 \sum_{k=0}^{n} p^k = a_0 \frac{1 - p^{n+1}}{1 - p}, \qquad n \geq 0, \tag{6}$$

$$E(s) = a_0/(1 - p), \tag{7}$$

and

$$E(r_n) = -a_0 p^{n+1}/(1 - p), \qquad n = 0, 1, 2, \ldots . \tag{8}$$

Now,

$$E(\bar{r}_n) = \sum_{k=0}^{n} \mu_{nk} E(r_k) = \frac{-a_0 p}{1 - p} \sum_{k=0}^{n} \mu_{nk} p^k$$

$$= \frac{-a_0 p}{1 - p} P_n(p), \qquad n \geq 0. \tag{9}$$

All methods will have the property that $E(\bar{r}_n) = 0$ for $n \geq 1$.

Definition. The summation method U is called *E-admissible* if the characteristic polynomial has the form

$$P_n(\lambda) = (\lambda - p)^{k_n} d_{n-k_n}(\lambda)/(1 - p)^{k_n} d_{n-k_n}(1), \tag{10}$$

where $k_1, k_2, \ldots$ are positive integers, $1 \leq k_n \leq n$, $d_j(\lambda)$ is a polynomial of degree j in λ, and $d_j(1) \neq 0$ for any j.

Clearly, $|E(\bar{r}_n)|$ is minimized if and only if U is E-admissible. Perhaps the simplest example of an E-admissible method is

$$P_n(\lambda) = \lambda^{n-1}(\lambda - p)/(1 - p), \qquad n \geq 1, \tag{11}$$

which leads to the following very simple choice.

Method I

$$\begin{aligned} \mu_{nk} &= 0, \qquad 0 \leq k \leq n - 2; \\ \mu_{n,n-1} &= -p/(1 - p), \qquad \mu_{nn} = 1/(1 - p). \end{aligned} \tag{12}$$

For this matrix U, $\bar{s}_n$ for each n is the expected value of s given $s_0, s_1, \ldots, s_n$.

Does there exist a U that minimizes *both* $|E(\bar{r}_n)|$ and $E(\bar{r}_n^2)$? The answer is yes: It can be found as follows.

Let

$$\omega_{nk} = \sum_{j=0}^{k-1} \mu_{nj}, \quad 1 \leq k \leq n, \qquad \omega_{n0} = 0. \tag{13}$$

Then

$$\bar{r}_n^2 = r_n^2 - 2r_n \sum_{k=1}^{n} \omega_{nk} a_k + \left(\sum_{k=1}^{n} \omega_{nk} a_k\right)^2. \tag{14}$$

Now

$$E(\bar{r}_n^2) = E(r_n^2) - 2 \sum_{k=1}^{n} \omega_{nk} E(r_n a_k) + E\left(\sum_{k=1}^{n} \omega_{nk} a_k\right)^2, \tag{15}$$

but for $k \leq n$,

$$\begin{aligned} E(r_n a_k) &= E\left(-\sum_{j=n+1}^{\infty} a_j a_k\right) = -a_0^2 \sum_{j=n+1}^{\infty} p^{j-k} q^k \\ &= -a_0^2 \left(\frac{q}{p}\right)^k \sum_{j=n+1}^{\infty} p^j = -a_0^2 \left(\frac{q}{p}\right)^k \frac{p^{n+1}}{1-p}, \end{aligned} \tag{16}$$

$$E(\bar{r}_n^2) = E(r_n^2) + \frac{2a_0 p^{n+1}}{1-p} \sum_{k=1}^{n} \omega_{nk} \left(\frac{q}{p}\right)^k + E\left(\sum_{k=1}^{n} \omega_{nk} a_k\right)^2. \tag{17}$$

For the last term,

$$\begin{aligned} E\left(\sum_{k=1}^{n} \omega_{nk} a_k\right)^2 &= E\left(\sum_{l=1}^{n} \sum_{k=1}^{n} \omega_{nk} \omega_{nl} a_k a_l\right) \\ &= E\left(\sum_{k=1}^{n} \omega_{nk}^2 a_k^2\right) + 2E\left(\sum_{k=1}^{n} \omega_{nk} \sum_{l=1}^{k-1} \omega_{nl} a_k a_l\right) \\ &= a_0^2 \sum_{k=1}^{n} \omega_{nk}^2 q^k + 2a_0^2 \sum_{k=1}^{n} \omega_{nk} \sum_{l=1}^{k-1} \omega_{nl} p^{k-l} q^l. \end{aligned} \tag{18}$$

Let

$$F = E(\bar{r}_n^2) - \lambda\left(\sum_{k=1}^{n} \omega_{nk} p^{k-1} + \frac{p^n}{1-p}\right). \tag{19}$$

Then

$$\frac{\partial F}{\partial \omega_{nj}} = \frac{2a_0^2 p^{n+1}}{1-p} \left(\frac{q}{p}\right)^j + 2a_0^2 q^j \omega_{nj} + 2a_0^2 \sum_{k=1}^{j-1} \omega_{nk} p^{j-k} q^k - \lambda p^{j-1}. \tag{20}$$

Setting $\partial F/\partial\omega_{nj} = 0$ for $1 \le j \le n$ gives

$$\omega_{nj} = \frac{\lambda p^{j-1}}{2a_0^2 q^j} - \frac{p^{n+1-j}}{1-p} - \sum_{k=1}^{j-1} \omega_{nk}\left(\frac{p}{q}\right)^{j-k} \tag{21}$$

or

$$\omega_{nk} = \frac{p^{n+1-k}}{1-p}\left(\frac{p^2}{q} - 1\right), \qquad 2 \le k \le n. \tag{22}$$

Since $|E(\bar{r}_n)|$ is a minimum,

$$\begin{aligned} P_n(p) = 0 = \omega_{n1} + (\omega_{n2} - \omega_{n1})p + \cdots \\ + (\omega_{nn} - \omega_{n,n-1})p^{n-1} + (1 - \omega_{nn})p^n, \end{aligned} \tag{23}$$

and this implies

$$\sum_{k=1}^{n} \omega_{nk} p^{k-1} = \frac{-p^n}{1-p}. \tag{24}$$

We can now determine $\omega_{n1} = \mu_{n0}$ and then, from (22), all the μ_{nk}:

$$\omega_{n1} = \mu_{n0} = \frac{-p^n}{1-p} - \sum_{k=2}^{n} \omega_{nk} p^{k-1}. \tag{25}$$

Method II. This method minimizes both $|E(\bar{r}_n)|$ and $E(\bar{r}_n^2)$:

$$\begin{aligned} \mu_{n0} &= \frac{-p^n}{1-p}\left[1 + (n-1)\left(\frac{p^2}{q} - 1\right)\right]; \\ \mu_{n1} &= \frac{p^{n-1}}{(1-p)}\left(\frac{p^2}{q} - 1\right)[1 + (n-1)p] + \frac{p^n}{1-p}; \\ \mu_{nk} &= \left(\frac{p^2}{q} - 1\right)p^{n-k}, \qquad 2 \le k \le n-1; \\ \mu_{nn} &= \frac{1}{1-p}\left(1 - \frac{p^3}{q}\right). \end{aligned} \tag{26}$$

We shall be concerned with yet a third E-admissible method, namely, that arising from the choice

$$P_n(\lambda) = (\lambda - p)^n/(1-p)^n, \qquad n \ge 0, \tag{27}$$

i.e., the choice that forces $P_n(\lambda)$ to vanish as strongly as possible at $x = p$.

Method III. This yields the weights

$$\mu_{nk} = \binom{n}{k}(-p)^{n-k}/(1-p)^n. \tag{28}$$

These are related to the Euler means of Section 2.3.2. (There $p < 0$.)

Theorem. Let $p \in \mathscr{C}$. Then Methods I–III define regular methods iff $p \neq 1$, $|p| < 1$, and $p \leq 0$, respectively.

Proof. Application of Theorem 2.2(1) to the weights of Methods I and II is trivial. For Method III, note that for fixed k

$$|\mu_{nk}| \sim M_k n^k \left|\frac{p}{1-p}\right|^n, \qquad M_k \text{ independent of } n, \tag{29}$$

so that $\lim_{n\to\infty} \mu_{nk} = 0$ iff $|p/(1-p)| < 1$, that is, $\operatorname{Re} p < \frac{1}{2}$. Further,

$$\sum_{k=0}^{n} |\mu_{nk}| = \frac{|p|^n}{|1-p|^n} \sum_{k=0}^{n} \binom{n}{k} |p|^{-k} = \left|\frac{|p|+1}{|1-p|}\right|^n. \tag{30}$$

Now, the triangle with vertices $\{0, 1, p\}$ in the complex plane has legs of length 1, $|p|$, and $|1-p|$. Thus $[|p|+1]/|1-p| > 1$ unless p is real and negative, in which case the ratio is 1 and $\sum_{k=0}^{n} |\mu_{nk}| = 1$. ■

To obtain the final summation formulas, note that in almost all applications the variables $\{x_k\}_{k=1}^{\infty}$ are identically distributed with the first moment p unknown. One could then estimate p by the method of moments

$$p \sim \frac{1}{n} \sum_{k=1}^{n} x_k = \frac{1}{n} \sum_{k=1}^{n} \frac{a_k}{a_{k-1}} = p_n, \qquad n \geq 1, \tag{31}$$

or by the maximum likelihood estimate. The first estimate provides the more useful results.

Thus, for instance, for Methods I and III one has

$$\bar{s}_0 = s_0, \qquad \bar{s}_n = (-p_n s_{n-1} + s_n)/(1 - p_n), \quad n \geq 1, \tag{32}$$

and

$$\bar{s}_n = \sum_{k=0}^{n} \binom{n}{k} (-p_n)^{n-k} s_k/(1 - p_n)^n, \tag{33}$$

respectively, and a similar formula is obtained for Method II with p in (26) replaced by p_n and q by $q_n = n^{-1} \sum_{k=1}^{n} a_k^2/a_{k-1}^2$.

The form of the sequence of expected errors (8) is a fortunate consequence of the derivation since many of the sequences encountered in practice are at least approximately a constant plus an exponentially decreasing term.

The above methods (like many nonlinear methods) cannot be applied to certain sequences. For instance, if $s_n = s_{n+1}$ for some value of n, then a_{n+2}/a_{n+1} is undefined and so is p_{n+1}. The problem in definition is not resolved by considering only subsequences of $\{s_n\}$ containing no adjacent duplicate members since the possibility that $p_n = 1$ is still not obviated.

In fact, it is possible to manufacture examples of convergent infinite series where $p_n = 1$ for an infinite number of n, for instance, by folding together the two absolutely convergent series

$$\lambda + \lambda^2 + \lambda^3 + \cdots, \quad c\lambda + c\lambda^2 + c\lambda^3 + \cdots, \qquad |\lambda| < 1, \tag{34}$$

to obtain

$$\lambda + c\lambda + \lambda^2 + c\lambda^2 + \cdots. \tag{35}$$

The sequence a_n/a_{n-1}, $n = 1, 2, 3, \ldots$, is then

$$c, \quad \lambda/c, \quad c, \quad \lambda/c, \quad \ldots, \tag{36}$$

so that

$$p_{2n} = \frac{1}{2n}\sum_{k=1}^{2n} \frac{a_k}{a_{k-1}} = \frac{1}{2n} n\left(c + \frac{\lambda}{c}\right) = \left(c + \frac{\lambda}{c}\right), \qquad n = 1, 2, 3, \ldots, \tag{37}$$

and for the choice

$$c = 1 + \sqrt{1 - \lambda}, \tag{38}$$

$p_{2n} = 1$, $n = 1, 2, 3, \ldots$.

12.3. Properties of the Methods

For Method I there is a simple necessary and sufficient condition for convergence. We have

$$\bar{s}_n = (s_n - p_n s_{n-1})/(1 - p_n) = s_{n-1} + a_n/(1 - p_n). \tag{1}$$

Theorem 1. Let $\mathbf{s} \in \mathscr{C}_C$. Then for Method I

$$\bar{s} = s \qquad \text{iff} \qquad \lim_{n\to\infty} a_n/(1 - p_n) = 0. \tag{2}$$

Proof. Trivial. ■

Method I is not regular because of its lack of definition for certain convergent sequences, but if both $\{s_n\}$ and $\{\bar{s}_n\}$ converge then $\lim_n s_n = \lim_n \bar{s}_n$ so long as $\{p_n\}$ is bounded away from 1.

Corollary. For any convergent alternating series $\sum_{n=0}^{\infty} a_n$, Method I preserves convergence. Further if $|a_n|$ is monotone decreasing, then $\bar{s}_n$ lies between s_{n-1} and s_n for all n.

Proof. Since $-1 < a_n/a_{n-1} < 0$, $-1 < p_n < 0$, and $\frac{1}{2} < (1 - p_n)^{-1} < 1$, Eq. (1) immediately yields the corollary. ■

Further, Methods I and II preserve convergence when $\{s_n\}$ is the sequence of partial sums of any series, $\sum_{k=0}^{\infty} a_k$, for which Raabe's test (Knopp, 1947, p. 285) is applicable.

Theorem 2. Let $a_n > 0$ for all n, and let $\mathbf{s} \in \mathscr{C}_C$ with

$$n(a_{n+1}/a_n - 1) \leq . -\gamma. \tag{3}$$

Then Method I is regular for $\mathbf{s}$ if $-\gamma < -1$ and Method II is regular for $\mathbf{s}$ if $-\gamma < -2$.

Proof. We may rewrite (3)

$$n(a_{n+1}/a_n - 1 + 1/n) \leq . -\beta < 0, \qquad \beta = \gamma - 1, \tag{4}$$

so that

$$na_{n+1} - (n-1)a_n \leq . -\beta a_n, \tag{5}$$

or

$$(n-1)a_n - na_{n+1} \geq . \beta a_n . > 0. \tag{6}$$

Thus ultimately $\{(n-1)a_n\}$ is monotone decreasing. Therefore

$$\lim_{n\to\infty} na_{n+1} = \lim_{n\to\infty} na_n$$

exists.

Now from

$$a_{n+1}/a_n - 1 \leq . -\gamma/n, \tag{7}$$

we can conclude

$$a_k/a_{k-1} \leq . 1 - \gamma/(k-1), \tag{8}$$

$$p_n = \frac{1}{n}\sum_{k=1}^{n} \frac{a_k}{a_{k-1}} \leq \frac{1}{n}\left[\sum_{k=1}^{N} \frac{a_k}{a_{k-1}} + \sum_{k=N+1}^{n}\left(1 - \frac{\gamma}{k-1}\right)\right], \tag{9}$$

or

$$p_n \leq [n - \gamma \ln n + M(n)]/n, \tag{10}$$

where $\{M(n)\}$ is a bounded sequence. Here we have used the fact that

$$\sum_{k=N+1}^{n} \frac{1}{k-1} = \ln n + O(1). \tag{11}$$

Thus

$$1 - p_n \geq [\gamma \ln n - M(n)]/n, \tag{12}$$

and for n large enough, the right-hand side is positive. Thus

$$\frac{a_n}{1 - p_n} \le. \frac{na_n}{\gamma \ln n - M(n)}, \tag{13}$$

or

$$\lim_{n\to\infty} a_n/(1 - p_n) = 0. \tag{14}$$

This gives the result for Method I.

Now

$$\begin{aligned} |p_n|^n &\le. \left[1 + \frac{M(n) - \gamma \ln n}{n}\right]^n, \\ &=. \exp\left[n \ln\left(1 + \frac{M(n) - \gamma \ln n}{n}\right)\right], \\ &=. \exp\left[n\left(\frac{M(n) - \gamma \ln n}{n} + \frac{\varepsilon(n)}{n}\right)\right], \end{aligned} \tag{15}$$

by a Taylor's series argument, where $\{\varepsilon(n)\}$ is a null sequence. Thus

$$\frac{n|p_n|^n}{1 - p_n} \le. \frac{\beta n^{2-\gamma}}{\gamma \ln n - M(n)}, \tag{16}$$

and since $2 - \gamma < 0$,

$$\lim_{n\to\infty} |p_n|^n/(1 - p_n) = 0 \tag{17}$$

and the result is established for Method II. ■

Both Methods I and II preserve convergence for series for which the ratio test shows convergence provided one further stipulation is added for Method II.

Theorem 3. Let $\mathbf{s} \in \mathscr{R}_C$ with

$$\overline{\lim_{n\to\infty}}\, a_n/a_{n-1} < 1. \tag{18}$$

Then Method I is regular for $\mathbf{s}$.

If in addition

$$\underline{\lim_{n\to\infty}}\, a_n/a_{n-1} > -1, \tag{19}$$

the same holds for Method II.

Proof. Since p_n is the Cesaro means of a_n/a_{n-1},

$$\varliminf_{n\to\infty} a_n/a_{n-1} \le \varliminf_{n\to\infty} p_n \le \varlimsup_{n\to\infty} p_n \le \varlimsup_{n\to\infty} a_n/a_{n-1}. \tag{20}$$

Thus, for every $\varepsilon > 0$,

$$\varliminf_{n\to\infty} a_n/a_{n-1} - \varepsilon <. \, p_n <. \varlimsup_{n\to\infty} a_n/a_{n-1} + \varepsilon. \tag{21}$$

By virtue of (18),

$$p_n <. \, r < 1, \tag{22}$$

or

$$(1 - p_n) >. \, 1 - r > 0, \tag{23}$$

or

$$(1 - p_n)^{-1} <. \, (1 - r)^{-1}. \tag{24}$$

Theorem 1 may be invoked to show regularity for Method I.

Using (19) and (22) now gives

$$-1 < r_1 <. \, p_n <. \, r_2 < 1, \tag{25}$$

so

$$\frac{|p_n|^n}{1 - p_n} <. \, \frac{[\max(|r_1|, r_2)]^n}{1 - r_2}. \tag{26}$$

Thus

$$\lim_{n\to\infty} n|p_n|^n/(1 - p_n) = 0. \quad \blacksquare \tag{27}$$

Many sequences encountered in practice can be expressed (at least approximately) as a constant plus a linear combination of exponential terms. More precisely, one can define a sequence space $\mathscr{H}$ as follows.

Let $\mathscr{S}$ be the space of complex sequences

$$\mathscr{S} = \{\mathbf{u} \,|\, u_n \ne 0, u_{n+1} = u_n[1 + o(1)]\}. \tag{28}$$

Define

$$\mathscr{H} = \left\{\mathbf{s} \,|\, s_n = s + \sum_{r=1}^{k} \lambda_r^n u_n^{(r)}, \lambda_r \in N, 1 > |\lambda_1| > |\lambda_j|, 2 \le j \le k, \mathbf{u}^{(r)} \in \mathscr{S}\right\}. \tag{29}$$

Note $\mathscr{H}$ is a generalization of $\mathscr{C}_{E^k}(N)$ [see Eq. 2.2(12)].

Theorem 4. Method II is regular for $\mathscr{H}$; Method I accelerates $\mathscr{H}$.

Proof. Let $\mathbf{u}^{(j)} \in \mathscr{S}$. By an application of Theorem 1.4(2),

$$|u_n^{(j)}| = |u_0^{(j)}| e^{n\varepsilon_n^{(j)}}, \qquad \boldsymbol{\varepsilon}^{(j)} \in \mathscr{R}_N. \tag{30}$$

Now,

$$a_{n+1} = \sum_{r=1}^{k} \lambda_r^{n+1} u_n^{(r)}[1 + o(1)] - \sum_{r=1}^{k} \lambda_r^n u_n^{(r)} = \sum_{r=1}^{k} \lambda_r^n u_n^{(r)}[(\lambda_r - 1) + o(1)]. \quad (31)$$

Thus

$$\lim_{n\to\infty} \left| \frac{\lambda_r^n u_n^{(r)}}{\lambda_1^n u_n^{(1)}} \right| = \lim_{n\to\infty} \left| \frac{u_0^{(r)}}{u_0^{(1)}} \right| \exp\left[n\left(\ln\left| \frac{\lambda_r}{\lambda_1} \right| + \varepsilon_n^{(r)} - \varepsilon_n^{(1)} \right) \right] = 0, \quad 2 \le r \le k, \quad (32)$$

by virtue of the fact that $\ln|\lambda_r/\lambda_1| < 0$. Thus

$$\lim_{n\to\infty} a_{n+1}/\lambda_1^n u_n^{(1)} = \lambda_1 - 1. \quad (33)$$

Also,

$$\lim_{n\to\infty} a_n/\lambda_1^{n-1} u_{n-1}^{(1)} = \lambda_1 - 1. \quad (34)$$

Dividing these two limits gives

$$\lim_{n\to\infty} a_{n+1}/a_n = \lambda_1. \quad (35)$$

Since $\mathbf{p}$ is the Cesaro mean sequence of the sequence $\{a_n/a_{n-1}\}$, it also converges to λ_1, and so (1) gives

$$\bar{s}_n = s + \sum_{r=2}^{k} \lambda_r^{n-1} u_{n-1}^{(r)} \left[\left(1 + \frac{\lambda_{r-1}}{1 - \lambda_1} \right) + o(1) \right] + \lambda_1^{n-1} u_{n-1}^{(1)} o(1). \quad (36)$$

Thus

$$\lim_{n\to\infty} [(\bar{s}_n - s)/\lambda_1^{n-1} u_{n-1}^{(1)}] = 0. \quad (37)$$

From (32),

$$\lim_{n\to\infty} [(s_n - s)/\lambda_1^n u_n^{(1)}] = 1. \quad (38)$$

Dividing these limits shows that

$$\lim_{n\to\infty} [(\bar{s}_n - s)/(s_n - s)] = 0, \quad (39)$$

and this is the desired result.

For the result for Method II, note that

$$\frac{p_n^n}{1 - p_n} = \frac{\lambda_1^n}{1 - \lambda_1} [1 + o(1)], \quad (40)$$

and this, used in (25), implies the method is regular for $\mathbf{s}$. ∎

Note that all these methods map the partial sums of $\sum_{n=0}^{\infty} x^n$ into

$$\bar{s}_0 = 1, \qquad \bar{s}_n = 1/(1 - x), \quad n \geq 1. \tag{41}$$

This property is shared by other nonlinear transformation, for example, the Shanks e_1^m transformation (Shanks, 1955). In fact, a transformation related to Method I was mentioned in passing in Shanks (1955, pp. 25–26) under the name "geometric extrapolation." This transformation is defined by

$$\bar{s}_n = \frac{s_n - \lim_n (a_n/a_{n-1}) s_{n-1}}{1 - \lim_n (a_n/a_{n-1})}, \qquad n \geq 1. \tag{42}$$

Method III is, for certain classes of sequences, the most effective method of all. We now make this more precise.

Lemma. Let $\mathbf{w} \in \mathscr{C}$ be bounded and belong to a bounded subset S of $(-\infty, 0]$.

$$\bar{s}_n = \sum_{k=0}^{n} \binom{n}{k} (-w_n)^{n-k} s_k \Big/ (1 - w_n)^n, \tag{43}$$

it follows that

$$\lim_{n \to \infty} \bar{s}_n = s. \tag{44}$$

Proof. Let

$$\sup_{z \in S} \left| \frac{z}{1 - z} \right| = d < 1. \tag{45}$$

Note also that

$$|1 - z|^{-1} \leq 1, \qquad z \in S. \tag{46}$$

Since

$$\sum_{k=0}^{n} \binom{n}{k} (-w_n)^{n-k} \Big/ (1 - w_n)^n = 1, \tag{47}$$

one can write

$$\bar{r}_n = \bar{s}_n - s = \sum_{k=0}^{n} \binom{n}{k} (s_k - s)(-w_n)^{n-k} \Big/ (1 - w_n)^n. \tag{48}$$

Choose N such that $w_n \in S$, $n > N$, and $|s_n - s| = |r_n| < \varepsilon$, $n > N$. Since r_k is bounded, $|r_k| < C$

$$
\begin{aligned}
|\bar{r}_n| &< C \sum_{k=0}^{N} \binom{n}{k} \left| \frac{-w_n}{1-w_n} \right|^{n-k} |1-w_n|^{-k} \\
&\quad + \varepsilon \sum_{k=N+1}^{n} \binom{n}{k} |w_n|^{n-k} \Big/ |1-w_n|^n, \\
&< C \sum_{k=0}^{N} \binom{n}{k} d^{n-k} + \varepsilon \sum_{k=0}^{n} \binom{n}{k} |w_n|^{n-k} \Big/ |1-w_n|^n, \\
&= C \sum_{k=0}^{N} \binom{n}{k} d^{n-k} + \varepsilon \left| \frac{1+|w_n|}{|1-w_n|} \right|^n,
\end{aligned} \tag{49}
$$

or

$$
|\bar{r}_n| <. C \sum_{k=0}^{N} \binom{n}{k} d^{n-k} + \varepsilon. \tag{50}
$$

Now

$$
\binom{n}{k} = \frac{n^k}{k!}[1 + O(n^{-1})], \tag{51}
$$

for fixed k, so

$$
\lim_{n\to\infty} \binom{n}{k} d^n = 0, \tag{52}
$$

taking lim sup term by term gives

$$
\overline{\lim_{n\to\infty}} |\bar{r}_n| < \varepsilon, \tag{53}
$$

or, since ε was arbitrary,

$$
\overline{\lim_{n\to\infty}} |\bar{r}_n| = 0 \qquad \text{or} \qquad \lim_{n\to\infty} \bar{r}_n = 0. \quad \blacksquare \tag{54}
$$

Theorem 5. Method III is regular for all alternating series

$$
\sum_{k=0}^{\infty} (-1)^k b_k, \qquad b_k > 0,
$$

for which

$$
\sup_{k\geq 1} b_k/b_{k-1} = M < \infty. \tag{55}
$$

Proof. In the lemma, let $a_k = (-1)^k b_k$,

$$
a_k/a_{k-1} = -b_k/b_{k-1} < 0, \tag{56}
$$

so that $p_n < 0$. Also,

$$|p_n| = \frac{1}{n} \sum_{k=1}^{n} \frac{b_k}{b_{k-1}} \leq M. \tag{57}$$

Thus p_n belongs to a bounded set and the proof of the theorem is complete. ■

12.4. Numerics

As the reader will see, the three methods derived in the previous section have very unusual properties.

We first discuss their application to probabilistic sequences. It must be kept in mind that, although the methods are used on individual random sequences, each method was designed for a *space* of random sequences. The space (and consequently the method) is identified by its parameters p (and perhaps q). The concept of applying one of the methods to an individual sequence is ambiguous. It really makes more sense to talk about the *average value* of the transformed sequence over a great number of trials s_n^i, in other words,

$$\frac{1}{N} \sum_{i=1}^{N} \bar{s}_n^i. \tag{1}$$

This should, in some sense, be close to the expected value of s_n.

Now suppose that $\{x_n\}_{n=1}^{\infty}$ is a sequence of independent beta-distributed random variables. Recall the random variable x is beta distributed when its probability distribution function is the incomplete beta integral

$$F(\alpha, \beta, x^*) = \Pr(x \leq x^*) = \frac{1}{B(\alpha, \beta)} \int_0^{x^*} t^{\alpha-1}(1-t)^{\beta-1}\, dt \tag{2}$$

with corresponding probability density function

$$f(\alpha, \beta, x) = \frac{1}{B(\alpha, \beta)} x^{\alpha-1}(1-x)^{\beta-1}. \tag{3}$$

Note that

$$E(x) = \int_0^1 tf(\alpha, \beta, t)\, dt = \frac{B(\alpha+1, \beta)}{B(\alpha, \beta)} = \frac{\alpha}{\alpha+\beta} \tag{4}$$

and

$$E(x^2) = \int_0^1 t^2 f(\alpha, \beta, t)\, dt = \frac{B(\alpha+2, \beta)}{B(\alpha, \beta)} = \frac{\alpha(\alpha+1)}{(\alpha+\beta)(\alpha+\beta+1)}. \tag{5}$$

Since each x_k is beta distributed with parameters (α, β), (4) and (5) hold for each x_k. When the x_k are independent

$$E\left(\prod_{k=1}^{n} x_k\right) = \prod_{k=1}^{n} E(x_k) = \left(\frac{\alpha}{\alpha + \beta}\right)^n, \qquad n \geq 1. \tag{6}$$

Taking $\beta = 1$ for the numerics, define

$$\begin{aligned} s_n(\alpha) &= 1 + x_1 + x_1 x_2 + \cdots + x_1 x_2 \cdots x_n, \\ t_n(\alpha) &= 1 - x_1 + x_1 x_2 - + \cdots + (-1)^n x_1, x_2, \ldots, x_n. \end{aligned} \tag{7}$$

From (6),

$$E(s_n) = \frac{1 - (\alpha/\alpha + 1)^n}{1 - (\alpha/\alpha + 1)}, \tag{8}$$

so

$$E(s) = \alpha + 1, \tag{9}$$

while

$$E(t_n) = \frac{1 + (-1)^n (\alpha/\alpha + 1)^{n+1}}{1 + (\alpha/\alpha + 1)}, \tag{10}$$

so

$$E(t) = (\alpha + 1)/(2\alpha + 1). \tag{11}$$

Computer generated values of x_k will be selected as follows. Pick $y_1, y_2, \ldots, y_n$ from a uniform density on $[0, 1]$; then for $k = 1, 2, \ldots, n$ the value x_k is defined by (2), i.e.,

$$x_k = y_k^{1/\alpha}. \tag{12}$$

It is possible that the sample values x_k are atypical of beta-distributed random variables since the distribution function controls only the *likelihood* (probability) that particular values are obtained and not the possibility that particular values occur. For instance, when $\alpha \gg 1$ the distribution function for x_k is skewed toward $x = 1$ (see Fig. 1), so that one "expects"

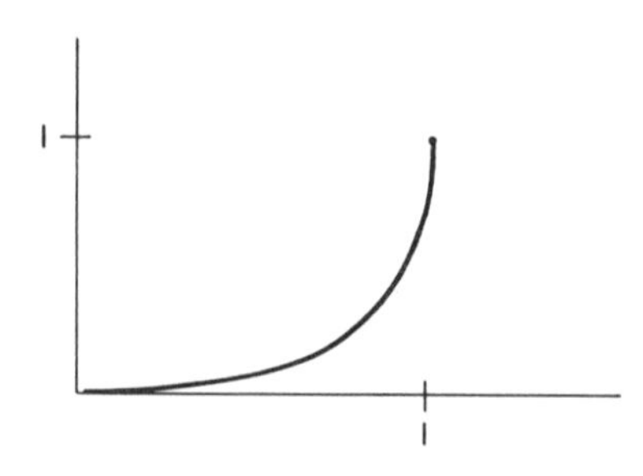

Fig. 1

Table I

Effect of Methods on Two Typical Probabilistic Sequences[a]

n	s_n	$\bar{s}_n$ I	$\bar{s}_n$ II	$\bar{s}_n$ II	n	t_n	$\bar{t}_n$ I	$\bar{t}_n$ II	$\bar{t}_n$ III
3	3.168	6.872	6.889	29.95	3	0.019	0.466	0.475	0.479
6	4.616	6.415	6.523	8333	6	0.881	0.500	0.487	0.471
9	5.495	7.294	7.347	—	9	0.160	0.504	0.515	0.483
12	6.071	6.971	7.022	—	12	0.695	0.438	0.473	0.492
15	6.461	7.324	7.355	—	15	0.228	0.455	0.468	0.493
18	6.754	7.437	7.457	—	18	0.596	0.420	0.406	0.484

[a] $\mathbf{s}(\alpha)$ and $\mathbf{t}(\alpha)$, $\alpha = 8$; $s = 7.2766$, $t = 0.3360$. Expected value $s = 9$; expected value $t = \frac{9}{17} = 0.4737$.

most simulated values for x_k to lie near 1 (and hence for the sequence s_n to converge more slowly than for smaller values of α). However, it is still *possible* to generate values for x_k anywhere in $[0, 1]$. Even when α is small, say, $0 < \alpha < 1$, so that any individual sequences **s** and **t** must converge rapidly, the inevitability of producing x_k arbitrarily close to 1 prevents one from concluding that $\overline{\lim}\, a_n/a_{n-1} < 1$ and thus using Theorem 12.3(3) to demonstrate regularity of the methods for such sequences; this occurs in spite of the fact that, for all practical purposes, the sequences *converge exponentially*. In fact, since t_n is composed of alternating monotone decreasing terms the difference $|t - t_n|$ may be bounded and t computed confidently to any desired accuracy. Table I elaborates on this strange phenomenon. For the given individual sequence t_n the methods seem to be summing **t** to its expected, rather than its actual, value. Of course, this is precisely what one would expect, and indeed demand, of a method to be applied to a sequence of experiments arising from a fixed probability distribution. But this bizarre

Table II

Effect of Methods on Various Analytic Sequences

n	$(\text{PI } 2)_n$ I	$(\text{PI } 2)_n$ II	$(\text{PI } 2)_n$ III	$(\text{LN } 2)_N$ I	$(\text{LN } 2)_N$ II	$(\text{LN } 2)_N$ III	$(\text{FAC})_n$ III
5	1.5213	1.5116	1.5323	0.68586	0.66994	0.69327	0.608398
10	1.5735	1.5671	2.9197	0.69619	0.70758	0.693147	0.61003
15	1.5945	1.5780	−1747.9	0.69143	0.68257	0.693147	0.60863
20	1.6058	1.5951	—	0.69427	0.70158	0.693147	0.60798
25	1.6130	1.6052	—	0.69235	0.68608	0.693147	

kind of nonregularity is clearly a hazard if the method is to be applied to any individual series, in particular, to an analytic sequence. The effect these methods have on analytic sequences is capricious; some examples are given in Table II. Methods I and II produced only so-so results on every analytic sequence on which they were tested. Method III is a disaster on monotone sequences but performs very well on certain sequences alternating about their limits. It even seems to sum to the usually ascribed value the sequence $(\mathrm{FAC})_n = 1 - 1! + 2! + - \cdots + (-1)^n n!$.

Chapter 13 | Multiple Sequences

13.1. Rectangular Transformations

Given a double sequence $\{s_{nk}\}$, define the transformed *double array* $\{\bar{s}_{nk}\}$ componentwise by

$$\bar{s}_{nk} = \sum_{i=0}^{n} \sum_{j=0}^{k} \mu_{ij}^{nk} s_{ij}, \qquad n, k \geq 0. \tag{1}$$

The transformation is completely characterized by the four-dimensional array of weights

$$U = [\mu_{ij}^{nk}], \qquad 0 \leq i \leq n, \quad 0 \leq j \leq k. \tag{2}$$

Obviously, convergence in the $\{s_{nk}\}$ array can be path dependent. Let s_{nk} be located at the point $(k, -n)$ of $J^0 \times -J$. We shall be concerned here with only two modes of convergence, *horizontal*, $\lim_{k\to\infty} s_{nk}$, and *vertical* $\lim_{n\to\infty} s_{nk}$. (This designation differs slightly from the convention previously used for array transformations $\mathbf{s} \to \mathbf{s}^{(k)}$, but here it is more useful to think of $\{s_{nk}\}$ as a rectangular, rather than a triangular, array.)

We shall assume that

$$\begin{aligned} \lim_{k\to\infty} s_{nk} &= \beta_n, \qquad n \geq 0, \\ \lim_{n\to\infty} s_{nk} &= \alpha_k, \qquad k \geq 0. \end{aligned} \tag{3}$$

Definition. The transformation defined by (1) is called *horizontally regular* if

$$\lim_{k\to\infty} \bar{s}_{nk} = \beta_n, \qquad n \geq 0, \tag{4}$$

and *vertically regular* if

$$\lim_{n\to\infty} \bar{s}_{nk} = \alpha_k, \qquad k \geq 0. \tag{5}$$

The material in the remainder of this section is due to Higgins (1976).

Theorem 1. The transformation defined by (1) is horizontally regular iff

(i) given $n \geq 0$,

$$\sum_{i=0}^{n} \sum_{j=0}^{k} |\mu_{ij}^{nk}| \leq R_n$$

for some positive number R_n independent of k;

(ii) given $n \geq 0$,

$$\lim_{k\to\infty} \sum_{j=0}^{k} \mu_{rj}^{nk} = \delta_{nr};$$

(iii) given $n \geq 0, j \geq 0, 0 \leq i \leq n$,

$$\lim_{k\to\infty} \mu_{ij}^{nk} = 0.$$

Proof

$\Leftarrow$: By virtue of (3), one can write $s_{ij} = \beta_i + \varepsilon_{ij}$, where $\lim_{j\to\infty} \varepsilon_{ij} = 0$. Thus

$$\begin{aligned}\bar{s}_{nk} - \beta_n &= \sum_{i=0}^{n} \sum_{j=0}^{k} \mu_{ij}^{nk}(\beta_i + \varepsilon_{ij}) - \beta_n \\ &= \sum_{i=0}^{n-1} \beta_i \sum_{j=0}^{k} \mu_{ij}^{nk} + \left(\sum_{j=0}^{k} \mu_{nj}^{nk} - 1\right)\beta_n + \sum_{i=0}^{n} \sum_{j=0}^{k} \mu_{ij}^{nk}\varepsilon_{ij}.\end{aligned} \tag{6}$$

For n fixed, condition (ii) guarantees that the first two terms on the right-hand side of (6) can be made arbitrarily small for large k. Now separate the remaining term as

$$\sum_{i=0}^{n} \sum_{j=0}^{k} \mu_{ij}^{nk}\varepsilon_{ij} = \sum_{i=0}^{n} \sum_{j=0}^{J} \mu_{ij}^{nk}\varepsilon_{ij} + \sum_{i=0}^{n} \sum_{j=J+1}^{k} \mu_{ij}^{nk}\varepsilon_{ij}. \tag{7}$$

Since n is fixed and $\lim_{j\to\infty} \varepsilon_{ij} = 0$ for $0 \leq i \leq n$, given $\varepsilon > 0$ it is possible to pick J such that $|\varepsilon_{ij}| < \varepsilon/2R_n$ for $0 \leq i \leq n$ and $j > J$. Thus for k sufficiently large, the second term on the right-hand side of (7) has modulus less than $\varepsilon/2$ when condition (i) applies.

Now with n and J fixed, define

$$M = \max_{\substack{0 \leq i \leq n \\ 0 \leq j \leq J}} |\varepsilon_{ij}|, \tag{8}$$

and pick K so large that $k > K$ implies

$$|\mu_{ij}^{nk}| < \varepsilon/2MnJ \tag{9}$$

for $0 \leq i \leq n$ and $0 \leq j \leq J$. Condition (iii) has obviously been used to guarantee (9).

Thus for k sufficiently large, both terms on the right-hand side of (7) can be made arbitrarily small so that the three conditions of the theorem imply $\lim_{k\to\infty} (\bar{s}_{nk} - \beta_n) = 0$ for each n.

$\Rightarrow$: First, let r be fixed and apply the transformation to the array

$$s_{nk} = \begin{cases} 1, & n = r, \quad k \geq 0 \\ 0, & \text{otherwise} \end{cases} \tag{10}$$

obtaining

$$\bar{s}_{nk} = \begin{cases} 0, & r > n \\ \displaystyle\sum_{j=0}^{k} \mu_{rj}^{nk}, & n \geq r. \end{cases} \tag{11}$$

By the horizontal regularity,

$$\delta_{nr} = \lim_{k\to\infty} \bar{s}_{nk} = \lim_{k\to\infty} \sum_{j=0}^{k} \mu_{rj}^{nk}, \qquad n \geq r. \tag{12}$$

Now let r vary to demonstrate the necessity of condition (ii).

Second, for the necessity of condition (iii), fix i and j and define the double arrays $D_{ij} = (d_{nk})$, where $d_{nk} = \delta_{in}\delta_{jk}$. Applying the transformation, we obtain

$$\bar{s}_{nk} = \begin{cases} \mu_{ij}^{nk}, & n \geq i, \quad k \geq j \\ 0, & \text{otherwise.} \end{cases} \tag{13}$$

Thus for $i \leq n$,

$$\lim_{k\to\infty} \mu_{ij}^{nk} = \lim_{k\to\infty} \bar{s}_{nk} = \lim_{k\to\infty} s_{nk} = 0 \tag{14}$$

by horizontal regularity. Varying i and j, we obtain the necessity of condition (iii).

Third, the necessity of condition (i) will be demonstrated by contradiction. Suppose there is an integer n for which

$$\overline{\lim_{k\to\infty}} \sum_{i=0}^{n} \sum_{j=0}^{k} |\mu_{ij}^{nk}| = +\infty. \tag{15}$$

Then there must be at least one integer i such that $0 \leq i \leq n$ and

$$\overline{\lim_{k\to\infty}} \sum_{j=0}^{k} |\mu_{ij}^{nk}| = +\infty. \tag{16}$$

If $i = n$, define the Toeplitz transformation with matrix (c_{kj}) by

$$c_{kj} = \mu_{nj}^{nk}, \qquad k \geq 0, \quad 0 \leq j \leq k. \tag{17}$$

This Toeplitz matrix represents a nonregular method since $\sum_{j=1}^{k} |c_{kj}|$ is unbounded. Thus, there is a sequence $\{x_j\}$ with $\lim_j x_j = x$ and $\sum_{j=0}^{k} c_{kj} x_j$ does not converge to x as k goes to infinity. Consider now the double array that has zero entries except for row n, wherein lies the sequence $\{x_j\}$, and apply the transformation of the theorem to obtain

$$\bar{s}_{nk} = \sum_{j=0}^{k} \mu_{nj}^{nk} x_j = \sum_{j=0}^{k} c_{kj} x_j \not\to x \qquad \text{as} \quad k \to \infty. \tag{18}$$

Therefore the method is not horizontally regular if $i = n$.

If $i < n$, consider the Toeplitz transformation whose matrix (c_{kj}) is given by

$$c_{kj} = \mu_{ij}^{nk}, \qquad k \geq 0, \quad 0 \leq j \leq k. \tag{19}$$

By Hardy (1956, p. 43),

$$\overline{\lim_{k\to\infty}} \sum_{j=0}^{k} |c_{kj}| = +\infty$$

implies the existence of a bounded sequence $\{y_j\}$ with the property that $\{\sum_{j=0}^{k} c_{kj} y_j\}_{k=1}^{\infty}$ does not converge as $k \to \infty$. Now apply the transformation of the theorem to the double array (s_{nk}), which has all zero entries except in row i, wherein lies the sequence $\{y_i\}$. Clearly,

$$\bar{s}_{nk} = \sum_{j=0}^{k} \mu_{ij}^{nk} y_j = \sum_{j=0}^{k} c_{kj} y_j, \tag{20}$$

but $\lim_{k\to\infty} s_{nk} = 0$; thus if $i < n$, the method is not horizontally regular.

The necessity of condition (i) is now established by contradiction. ∎

The proof actually substantiates some stronger results regarding the transformation (1). For example, note that with n fixed, the three conditions of the theorem are necessary and sufficient for $\lim_{k\to\infty} s_{nk} = \beta_n$. Also, conditions (i) and (iii) are sufficient conditions that the transformation (1) map a double array with all row limits zero to a double array with all row limits zero.

The three corresponding necessary and sufficient conditions for vertical regularity can be obtained by analogy. The existence of higher-dimensional analogs of these methods and of this theorem also is clear.

An extension of these ideas that has practical application but will not be pursued in this book is the following.

Suppose $f: J^0 \to J^0$ and $g: J^0 \to J^0$ with either $\lim_n f(n) = \infty$ or $\lim_n g(n) = \infty$ (or both). Call the double array *(f, g)-convergent to s* if

$$\lim_{n \to \infty} s_{f(n), g(n)} = s. \tag{21}$$

The natural question is, What are necessary and sufficient conditions on the weights μ_{ij}^{nk} in (1) to ensure that if (s_{nk}) is (f, g)-convergent to s then $(\bar{s}_{nk})$ is (f, g)-convergent to s?

If the double array (s_{nk}) has

$$\lim_{k \to \infty} s_{nk} = \beta, \qquad n \geq 0, \tag{22}$$

then condition (ii) of Theorem 1 can be somewhat relaxed and still maintain horizontal regularity of the transformation (1).

Theorem 2. Suppose that the double array (s_{nk}) enjoys property (22) and that conditions (i) and (iii) of Theorem 1 are satisfied. If

$$\lim_{k \to \infty} \sum_{i=0}^{n} \sum_{j=0}^{k} \mu_{ij}^{nk} = 1, \qquad n \geq 0, \tag{23}$$

then

$$\lim_{k \to \infty} \sum_{i=0}^{n} \sum_{j=0}^{k} \mu_{ij}^{nk} s_{ij} = \beta, \qquad n > 0. \tag{24}$$

Proof. Obvious in view of the proof of Theorem 1. ∎

This theorem aids in the design of transformations of the double arrays. If we assume that the better approximations to the limit β appear for the larger indices n and k, the weights μ_{ij}^{nk} should put more mass on the larger indices i and j than on the smaller indices.

Therefore, let f be a function from $J^0 \times J^0$ to $\mathscr{R}^+$ satisfying

$$f(\mu, j) > f(\nu, j) \qquad \text{for} \quad \mu > \nu \quad \text{and all} \quad j, \tag{25}$$

$$f(i, \mu) > f(i, \nu) \qquad \text{for} \quad \mu > \nu \quad \text{and all} \quad i, \tag{26}$$

and $\sum_{j=0}^{\infty} f(0, j)$ diverges. We choose the weights μ_{ij}^{nk} by

$$\mu_{ij}^{nk} = f(i, j) \Big/ \sum_{\mu=0}^{n} \sum_{\nu=0}^{k} f(\mu, \nu), \tag{27}$$

which leads to $\sum_{i=0}^{n} \sum_{j=0}^{k} \mu_{ij}^{nk} = 1$ for all n and k, so that (23) is satisfied.

Condition (iii) of Theorem 1 is satisfied in view of the divergence of the sum $\sum_{j=0}^{\infty} f(1, j)$. Condition (i) is satisfied because of the positivity of f.

Thus, the weights (27) define a transformation for which $\lim_{k \to \infty} s_{nk} = \beta$ for all n implies $\lim_{k \to \infty} \bar{s}_{nk} = \beta$ for all n. These weights will be computationally useful only when the function f reflects the input double array.

13.2. Crystal Lattice Sums

An important class of multidimensional sums arises in the theory of crystal lattices, specifically in the computation of the lattice energy per atom of a given crystalline material. Let $f: J^p \to \mathscr{R}$, $J^p = J \times \cdots \times J$, $M_p = (m_1, \ldots, m_p)$, $A_p = (a_1, \ldots, a_p)$, $0_p = (0, 0, \ldots, 0)$, $\|M_p\| = (m_1^2 + \cdots + m_p^2)^{1/2}$. [Where there is no chance of misunderstanding, we omit the subscript p.) The sums of interest are generally of the form

$$S = \sum_{M \neq A} \frac{f(M)}{\|M - A\|^{2s}}, \qquad s \in \mathscr{C}. \tag{1}$$

f is usually quite simple, typical examples being

$$f = 1, \qquad f = (-1)^{m_1 + m_2 + \cdots + m_p}, \tag{2}$$

although in so-called phase modulated sums (Glasser, 1974) more complicated functions occur.

Sometimes the m_j range over only even or odd numbers, but it is not useful to develop a special notation to deal with such cases.

It is not at all obvious when (1) converges. The following theorem is often applicable.

Theorem. Let $M - A \neq 0$ and let f be bounded. Then (1) converges and represents an analytic function of s for $\operatorname{Re} s > p/2$, convergence being obtained regardless of the order in p-space in which the terms are added up.

Proof. The most elegant demonstration uses the theory of theta functions. This proof is given in Section 13.2.2. ■

For a discussion of the physical context in which such sums arise, see the classic treatise by Born and Huang (1954).

We shall take an approach with these sums that is fundamentally different from the procedures used previously in this book to accelerate the convergence of series or sequences. The techniques given here will not be general, but will very much depend on the specific character of f. This is, of course, very much in contrast to the previous work—for instance, the fact that the remainder sequence possessed an asymptotic series of Poincaré type—where only the *general form* of the sequence or series was of interest. The present kind of endeavor might be called the *analytic* approach to sequence transformations. The arguments used will depend on known properties of mathematical functions, such as theta functions, and on the application of a powerful formula from classical analysis, the Poisson summation formula.

13.2.1. Exact Methods

Definition. Let f be locally $L(0, \infty)$ and let the integral

$$\mathcal{M}(f; s) = \int_0^\infty x^{s-1} f(x)\, dx \tag{1}$$

converge for $\operatorname{Re} s = t_0$, $\operatorname{Re} s = t_1$, $t_0 < t_1$. $\mathcal{M}$ is called the *Mellin transform* of f.

Clearly the integral converges for

$$\alpha < \operatorname{Re} s < \beta \tag{2}$$

where $\alpha = \inf t_0$ and $\beta = \sup t_1$. (2) is called the strip of absolute convergence of (1).

The *Mellin inversion theorem* states that if f is of bounded variation in a neighborhood of $x \in (0, \infty)$, then, for any $\alpha < c < \beta$,

$$\frac{f(x^+) + f(x^-)}{2} = \frac{1}{2\pi i} \lim_{R\to\infty} \int_{c-iR}^{c+iR} \mathcal{M}(f; s) x^{-s}\, ds. \tag{3}$$

Usually $\mathcal{M}$ may be continued analytically into a larger region $\mathcal{D}$ of the complex s plane, for instance,

$$\frac{\Gamma(s)}{a^s} = \int_0^\infty x^{s-1} e^{-ax}\, dx, \qquad \operatorname{Re} a > 0. \tag{4}$$

Here $\alpha = 0$, $\beta = \infty$, $\mathcal{D} = \mathcal{C} - \{0, -1, -2, \ldots\}$.

The theta functions for $x > 0$ are defined as follows.

$$\begin{aligned} \theta_2\left(0 \,\middle|\, \frac{ix}{\pi}\right) &= 2 \sum_{n=0}^{\infty} e^{-(n+1/2)^2 x}, \\ \theta_3\left(0 \,\middle|\, \frac{ix}{\pi}\right) &= 1 + 2 \sum_{n=1}^{\infty} e^{-n^2 x}, \\ \theta_4\left(0 \,\middle|\, \frac{ix}{\pi}\right) &= 1 + 2 \sum_{n=1}^{\infty} (-1)^n e^{-n^2 x}. \end{aligned} \tag{5}$$

For some of the many beautiful properties of these now almost forgotten functions, consult Whittaker and Watson (1962), Hancock (1909, Vol. I), or Bellman's more recent book (1961), which is compulsively readable. A good collection of formulas is in Abramowitz and Stegun (1964).

The following notation is standard:

$$\theta_i(0, q) = \theta_i(0 | \tau), \qquad q = e^{i\pi\tau}. \tag{6}$$

Thus $\theta_i(0|ix/\pi)$ corresponds to taking $q = e^{-x}$ in $\theta_i(0, q)$. Formulas such as the following can be found in Hancock (1909, Chapter XVIII):

$$\begin{aligned}\theta_2^2\left(0\left|\frac{ix}{\pi}\right.\right) &= 4\sum_{n=0}^{\infty}\frac{e^{-(n+1/2)x}}{(1+e^{-(2n+1)x})},\\ \theta_3^2\left(0\left|\frac{ix}{\pi}\right.\right) &= 1+4\sum_{n=0}^{\infty}\frac{(-1)^n e^{-(2n+1)x}}{(1-e^{-(2n+1)x})},\\ \theta_4^2\left(0\left|\frac{ix}{\pi}\right.\right) &= 1+4\sum_{n=0}^{\infty}\frac{(-1)^{n+1} e^{-(n+1)x}}{(1+e^{-(2n+2)x})}.\end{aligned} \tag{7}$$

Similar formulas exist for θ_j^4, etc.; see Hancock or Jacobi (1829), who gives a list of 47 such relationships.

It can be shown that $\theta_j(0|ix/\pi)$ has an algebraic singularity at $x = 0$; hence Mellin transforms of θ_2, $\theta_3 - 1$, $\theta_4 - 1$, etc., have a half-plane of convergence.

The Mellin transforms of theta functions generally involve meromorphic functions such as Riemann's zeta function, defined for $\operatorname{Re} s > 1$ by

$$\zeta(s) = \sum_{n=1}^{\infty}\frac{1}{n^s}. \tag{8}$$

We shall need the formulas

$$\begin{aligned}(1-2^{1-s})\zeta(s) &= \sum_{n=1}^{\infty}\frac{(-1)^{n-1}}{n^s}, &\quad \operatorname{Re} s > 1;\\ (1-2^{-s})\zeta(s) &= \sum_{n=0}^{\infty}\frac{1}{(2n+1)^s}, &\quad \operatorname{Re} s > 1.\end{aligned} \tag{9}$$

Another useful function is

$$L(s) = \sum_{n=0}^{\infty}\frac{(-1)^n}{(2n+1)^s}, \qquad \operatorname{Re} s > 1, \tag{10}$$

which satisfies the relationship

$$L(1-s) = (2/\pi)^s\Gamma(s)\sin(\pi s/2)L(s). \tag{11}$$

Obviously, $L(s)$ can be expressed in terms of the *generalized zeta function*

$$\zeta(s, a) = \sum_{n=0}^{\infty}\frac{1}{(n+a)^s}, \qquad -a \notin J^0, \quad \operatorname{Re} s > 1. \tag{12}$$

The Mellin transforms of powers of the theta functions can be found from such formulas as (7). For instance,

$$\begin{aligned}\mathscr{M}\left[\theta_2^2\left(0\,\middle|\,\frac{ix}{\pi}\right)\right] &= 4\mathscr{M}\left[\sum_{n,k=0}^{\infty}(-1)^k e^{-(n+1/2)(2k+1)x}\right]\\ &= 4\Gamma(s)\sum_{n,k=0}^{\infty}\frac{(-1)^k}{(n+\frac{1}{2})^s(2k+1)^s}\\ &= 4\Gamma(s)L(s)\sum_{n=0}^{\infty}\frac{2^s}{(2n+1)^s} = 4(2^s-1)\Gamma(s)\zeta(s)L(s). \qquad (13)\end{aligned}$$

Mellin transforms of *products* of theta functions can be found by using the *Landen transformations*,

$$\begin{aligned}\theta_2(0,q)\theta_3(0,q) &= \tfrac{1}{2}\theta_2^2(0,q^{1/2}),\\ \theta_2(0,q)\theta_4(0,q) &= \tfrac{1}{2}e^{-\pi i/4}\theta_2^2(0,iq^{1/2}),\\ \theta_3(0,q)\theta_4(0,q) &= \theta_4^2(0,q^2),\end{aligned} \qquad (14)$$

and the formula

$$\mathscr{M}\{f(ax);s\} = a^{-s}\mathscr{M}\{f(x);s\}. \qquad (15)$$

Table I gives some of the Mellin transforms that can be found this way.

Table I

Mellin Transforms Involving $\theta_j = \theta_j(0|ix/\pi)$

$f(x)$	$\mathscr{M}(f;s)$
θ_2	$2(2^{2s}-1)\Gamma(s)\zeta(2s)$
θ_3	$2\Gamma(s)\zeta(2s)$
θ_4	$2(2^{1-2s}-1)\Gamma(s)\zeta(2s)$
θ_2^2	$4(2^s-1)\Gamma(s)\zeta(s)L(s)$
θ_3^2-1	$4\Gamma(s)\zeta(s)L(s)$
θ_4^2-1	$4(2^{1-s}-1)\Gamma(s)\zeta(s)L(s)$
$(\theta_3-1)^2$	$4\Gamma(s)[L(s)\zeta(s)-\zeta(2s)]$
$(\theta_4-1)^2$	$4(1-2^{1-2s})\Gamma(s)[\zeta(2s)-L(s)\zeta(s)]$
$\theta_2\theta_3$	$2^{s+1}(2^s-1)\Gamma(s)\zeta(s)L(s)$
$\theta_3\theta_4-1$	$2^{2-s}(2^{1-s}-1)\Gamma(s)\zeta(s)L(s)$
$(\theta_3-1)(\theta_4-1)$	$-2^{2-s}\Gamma(s)[2^{-s}\zeta(2s)+(1-2^{1-s})\zeta(s)L(s)]$
$\theta_2\theta_3\theta_4$	$-2^{s+1}\Gamma(s)L(2s-1)$
θ_2^4	$16(1-2^{1-s})(1-2^{-s})\Gamma(s)\zeta(s)\zeta(s-1)$
θ_3^4-1	$8(1-2^{2-2s})\Gamma(s)\zeta(s)\zeta(s-1)$
θ_4^4-1	$-8(1-2^{1-s})(1-2^{2-s})\Gamma(s)\zeta(s)\zeta(s-1)$
$\theta_2^2\theta_4^2$	$2^{s+2}\Gamma(s)L(s)L(s-1)$
$\theta_3^2\theta_4^2-1$	$-2^{3-s}(1-2^{2-s})(1-2^{1-s})\Gamma(s)\zeta(s)\zeta(s-1)$
$\theta_2^2\theta_3^2$	$2^{2+s}(1-2^{1-s})(1-2^{-s})\Gamma(s)\zeta(s)\zeta(s-1)$

To see how these formulas can be used to obtain closed-form expressions for lattice sums, consider

$$S = \sum_{-\infty}^{\infty}{}' \frac{1}{(m_1^2 + m_2^2 + m_3^2 + m_4^2)^s}$$

$$= \sum_{-\infty}^{\infty}{}' \frac{1}{\Gamma(s)} \int_0^\infty x^{s-1} e^{-(m_1^2+m_2^2+m_3^2+m_4^2)x}\, dx$$

$$= \frac{1}{\Gamma(s)} \int_0^\infty x^{s-1} [\theta_3^4(0 \mid ix/\pi) - 1]\, dx$$

$$= 8(1 - 2^{2-2s})\zeta(s)\zeta(s-1). \tag{16}$$

[Later it is shown that this sum converges for Re $s > p/2 = 2$. Since $\zeta(s-1)$ has a pole at $s = 2$, the result is sharp.]

As another example, consider

$$S = \sum_{m,n=1}^{\infty} \frac{1}{(m^2 + n^2)^s}$$

$$= \frac{1}{\Gamma(s)} \sum_{m,n=1}^{\infty} \int_0^\infty x^{s-1} e^{-(m^2+n^2)x}\, dx$$

$$= \frac{1}{4\Gamma(s)} \int_0^\infty x^{s-1} \left[\theta_3\left(0 \middle| \frac{ix}{\pi}\right) - 1\right]^2 dx$$

$$= L(s)\zeta(s) - \zeta(2s). \tag{17}$$

A short table (Table II) lists two-dimensional sums determined by Glasser.

Table II

$$S = \sum_{-\infty}^{\infty}{}' f(m, n)$$

f	S
$(m^2 + n^2)^{-s}$	$4\zeta(s)L(s)$
$(-1)^{m+n}(m^2 + n^2)^{-s}$	$-4(1 - 2^{1-2s})\zeta(s)\mathrm{L}(s)$
$(-1)^{n+1}(m^2 + n^2)^{-s}$	$2^{2-s}(1 - 2^{1-s})\zeta(s)\mathrm{L}(s)$
$[(2m + 1)^2 + (2n + 1)^2]^{-s}$, $m, n \geq 0$	$2^{-s}(1 - 2^{-s})\zeta(s)\mathrm{L}(s)$
$(m^2 + 4n^2)^{-s}$	$2(1 - 2^{-s} + 2^{1-2s})\zeta(s)\mathrm{L}(s)$

Certain other related sums have been obtained, i.e.,

$$\sum_{-\infty}^{\infty}{}' (m^2 + mn + n^2)^{-s} = 6\zeta(s)g(s), \qquad g(s) = \sum_{n=0}^{\infty} [(3n+1)^{-s} - (3n+2)^{-s}] \tag{18}$$

(Fletcher *et al.*, 1962, p. 95), and

$$\sum_{m_1, m_2, m_3 \geq 1} (-1)^{m_2}[(m_1 - \tfrac{1}{2})^2 + m_2^2 + m_3^2]^{-s}, \tag{19}$$

whose derivation is rather complicated (Glasser, 1973b). Obviously, the following case can be expressed by a single sum:

$$\sum_{m_j \geq 1} (m_1 + m_2 + \cdots + m_p)^{-s} = \sum_{k=0}^{\infty} \binom{k+p-1}{k} \frac{1}{(k+p)^s}. \tag{20}$$

The difficulty in computing odd-dimensional sums by the use of theta functions is that most of the known theta function identities involve an even number of theta functions. Glasser (1937b) uses a number-theoretic approach to obtain additional sums, and the theory of basic hypergeometric series (Glasser, 1975) can be used to deduce the five-dimensional sum

$$\sum_{m_1 \geq 0; m_2, \cdots; m_5 \geq 1} (m_1 m_2 + m_2 m_3 + m_3 m_4 + m_4 m_1 + m_2 m_5)^{-s}$$
$$= \zeta(s)\zeta(s-2) - \zeta^2(s-1). \tag{21}$$

(The region of convergence of this sum cannot be deduced from the theorem of Section 13.2.)

13.2.2. Approximate Methods: The Poisson Summation Formula

Many approximation techniques have been developed to deal with lattice sums, beginning, perhaps, with Born's and Huang's approach, which uses values of the incomplete gamma function. That approach is not very adaptable to general values of s. Other approaches (van der Hoff and Benson, 1953; Benson and Schreiber, 1955; Hautot, 1974) use methods that convert the sum to a multidimensional sum involving the modified Bessel functions K_ν. This might, at first glance, seem to be compounding the problems. However, the transformed sums converge with extraordinary rapidity, and often the contributions at just a few lattice points serve to give six- or eight-place accuracy. Several approaches are possible, including one (Hautot, 1974) using Schlömilch series. My own preference is to begin with the following striking result, which can be found in any book on Fourier methods [e.g., Butzer and Nessel (1971, p. 202)].

Theorem. Let $f \in L(-\infty, \infty)$,

$$F(x) = \int_{-\infty}^{\infty} e^{-ixt} f(t)\, dt, \qquad x \in \mathscr{R}. \tag{1}$$

Then, if f is of bounded variation,

$$2\pi \sum_{k=-\infty}^{\infty} f(x + 2k\pi) = \lim_{n\to\infty} \sum_{k=-n}^{n} e^{ikx} F(k), \qquad x \in \mathscr{R}, \tag{2}$$

where, at points of discontinuity, $f(a) = \frac{1}{2}[f(a^+) + f(a^-)]$.

Proof. See Butzer and Nessel (1971). ■

There follows a list of formulas that will subsequently be of use. For the computation of the integrals involved, consult Erdélyi *et al.* (1954, Vol. I).

$$f(t) = e^{-at^2} \cos bt, \qquad a \in \mathscr{R}^+, \quad b \in \mathscr{R};$$

$$\sum_{-\infty}^{\infty} e^{-a(x+2k\pi)^2} \cos[b(x + 2k\pi)] = \frac{1}{2\sqrt{\pi a}} \sum_{-\infty}^{\infty} e^{ikx} e^{-(k^2+b^2)/4a} \cosh\left(\frac{bk}{2a}\right). \tag{S-1}$$

$$f(t) = |t|^{\pm\mu} K(a|t|), \qquad a \in \mathscr{R}^+;$$

$$\sum_{-\infty}^{\infty} e^{ikx}(k^2 + a^2)^{\mp\mu-1/2} = \frac{2\sqrt{\pi}}{(2a)^{\pm\mu}\Gamma(\pm\mu + 1/2)} \times \sum_{-\infty}^{\infty} |x + 2k\pi|^{\pm\mu} K_\mu(a|x + 2k\pi|). \tag{S-2}$$

(By analytic continuation and use of the well-known asymptotic properties of K_μ, one finds that these sums are convergent and equal when $\mathrm{Re}(\pm\mu) > 0$.)

$$f(t) = \sqrt{t^2 + a^2}^{\,-1} e^{-b\sqrt{t^2+a^2}}, \qquad a, b \in \mathscr{R}^+;$$

$$\pi \sum_{-\infty}^{\infty} [(x + 2k\pi)^2 + a^2]^{-1/2} e^{-b\sqrt{(x+2k\pi)^2+a^2}} = \sum_{-\infty}^{\infty} e^{ikx} K_0(a\sqrt{b^2 + k^2}); \tag{S-3}$$

$$f(t) = e^{-b\sqrt{t^2+a^2}}, \qquad a, b \in \mathscr{R}^+;$$

$$\frac{\pi}{ab} \sum_{-\infty}^{\infty} e^{-b\sqrt{(x+2k\pi)^2+a^2}} = \sum_{-\infty}^{\infty} e^{ikx} \sqrt{b^2 + k^2}^{\,-1} \times K_1(a\sqrt{b^2 + k^2}); \tag{S-4}$$

$$f(t) = (t^2 + b^2)^{\mp\mu/2} K_\mu(a\sqrt{t^2 + b^2}), \qquad a, b \in \mathscr{R}^+;$$

$$\sqrt{2\pi}^3 a^{\pm\mu} b^{1/2\mp\mu} \sum_{-\infty}^{\infty} [(x + 2k\pi)^2 + a^2]^{\pm\mu/2-1/4} K_{\pm\mu-1/2}(b\sqrt{(x+2k\pi)^2+a^2})$$
$$= \sum_{-\infty}^{\infty} e^{ikx}(k^2 + b^2)^{\mp\mu/2} K_\mu(a\sqrt{b^2 + k^2}), \qquad \mu \in \mathscr{C}. \tag{S-5}$$

We are now in a position to complete the proof of the theorem in Section 13.2. Let

$$S = \sum_{\substack{|m_j|<N_j \\ M \neq 0}} \frac{f(M)}{\|M\|^{2s}}. \tag{3}$$

(Without loss of generality we may assume that $A = 0$.) Then

$$S = \frac{1}{\Gamma(s)} \int_0^\infty g\, dx, \tag{4}$$

$$g = x^{s-1} \sum_{\substack{|m_j|<N_j \\ M \neq 0}} f(M) e^{-\|M\|^2 x}. \tag{5}$$

We get

$$g \leq C x^{\mathrm{Re}\, s-1}[\theta_3(0|ix/\pi)^p - 1] = h, \tag{6}$$

and, by (S-1) with $b = x = 0$

$$h = O(x^{\mathrm{Re}\, s-1-p/2}), \qquad x \to 0^+, \tag{7}$$

while as $x \to \infty$, $h = O(e^{-\alpha x})$, $\alpha > 0$. Thus, under the stated conditions, h is integrable and, by dominated convergence, $\lim_{N_j \to \infty} S$ exists. ■

Expansion (S-2) will be the principal tool. Let $k \to m_p$,

$$a = (m_1^2 + \cdots + m_{p-1}^2 + \delta^2)^{1/2},$$

multiply both sides by $e^{ix(m_1+\cdots+m_{p-1})}$, take the upper sign on μ, $\mu \to s - \frac{1}{2}$, and sum:

$$\sum_{-\infty}^{\infty} \frac{e^{ix(m_1+\cdots+m_p)}}{[\|M_p\|^2 + \delta^2]^s} = \frac{2^{3/2-s}\sqrt{\pi}}{\Gamma(s)} \sum_{-\infty}^{\infty} \frac{|x + 2m_p\pi|^{s-1/2} e^{ix(m_1+\cdots+m_{p-1})}}{[\|M_{p-1}\|^2 + \delta^2]^{s/2-1/4}}$$
$$\times K_{s-1/2}(\sqrt{\|M_{p-1}\|^2 + \delta^2}\,|x + 2m_p\pi|). \tag{8}$$

Now let $\delta \to 0$. The result can be written

$$\sum_{-\infty}^{\infty}{}' \frac{e^{ix(m_1+\cdots+m_p)}}{\|M_p\|^{2s}} = \frac{2^{3/2-s}\sqrt{\pi}}{\Gamma(s)} \sum_{\substack{-\infty \\ M_{p-1} \neq 0}}^{\infty} \frac{|x + 2m_p\pi|^{s-1/2}}{\|M_{p-1}\|^{s-1/2}} e^{ix(m_1+\cdots+m_{p-1})}$$
$$\times K_{s-1/2}(\|M_{p-1}\|\,|x + 2m_p\pi|) + 2\sum_{k=1}^{\infty} \frac{\cos kx}{k^{2s}}. \tag{9}$$

As it stands, this holds only for $x \neq 2j\pi$, $j \in J$. For $x \to 0^+$, the $m_p = 0$ term must be peeled off and the relationship

$$\varepsilon^\nu K_\nu(\varepsilon) \approx (\tfrac{1}{2})^{1-\nu}\Gamma(\nu), \qquad \varepsilon \to 0^+, \tag{10}$$

used. The result gives the original sum as a sum over one lower dimension plus a rapidly convergent series of Bessel functions.

$$\sum_{-\infty}^{\infty}{}' \|M_p\|^{-2s} = 2\zeta(2s) + \frac{\sqrt{\pi}\,\Gamma(s - 1/2)}{\Gamma(s)} \sum_{-\infty}^{\infty}{}' \|M_{p-1}\|^{1-2s}$$

$$+ \frac{2\pi^s}{\Gamma(s)} \sum_{-\infty}^{\infty}{}' \frac{|m_p|^{s-1/2}}{\|M_{p-1}\|^{s-1/2}} K_{s-1/2}(\|M_{p-1}\|\,|2m_p\pi|). \quad (11)$$

The Bessel function expansions on the right of (9) and (11), still expansions over p-space, converge with great rapidity. Also, for the values of x of greatest interest, the cosine series on the right of (9) can be evaluated in terms of zeta and related functions. For other values of s, it can be dealt with by the asymptotic techniques of Section 1.6. In many cases, s is an integer. The series on the right then becomes a series of exponentials. (An example is given later on.) In any event, the Bessel function K_ν can be considered a known quantity, its computation today being standard software.

For $s = \frac{1}{2}$ in the case of a three-dimensional sum, there is convergence provided $x \neq 2j\pi$, $j \in J$. The (m_1, m_2) sum can be expressed in terms of exponentials by (S-3), i.e.,

$$\sum_{m_1=1}^{\infty} \sum_{m_2=-\infty}^{\infty} e^{ix(m_1+m_2)} K_0(a\sqrt{m_1^2 + m_2^2}) = \pi \sum_{m_2=-\infty}^{\infty} [(x + 2m_2\pi)^2 + a^2]^{-1/2}$$

$$\times (e^{\sqrt{(x+2m_2\pi)^2+a^2} - ix} - 1)^{-1}, \quad (12)$$

and this can be used in an obvious way in (11). The same applies, with (S-4), when $p = 3$ and $s = \frac{3}{2}$.

As an example of how an error analysis of these sums proceeds, let us examine (11). Assume the Bessel function sum is truncated, with all points inside the hypercube

$$\sup_{0 \le j \le p} |x_j| = N \quad (13)$$

included. Let

$$R_N = \sum_{|m_j|=N+1} \frac{|m_p|^{s-1/2}}{\|M_{p-1}\|^{s-1/2}} K_{s-1/2}(\|M_{p-1}\|\,|2m_p\pi|). \quad (14)$$

For an analysis of R_N, we shall need several preliminary results useful with sums of this kind.

Lemma 1. Let $\alpha, n > 0$, $\beta > \alpha/n$. Then

$$\sum_{k=n}^{\infty} k^\alpha e^{-\beta k} \le n^\alpha e^{-\beta n}(1 - e^{\alpha/n - \beta})^{-1}. \quad (15)$$

Proof. By calculus one finds that

$$x^\alpha \leq (\alpha/\delta)^\alpha e^{-\alpha} e^{\delta x}, \qquad \alpha, \delta, x > 0. \tag{16}$$

Letting $x \to k$, $\delta \to \alpha/n$, and substituting the result in (15) proves the lemma. ■

Lemma 2. For $\lambda \geq 1$, $\operatorname{Re} \nu > -\frac{1}{2}$,

$$\left| \frac{K_\nu(\lambda)e^\lambda}{\lambda^\nu} \right| \leq \frac{\Gamma(\operatorname{Re}\,\nu + \frac{1}{2})}{|\Gamma(\nu + \frac{1}{2})|} K_{\operatorname{Re}\,\nu}(1) = c_\nu. \tag{17}$$

Proof. This follows immediately from the integral

$$\frac{K_\nu(z)e^z}{z^\nu} = \sqrt{\frac{\pi}{2}}\,\Gamma\left(\nu + \frac{1}{2}\right)^{-1} \int_0^\infty e^{-zt}\left[t\left(1 + \frac{t}{2}\right)\right]^{\nu - 1/2} dt,$$

$$\operatorname{Re} \nu > -\tfrac{1}{2}, \quad \operatorname{Re} z > 0. \quad ■ \tag{18}$$

Lemma 3

$$e^{-\|M_p\|\alpha} \leq e^{-(\alpha/p)(m_1 + m_2 + \cdots + m_p)}, \qquad \alpha, m_j \geq 0. \tag{19}$$

Proof

$$(m_1^2 + \cdots + m_p^2)^{1/2} \geq \max |m_j|, \tag{20}$$

so

$$(m_1^2 + \cdots + m_p^2)^{1/2} \geq (1/p)(m_1 + \cdots + m_p) \tag{21}$$

and the lemma follows. ■

A straightforward application of all these results shows that for $s \geq \frac{1}{2}$

$$\begin{aligned} |R_N| \leq\; & \frac{2^{s+p+1/2}\pi^{2s-1/2}c_{s-1/2}(N+1)^{2s-1}\exp\{-[2\pi(p-1)/p](N+1)^2\}}{\Gamma(s)} \\ & \times \{1 - \exp[-2\pi(N+1)/p]\}^{1-p} \\ & \times \{1 - \exp[(2s-1)/(N+1) - 2\pi(p-1)(N+1)/p]\}^{-1} \\ \approx\; & \frac{2^{s+p+1/2}\pi^{2s-1/2}}{\Gamma(s)} K_{s-1/2}(1)\exp\{-[2\pi(p-1)/p](N+1)^2\}, \quad N \to \infty. \end{aligned} \tag{22}$$

For instance, if $N = 2$, the truncated sum will contain 26 terms if $p = 3$. The exponential term above is 4.2×10^{-17}. If only seven terms are taken ($N = 1$), the exponential term is still only 5.3×10^{-8}.

The case $s = 1$ of (9) is particularly important. It gives

$$\sum_{-\infty}^{\infty}{}' \frac{e^{ix(m_1 + \cdots + m_p)}}{m_1^2 + \cdots + m_p^2} = \pi \sum_{\substack{-\infty \\ (m_1, \ldots, m_{p-1}) \neq 0}}^{\infty} e^{ix(m_1 + \cdots + m_{p-1})} \times \frac{e^{-\sqrt{m_1^2 + \cdots + m_{p-1}^2}|x + 2m_p\pi|}}{\sqrt{m_1^2 + \cdots + m_{p-1}^2}} + 2\sum_{k=1}^{\infty} \frac{\cos kx}{k^2}, \tag{23}$$

a rapidly convergent series of exponentials.

Obviously the forgoing procedure is easily modified to account for sums with denominator $\|M - A\|^s$, $A = (a_1, \ldots, a_p)$. For many special cases, see Hautot's paper.

13.2.3. Laguerre Quadrature

This is an elementary but very accurate method for hand computations. It can be applied for certain functions f when $s - 1 - \frac{1}{2}p$ is a value β for which the abscissas and weights for the Laguerre quadrature formula for $x^{\beta}e^{-x}$ have been tabulated, e.g., $\beta = 0, -\frac{1}{4}, -\frac{1}{2}, -\frac{3}{4}$, etc (Concus *et al.*, 1963.) This is illustrated for $f \equiv 1$.

$$\sum_{-\infty}^{\infty}{}' \frac{1}{\|M\|^{2s}} = \frac{1}{\Gamma(s)} \int_0^{\infty} x^{s-1-p/2} e^{-x} h(x)\, dx,$$
$$h(x) = e^x x^{p/2} [\theta_3(0 | ix/\pi)^p - 1]. \tag{1}$$

The integral on the right is easily evaluated by Laguerre quadrature, since the series for θ_3 converges with great rapidity.

For example, let $p = 2$, $s = \frac{3}{2}$.

$$S = \sum_{-\infty}^{\infty}{}' \frac{1}{(m_1^2 + m_2^2)^{3/2}} = \frac{2}{\sqrt{\pi}} \int_0^{\infty} x^{-1/2} e^{-x} h(x)\, dx,$$
$$h(x) = e^x x \left[\theta_3\left(0 \left| \frac{ix}{\pi} \right.\right)^2 - 1 \right]. \tag{2}$$

Laguerre quadrature with just three abcissas yields $S = 9.0352$, while the true value is 9.0336.

Appendix

A.1. Lagrangian Interpolation

Let $\mathbf{x}, \mathbf{y} \in \mathscr{C}_S$, and denote by $p_n^{(k)}(z)$ the polynomial of degree k that at $x_n, x_{n-1}, \ldots, x_{n+k}$ assumes the values $y_n, y_{n+1}, \ldots, y_{n+k}$, respectively. (It is assumed the x_j are distinct.) Then

$$p_n^{(k)}(z) = \sum_{m=0}^{k} y_{n+m} \prod_{\substack{i=0 \\ i \neq m}}^{k} \left(\frac{z - x_{n+i}}{x_{n+m} - x_{n+i}} \right). \tag{1}$$

It is easily shown that $p_n^{(k)}$ satisfies the recursion relationship

$$p_n^{(k+1)} = \frac{(x_n - z)p_{n+1}^{(k)} - (x_{n+k+1} - z)p_n^{(k)}}{x_n - x_{n+k+1}}, \quad n, k \geq 0, \quad p_n^{(0)} = y_n, \quad n \geq 0, \tag{2}$$

by putting $z = x_i$, $n \leq i \leq n + k + 1$.

Another useful expression for $p_n^{(k)}$ comes from expanding the determinant

$$\begin{vmatrix} p_n^{(k)} & 1 & z & z^2 & \cdots & z^k \\ y_n & 1 & x_n & x_n^2 & \cdots & x_n^k \\ y_{n+1} & 1 & x_{n+1} & x_{n+1}^2 & \cdots & x_{n+1}^k \\ \vdots & \vdots & \vdots & \vdots & & \vdots \\ y_{n+k} & 1 & x_{n+k} & x_{n+k}^2 & \cdots & x_{n+k}^k \end{vmatrix} = 0. \tag{3}$$

Let $u_j \in \mathscr{C}$, and denote the Vandermonde determinant V_m by

$$V_m(u_1, u_2, \ldots, u_m) = \begin{vmatrix} 1 & u_1 & u_1^2 & \cdots & u_1^m \\ 1 & u_2 & u_2^2 & \cdots & u_2^m \\ \vdots & \vdots & \vdots & & \vdots \\ 1 & u_m & u_m^2 & \cdots & u_m^m \end{vmatrix} = \prod_{i=0}^{m-1} \prod_{j=i+1}^{m} (u_j - u_i). \quad (4)$$

Expanding the determinants (3) by minors of the first column and using (4) shows that the determinantal expression is the same as the sum (1).

A.2. The Formula for the ε-Algorithm

The proof of Eqs. 6.7(1)–6.7(3) depends on two determinantal identities. It will be very useful to use Aitken's shorthand notation for determinants, writing only diagonal elements. For instance,

$$|a_1 b_3 d_4 e_7| = \begin{vmatrix} a_1 a_3 a_4 a_7 \\ b_1 b_3 b_4 b_7 \\ d_1 d_3 d_4 d_7 \\ e_1 e_3 e_4 e_7 \end{vmatrix}, \quad (1)$$

and so forth.

The two identities are the obvious generalizations to $n \times n$ determinants of

$$|h_1 b_2 c_3 d_4|\,|a_1 b_2 c_3| - |a_1 b_2 c_3 d_4|\,|h_1 b_2 c_3| = |h_1 a_2 b_3 c_4|\,|b_1 c_2 d_3|, \quad (2)$$

which relates determinants with different first rows, and

$$|a_1 b_2 c_3 d_4|\,|a_1 b_2 c_3 e_5| - |a_1 b_2 c_3 d_5|\,|a_1 b_2 c_3 e_4| = |a_1 b_2 c_3 d_4 e_5|\,|a_1 b_2 c_3|, \quad (3)$$

which is an expression of the cross product of determinants whose last rows and columns differ in a certain way [see Aitken (1956, p. 108, No. 2; p. 49, No. 8)].

First, Eq. 6.8(1) is true when $k = 1$ for

$$\varepsilon_0^{(n+1)} + [\varepsilon_1^{(n+1)} - \varepsilon_1^{(n)}]^{-1} = s_{n+1} + [(\Delta s_{n+1})^{-1} - (\Delta s_n)^{-1}]^{-1}$$

$$= \frac{s_n \Delta s_{n+1} - s_{n+1} \Delta s_n}{\Delta^2 s_n} = \varepsilon_2^{(n)}. \quad (4)$$

Next consider the case $k = 2m$, $m \geq 1$. Let

$$\Omega_n = \frac{\begin{vmatrix} 1 & \cdots & 1 \\ \Delta^2 s_n & \cdots & \Delta^2 s_{n+m} \\ \vdots & & \vdots \\ \Delta^2 s_{n+m-1} & \cdots & \Delta^2 s_{n+2m-1} \end{vmatrix}}{\begin{vmatrix} \Delta s_n & \cdots & \Delta s_{n+m} \\ \vdots & & \vdots \\ \Delta s_{n+m} & \cdots & \Delta s_{n+2m} \end{vmatrix}} - \frac{\begin{vmatrix} 1 & \cdots & 1 \\ \Delta^2 s_{n+1} & \cdots & \Delta^2 s_{n+m} \\ \vdots & & \vdots \\ \Delta^2 s_{n+m-1} & \cdots & \Delta^2 s_{n+2m-1} \end{vmatrix}}{\begin{vmatrix} \Delta s_{n+1} & \cdots & \Delta s_{n+m} \\ \vdots & & \vdots \\ \Delta s_{n+m} & \cdots & \Delta s_{n+2m-1} \end{vmatrix}} \tag{5}$$

and

$$\Omega_n' = \left[\frac{\begin{vmatrix} s_{n+1} & \cdots & s_{n+m+1} \\ \Delta s_{n+1} & \cdots & \Delta s_{n+m+1} \\ \vdots & & \vdots \\ \Delta s_{n+m} & \cdots & \Delta s_{n+2m} \end{vmatrix}}{\begin{vmatrix} 1 & \cdots & 1 \\ \Delta s_{n+1} & \cdots & \Delta s_{n+m+1} \\ \vdots & & \vdots \\ \Delta s_{n+m} & \cdots & \Delta s_{n+2m} \end{vmatrix}} - \frac{\begin{vmatrix} s_n & \cdots & s_{n+m} \\ \Delta s_n & \cdots & \Delta s_{n+m} \\ \vdots & & \vdots \\ \Delta s_{n+m-1} & \cdots & \Delta s_{n+2m-1} \end{vmatrix}}{\begin{vmatrix} 1 & \cdots & 1 \\ \Delta s_n & \cdots & \Delta s_{n+m} \\ \vdots & & \vdots \\ \Delta s_{n+m-1} & \cdots & \Delta s_{n+2m-1} \end{vmatrix}} \right]^{-1}. \tag{6}$$

We must show these are the same. Rearranging the elements of the first gives

$$\Omega_n = \frac{\begin{vmatrix} 1 & \cdots & 1 & 1 \\ \Delta^2 s_{n+1} & \cdots & \Delta^2 s_{n+m} & \Delta^2 s_n \\ \vdots & & \vdots & \vdots \\ \Delta^2 s_{n+m} & \cdots & \Delta^2 s_{n+2m-1} & \Delta^2 s_{n+m-1} \end{vmatrix}}{\begin{vmatrix} \Delta s_{n+1} & \cdots & \Delta s_{n+m} & \Delta s_n \\ \Delta^2 s_{n+1} & \cdots & \Delta^2 s_{n+m} & \Delta^2 s_n \\ \vdots & & \vdots & \vdots \\ \Delta^2 s_{n+m} & \cdots & \Delta^2 s_{n+2m-1} & \Delta^2 s_{n+m-1} \end{vmatrix}}$$

$$- \frac{\begin{vmatrix} 1 & \cdots & 1 \\ \Delta^2 s_{n+1} & \cdots & \Delta^2 s_{n+m} \\ \vdots & & \vdots \\ \Delta^2 s_{n+m-1} & \cdots & \Delta^2 s_{n+2m-2} \end{vmatrix}}{\begin{vmatrix} \Delta s_{n+1} & \cdots & \Delta s_{n+m} \\ \Delta^2 s_{n+1} & \cdots & \Delta^2 s_{n+m} \\ \vdots & & \vdots \\ \Delta^2 s_{n+m-1} & \cdots & \Delta^2 s_{n+2m-2} \end{vmatrix}}, \tag{7}$$

and using the determinantal identity (2) above one gets Eq. (8).

$$\Omega_n = \frac{\begin{vmatrix} 1 & \cdots & 1 & 1 \\ \Delta s_{n+1} & \cdots & \Delta s_{n+m} & \Delta s_n \\ \Delta^2 s_{n+1} & \cdots & \Delta^2 s_{n+m} & \Delta^2 s_n \\ \vdots & & \vdots & \vdots \\ \Delta^2 s_{n+m-1} & \cdots & \Delta^2 s_{n+2m-1} & \Delta^2 s_{n+m-1} \end{vmatrix} \begin{vmatrix} \Delta^2 s_{n+1} & \cdots & \Delta^2 s_{n+m} \\ \vdots & & \vdots \\ \Delta^2 s_{n+m} & \cdots & \Delta^2 s_{n+2m-1} \end{vmatrix}}{\begin{vmatrix} \Delta s_{n+1} & \cdots & \Delta s_{n+m} & \Delta s_n \\ \Delta^2 s_{n+1} & \cdots & \Delta^2 s_{n+m} & \Delta^2 s_n \\ \vdots & & \vdots & \vdots \\ \Delta^2 s_{n+m} & \cdots & \Delta^2 s_{n+2m-1} & \Delta^2 s_{n+m-1} \end{vmatrix} \begin{vmatrix} \Delta s_{n+1} & \cdots & \Delta s_{n+m} \\ \Delta^2 s_{n+1} & \cdots & \Delta^2 s_{n+m} \\ \vdots & & \vdots \\ \Delta^2 s_{n+m-1} & \cdots & \Delta^2 s_{n+2m-2} \end{vmatrix}}$$

$$= \frac{\begin{vmatrix} 1 & \cdots & 1 \\ \Delta s_n & \cdots & \Delta s_{n+m} \\ \vdots & & \vdots \\ \Delta s_{n+m-1} & \cdots & \Delta s_{n+2m-1} \end{vmatrix} \begin{vmatrix} 1 & \cdots & 1 \\ \Delta s_{n+1} & \cdots & \Delta s_{n+m+1} \\ \vdots & & \vdots \\ \Delta s_{n+m} & \cdots & \Delta s_{n+2m} \end{vmatrix}}{\begin{vmatrix} \Delta s_n & \cdots & \Delta s_{n+m} \\ \Delta s_{n+1} & \cdots & \Delta s_{n+m+1} \\ \vdots & & \vdots \\ \Delta s_{n+m} & \cdots & \Delta s_{n+2m} \end{vmatrix} \begin{vmatrix} \Delta s_{n+1} & \cdots & \Delta s_{n+m} \\ \vdots & & \vdots \\ \Delta s_{n+m} & \cdots & \Delta s_{n+2m-1} \end{vmatrix}}. \tag{8}$$

The second quantity [Eq. (6)] may be written

$$\Omega'_n = \frac{(-1)^k}{D_n} \begin{vmatrix} 1 & \cdots & 1 \\ \Delta s_n & \cdots & \Delta s_{n+m} \\ \vdots & & \vdots \\ \Delta s_{n+m-1} & \cdots & \Delta s_{n+2m-1} \end{vmatrix} \begin{vmatrix} 1 & \cdots & 1 \\ \Delta s_{n+1} & \cdots & \Delta s_{n+m+1} \\ \vdots & & \vdots \\ \Delta s_{n+m} & \cdots & \Delta s_{n+2m} \end{vmatrix},$$

$$D_n = \begin{vmatrix} \Delta s_{n+1} & \cdots & \Delta s_{n+m+1} \\ \vdots & & \vdots \\ \Delta s_{n+m} & \cdots & \Delta s_{n+2m} \\ s_{n+1} & \cdots & s_{n+m-1} \end{vmatrix} \begin{vmatrix} \Delta s_{n+1} & \cdots & \Delta s_n \\ \vdots & & \vdots \\ \Delta s_{n+m} & \cdots & \Delta s_{n+m-1} \\ 1 & \cdots & 1 \end{vmatrix} \tag{9}$$

$$- \begin{vmatrix} \Delta s_{n+1} & \cdots & \Delta s_{n+m} & \Delta s_n \\ \vdots & & \vdots & \vdots \\ \Delta s_{n+m} & \cdots & \Delta s_{n+2m-1} & \Delta s_{n+m-1} \\ s_{n+1} & \cdots & s_{n+m} & s_n \end{vmatrix} \begin{vmatrix} \Delta s_{n+1} & \cdots & \Delta s_{n+m} \\ \vdots & & \vdots \\ \Delta s_{n+m} & \cdots & \Delta s_{n+2m} \\ 1 & \cdots & 1 \end{vmatrix}$$

On D_n we use the second identity to find

$$D_n = \begin{vmatrix} \Delta s_{n+1} & \cdots & \Delta s_{n+m+1} & \Delta s_n \\ \vdots & & \vdots & \vdots \\ \Delta s_{n+m} & \cdots & \Delta s_{n+2m} & \Delta s_{n+m-1} \\ 1 & \cdots & 1 & 1 \end{vmatrix} \begin{vmatrix} \Delta s_{n+1} & \cdots & \Delta s_{n+m} \\ \vdots & & \vdots \\ \Delta s_{n+m} & \cdots & \Delta s_{n+2m-1} \end{vmatrix}. \tag{10}$$

Elementary determinant manipulations show that the first factor above is $(-1)^k$ times the first factor in the denominator of Ω_n. Thus $\Omega_n = \Omega_n'$.

The proof for $k = 2m + 1$ is similar.

A.3. Sylvester's Expansion Theorem

Let A be an $n \times n$ determinant, $n \geq 3$, with elements a_{ij} and denote the minor of element a_{ij} by M_{ij}. Let

$$D = \begin{vmatrix} a_{22} & \cdots & a_{2,n-1} \\ \vdots & & \vdots \\ a_{n-1,2} & \cdots & a_{n-1,n-1} \end{vmatrix}. \tag{1}$$

Then

$$AD = M_{11}M_{nn} - M_{1n}M_{n1} \tag{2}$$

[see Muir (1960, p. 132)].

Bibliography

Abramowitz, M., and Stegun, I. A. (eds.) (1964). "Handbook of Mathematical Functions." National Bureau of Standards Applied Mathematics Series #55, Washington, D.C.

Agnew, R. P. (1952). *Proc. Amer. Math. Soc.* **3**, 550–556.

Agnew, R. P. (1957). *Michigan Math. J.* **4**, 105–128.

Aitken, A. C. (1926). *Proc. Roy. Soc. Edinburgh Sect. A* **46**, 289–305.

Aitken, A. C. (1931). *Proc. Roy. Soc. Edinburgh Sect. A* **51**, 80–90.

Aitken, A. C. (1956). "Determinants and Matrices." Oliver & Boyd, London.

Allen, G. D., Chui, C. K., Madych, W. R., Narcowich, F. J., and Smith, P. W. (1975). *J. Approx. Theory* **14**, 302–316.

Atchison, T. A., and Gray, H. L. (1968). *SIAM J. Numer. Anal.* **5**, 451–459.

Bajšanski, B., and Karamata, J. (1960). *Acad. Serbe Sci. Publ. Inst. Math.* **14**, 109–114.

Baker, G. A., Jr., and Gammel, J. L. (eds.) (1970). "The Padé Approximant in Theoretical Physics." Academic Press, New York.

Banach, S. (1932). "Theorie des Opérations Linéaires." Chelsea, New York.

Baranger, J. (1970). *C.R. Acad. Sci. Paris Sér. A* **271**, 149–152.

Bauer, F. L. (1959). *In* "On Numerical Approximation" (R. E. Langer ed.). Univ. of Wisconsin Press, Madison, Wisconsin.

Bauer, F. L. (1965). *In* "Approximation of Functions" (H. Garabedian, ed.). American Elsevier, New York.

Bauer, F. L. *et al.* (1963). *Proc. Symp. Appl. Math. Amer. Math. Soc.* **15**.

Beardon, A. F. (1968). *J. Math. Anal. Appl.* **21**, 344–346.

Beckenbach, E. F., and Bellman, R. (1961). "Inequalities." Springer-Verlag, Berlin and New York.

Bellman, R. (1961). "A Brief Introduction to Theta Functions." Holt, New York.

Bellman, R. (1970). "Introduction to Matrix Analysis," 2nd ed. McGraw-Hill, New York.

Bellman, R., and Cooke, K. L. (1963). "Differential-Difference Equations." Academic Press, New York.

Benson, G. C., and Schreiber, H. P. (1955). *Canad. J. Phys.* **33**, 529–540.

Birkoff, G. D. (1930). *Acta Math.* **54**, 205–246.

Birkhoff, G. D., and Trjitzinsky, W. J. (1932). *Acta Math.* **60**, 1–89.

Born, M., and Huang, K. (1954). "Dynamical Theory of Crystal Lattices," Oxford Univ. Press, London and New York.

Brezinski, C. (1970). *C.R. Acad. Sci. Paris Sér. A* **270**, 1252–1253.

Brezinski, C. (1972). *RAIRO* **R1**, 61–66.

Brezinski, C. (1975). *Calcolo* **12**, 317–360.

Brezinski, C. (1976). *J. Comput. Appl. Math.* **2**, 113–123.

Brezinski, C. (1977). "Accélération de la Convergence en Analyse Numérique." Springer-Verlag, Berlin and New York.

Brezinski, C. (1978). "Algorithmes d'Accélération de la Convergence: Étude Numérique." Éditions Technip, Paris.

Bulirsch, R., and Stoer, J. (1966). *Numer. Math.* **8**, 1–13.

Bulirsch, R., and Stoer, J. (1967). *Numer. Math.* **9**, 271–278.

Butzer, P. L., and Nessel, R. J. (1971). "Fourier Analysis and Approximation." Academic Press, New York.

Chisolm, J. S. R. (1966). *J. Math. Phys.* **7**, 39.

Chow, Y. S., and Teicher, H. (1971). *Ann. Math. Statist.* **42**, 401–404.

Chrystal, G. (1959). "Algebra." Chelsea, New York.

Concus, P., Cassatt, D., Jaehnig, G., and Melby, E. (1963). *Math. Comp.* **17**, 245–256.

Cooke, R. G. (1955). "Infinite Matrices and Sequence Spaces." Dover, New York.

Cordellier, F. (1977). *C.R. Acad. Sci. Paris Sér. A* **284**, 389–392.

Cornyn, J. J., Jr. (1974). Direct Methods for Solving Systems of Linear Equations Involving Toeplitz or Hankel Matrices, NRL Memorandum Rep. 2920. Naval Research Laboratory, Washington, D.C.

Cowling, V. F., and King, J. P. (1962/1963). *J. Analyse Math.* **10**, 139–152.

Davis, P. (1963). "Interpolation and Approximation." Ginn (Blaisdell), Waltham, Massachusetts.

Dieudonné, J. (1969). "Foundations of Modern Analysis." Academic Press, New York.

Erdélyi, A. (1956). "Asymptotic Expansions." Dover, New York.

Erdélyi, A., Magnus, W., Oberhettinger, F., and Tricomi, F. G. (1953). "Higher Transcendental Functions," Vols. 1, 2, and 3. McGraw-Hill, New York.

Erdélyi, A., Magnus, W., Oberhettinger, F., and Tricomi, F. G. (1954). "Tables of Integral Transforms," Vols. 1 and 2. McGraw-Hill, New York.

Esser, H. (1975). *Computing* **14**, 367–369.

Fletcher, A., Miller, J. C. P., Rosenhead, L., and Comrie, L. J. (1962). "An Index of Mathematical Tables." Vol. 1. Oxford Univ. Press (Blackwell), London and New York.

Freud, G. (1966). "Orthogonal Polynomials." Pergamon, Oxford.

Gantmacher, F. R. (1959). "The Theory of Matrices," Vols. 1 and 2, Chelsea, New York.

Garreau, G. A. (1952). *Nederl. Akad. Wetensch. Proc. Ser. A* **14**, 237–244.

Gekeler, E. (1972). *Math. Comp.* **26**, 427–435.

Germain-Bonne, B. (1973). *RAIRO* **R1**, 84–90.

Germain-Bonne, B. (1978). Thesis, Univ. des Sciences et Techniques de Lille.

Gilewicz, J. (1978). "Approximants de Padé," Lecture Notes in Mathematics #667. Springer-Verlag, Berlin and New York.

Glasser, M. L. (1973a). *J. Math. Phys.* **14**, 409–414.

Glasser, M. L. (1973b). *J. Math. Phys.* **14**, 701–703.

Glasser, M. L. (1974). *J. Math. Phys.* **15**, 188–189.

Glasser, M. L. (1975). *J. Math. Phys.* **16**, 1237–1238.

Goldsmith, D. L. (1965). *Amer. Math. Monthly* **72**, 523–525.

Golomb, M. (1943). *Bull. Amer. Math. Soc.* **49**, 581–592.

Gordon, P. (1975). *SIAM J. Math. Anal.* **6**, 860–867.

Gray, H. L., and Atchison, T. A. (1967). *SIAM J. Numer. Anal.* **4**, 363–371.

Gray, H. L., and Atchison, T. A. (1968a). *J. Res. Nat. Bur. Standards* **72B**, 29–31.

Gray, H. L., and Atchison, T. A. (1968b). *Math. Comp.* **22**, 595–606.

Gray, H. L., and Clark, W. D. (1969). *J. Res. Nat. Bur. Standards* **73B**, 251–273.

Greville, T. N. E. (1968). Univ. of Wisconsin Math. Res. Center Rep. #877.
Haber, S. (1977). *SIAM J. Numer. Anal.* **14**, 668–685.
Hadamard, J. (1892). *J. Math. Pures. Appl.* **8**, 101–186.
Hancock, H. (1909). "Lectures on the Theory of Elliptic Functions," Vol. I. Dover, New York.
Hardy, G. H. (1956). "Divergent Series." Oxford Univ. Press, London and New York.
Hardy, G. H., and Rogosinski, W. W. (1956). "Fourier Series." Cambridge Univ. Press, London and New York.
Hardy, G. H., and Wright, E. M. (1954). "An Introduction to the Theory of Numbers." Oxford Univ. Press, London and New York.
Hautot, A. (1974). *J. Math. Phys.* **15**, 1722–1727.
Håvie, T. (1979). *BIT* **19**, 204–213.
Henrici, P. (1977). "Applied and Computational Complex Analysis," Vols. 1 and 2. Wiley (Interscience), New York.
Higgins, R. L. (1976). Thesis, Drexel Univ.
Householder, A. S. (1953). "Principles of Numerical Analysis." McGraw-Hill, New York.
Iguchi, K. (1975). *Inform. Process. Japan* **15**, 36–40.
Iguchi, K. (1976). *Inform. Process. Japan* **16**, 89–93.
Isaacson, E., and Keller, H. B. (1966). "Analysis of Numerical Methods." Wiley, New York.
Jacobi, C. G. I. (1829). "Fundamenta Nova Theoriae Functionum Ellipticarum." Königsberg.
Jacobi, C. G. I. (1846). *J. Reine Angew. Math.* **30**, 127–156.
Jakimouski, A. (1959). *Michigan Math. J.* **6**, 277–290.
Jameson, G. J. O. (1974). "Topology and Normed Spaces." Chapman & Hall, London.
Jones, B. (1970). *J. Inst. Math. Appl.* **17**, 27–36.
Kantorovich, L. V., and Akilov, G. P. (1964). "Functional Analysis in Normed Spaces" (transl. by D. E. Brown and A. P. Robertson). Pergamon, Oxford.
King, R. F. (1979). *SIAM J. Numer. Anal.* **16**, 719–725.
Knopp, K. (1947). "Theory and Application of Infinite Series." Hafner, New York.
Kress, R. (1971). *Computing* **6**, 274–288.
Kress, R. (1972). *Math. Comp.* **26**, 925–933.
Krylov, V. I. (1962). "Approximate Calculation of Integrals." Macmillan, New York.
Kummer, E. E. (1837). *J. Reine Angew. Math.* **16**, 206–214.
Lambert, J. D., and Shaw, B. (1965). *Math. Comp.* **19**, 456–462.
Lambert, J. D., and Shaw, B. (1966). *Math. Comp.* **20**, 11–20.
Laurent, P. J. (1964). Thesis, Grenoble.
Levin, D. (1973). *Internat. J. Comput. Math.* **B3**, 371–388.
Livingston, A. E. (1954). *Duke Math. J.* **21**, 309–314.
Lorch, L., and Newman, D. J. (1961). *Canad. J. Math.* **13**, 283–298.
Lorch, L., and Newman, D. J. (1962). *Comm. Pure Appl. Math.* **15**, 109–118.
Lotockiĭ, A. V. (1953). *Ivanov. Gos. Ped. Inst. Uč. Zap. Fiz-Mat. Nauki* **4**, 61–91.
Lubkin, S. (1952). *J. Res. Nat. Bur. Standards Sect. B* **48**, 228–254.
Luke, Y. L. (1969). "The Special Functions and Their Approximations," Vols. 1 and 2. Academic, New York.
Luke, Y. L. (1979). On a Summability Method, notes, Univ. of Missouri, Kansas City, Missouri.
Luke, Y. L., Fair, W., and Wimp, J. (1975). *Comp. Math. Appl.* **1**, 3–12.
Lyness, J. N. (1970). *Math. Comp.*, **24**, 101–135.
Lyness, J. N. (1971). *Math. Comp.*, **25**, 59–78.
McLeod, J. B. (1971). *Computing* **7**, 17–24.
McNamee, J., Stenger, F., and Whitney, E. L. (1971). *Math. Comp.* **25**, 141–154.
Miller, K. S. (1974). "Complex Stochastic Processes." Addison-Wesley, Reading, Massachusetts.
Milne-Thomson, L. M. (1960). "The Calculus of Finite Differences." Macmillan, London.

de Montessus de Balloire, R. (1902). *Bull. Soc. Math. France* **30**, 28–36.
Moore, E. H. (1920). *Bull. Amer. Math. Soc.* **26**, 394–395.
Muir, T. (1960). "A Treatise on the Theory of Determinants." Dover, New York.
Nikolskii, S. M. (1948). *Izv. Akad. Nauk SSSR Ser. Mat.* **12**, 259–278.
Olevskiĭ, A. M. (1975). "Fourier Series with Respect to General Orthogonal Systems," Springer-Verlag, Berlin and New York.
Olver, F. W. J. (1974). "Asymptotics and Special Functions." Academic Press, New York.
Ortega, J. M., and Rheinboldt, W. C. (1970). "Iterative Solution of Nonlinear Equations in Several Variables." Academic Press, New York.
Ostrowski, A. M. (1966). "Solutions of Equations and Systems of Equations," Academic Press, New York.
Ostrowski, A. M. (1973). "Solutions of Equations in Euclidean and Banach Spaces." Academic Press, New York.
Overholt, K. J. (1965). *BIT* **5**, 122–132.
Papoulis, A. (1965). "Probability, Random Variables, and Stochastic Processes." McGraw-Hill, New York.
Pennacchi, R. (1968). *Calcolo* **5**, 37–50.
Penrose, R. (1955). *Proc. Cambridge Philos. Soc.* **51**, 406–413.
Perron, O. (1929). "Die Lehre von den Kettenbrüchen." Chelsea, New York.
Perron, O. (1957). "Die Lehre von den Kettenbrüchen," 3rd ed., Vols. 1 and 2. Teubner, Stuttgart.
Petersen, G. M. (1966). "Regular Matrix Transformations." McGraw-Hill, New York.
Peyerimhoff, A. (1969). "Lecture Notes of Summability," Lecture Notes in Mathematics #107. Springer-Verlag, Berlin and New York.
Pollaczek, F. (1956). "Sur une Généralisation des Polynomes de Jacobi." Gauthiers-Villars, Paris.
Pyle, L. D. (1967). *Number. Math.* **10**, 86–102.
Rainville, E. D. (1960). "Special Functions," Macmillan, New York.
Reich, S. (1970). *Amer. Math. Monthly* **77**, 283–284.
Richtmeyer, R. D. (1957). "Difference Methods for Initial Value Problems." Wiley (Interscience), New York.
Rutishauser, H. (1954). *Z. Angew. Math. Phys.* **5**, 233–251.
Rutishauser, H. (1957). "Der Quotienten-Differenzen Algorithmus." Birkhäuser-Verlag, Basel.
Salzer, H. E. (1955). *J. Math. Phys.* **33**, 356–359.
Salzer, H. E. (1956). *MTAC* **10**, 149–156.
Salzer, H. E., and Kimbro, G. M. (1961). *Math. Comp.* **15**, 23–29.
Samuelson, P. A. (1945). *J. Math. Phys.* **24**, 131–134.
Scheid, F. (1968). "Numerical Analysis, Schàum's Outline Series." McGraw-Hill, New York.
Schmidt, J. R. (1941). *Philos. Mag.* **32**, 369–383.
Schur, I. (1921). *J. Reine Angew. Math.* **151**, 79–111.
Schwartz, C. (1969). *J. Comput. Phys.* **4**, 19–29.
Schwartz, L. (1961/1962). *In* "Séminaire Bourbaki," fasc. 3. Benjamin, New York.
Shanks, D. (1955). *J. Math. Phys.* **34**, 1–42.
Shaw, B. (1967). *J. Assoc. Comput. Math.* **14**, 143–154.
Shohat, J. A., and Tomarkin, J. D. (1943). "The Problems of Moments." American Mathematical Society, Providence, Rhode Island.
Shoop, R. A. (1979). *Pacific J. Math.* **80**, 255–262.
Slater, L. J. (1960). "Confluent Hypergeometric Functions." Cambridge Univ. Press, London and New York.
Smith, A. C. (1978). *Utilitas Math.* **13**, 249–269.
Smith, D. A., and Ford, W. F. (1979). *SIAM J. Numer. Anal.* **16**, 223–240.

Szegö, G. (1959). "Orthogonal Polynomials." American Mathematical Society, Providence, Rhode Island.
Titchmarsh, E. C. (1939). "The Theory of Functions." Oxford Univ. Press, London and New York.
Todd, J. (ed.) (1962). "Survey of Numerical Analysis." McGraw-Hill, New York.
Traub, J. F. (1964). "Iterative Methods for the Solution of Equations." Prentice-Hall, Englewood Cliffs, New Jersey.
Trench, W. F. (1964). *SIAM J. Appl. Math.* **12**, 515–522.
Trench, W. F. (1965). *SIAM J. Appl. Math.* **13**, 1102–1107.
Tucker, R. R. (1967). *Pacific J. Math.* **22**, 349–359.
Tucker, R. R. (1969). *Pacific J. Math.* **28**, 455–463.
Tucker, R. R. (1973). *Faculty Rev. Bull. N.C. A and T State Univ.* **65**, 60–63.
Uspensky, J. V. (1928). *Trans. Amer. Math. Soc.* **30**, 542–559.
van der Hoff, B. M. E., and Benson, G. C. (1953). *Canad. J. Phys.* **31**, 1087–1091.
Vučković, V. (1958). *Acad. Serbe. Sci. Publ. Inst. Math.* **12**, 125–136.
Walls, H. S. (1948). "Analytic Theory of Continued Fractions." Chelsea, New York.
Wasow, W. (1965). "Asymptotic Expansions For Ordinary Differential Equations." Wiley, New York.
Whittaker, E. T., and Watson, G. N. (1962). "A Course of Modern Analysis." Cambridge Univ. Press, London and New York.
Wilansky, A., and Zeller, K. (1957). *J. London Math. Soc.* **32**, 397–408.
Wimp, J. (1970). *SIAM J. Numer. Anal.* **7**, 329–334.
Wimp, J. (1972). *Math. Comp.* **26**, 251–254.
Wimp, J. (1974a). *Computing* **13**, 195–203.
Wimp, J. (1974b). *J. Approx. Theory* **10**, 185–198.
Wimp, J. (1974c). *Numer. Math.* **23**, 1–17.
Wimp, J. (1975). *Acceleration methods*, In "Encyclopedia of Computer Science and Technology," Vol. I. Dekker, New York.
Wright, E. M. (1955). *J. Reine Angew. Math.* **194**, 66–87.
Wynn, P. (1956a). *J. Math. Phys.* **35**, 318–320.
Wynn, P. (1956b). *Math. Tables Aids Comput.* **10**, 91–96.
Wynn, P. (1956c). *Proc. Cambridge Philos. Soc.* **52**, 663–671.
Wynn, P. (1959). *Numer. Math.* **1**, 142–149.
Wynn, P. (1961). *Nieuw Arch. Wisk* **9**, 117–119.
Wynn, P. (1962). *Math. Comp.* **16**, 301–322.
Wynn, P. (1963). *Nordisk Tidskr. Informat.-Behandl.* **3**, 175–195.
Wynn, P. (1966). *SIAM J. Numer. Anal.* **3**, 91–122.
Wynn, P. (1966). Univ. of Wisconsin Math. Res. Center Rep. #626.
Wynn, P. (1967). Univ. of Wisconsin Math. Res. Center Rep. #750.
Wynn, P. (1972). *C. R. Acad. Sci. Paris Sér. A* **275**, 1065–1068.
Zeller, K. (1952). *Math. Z.* **56**, 18–20.
Zeller, K. (1958). "Theorie du Limitierungsverfahren." Springer-Verlag, Berlin and New York.
Zemansky, M. (1949). *C.R. Acad. Sci. Paris* **228**, 1838–1840.

Index